Aquaculture Microbiology and Biotechnology

Volume 2

Aquaculture
Microbiology and
Biotechnology

Volume 2

Editors

Didier Montet

Centre International de Recherche en Agronomie
pour le Développement (CIRAD)
Montpellier
France

Ramesh C. Ray

Central Tuber Crops Research Institute
Bhubaneswar
India

CRC Press
Taylor & Francis Group
Boca Raton London New York

CRC Press is an imprint of the
Taylor & Francis Group, an **informa** business

A SCIENCE PUBLISHERS BOOK

CRC Press
Taylor & Francis Group
6000 Broken Sound Parkway NW, Suite 300
Boca Raton, FL 33487-2742

First issued in paperback 2017

© 2011 by Taylor & Francis Group, LLC
CRC Press is an imprint of Taylor & Francis Group, an Informa business

ISBN 13: 978-1-138-11375-6 (pbk)
ISBN 13: 978-1-57808-711-2 (hbk)

Library of Congress Cataloging-in-Publication Data

Aquaculture microbiology and biotechnology/editors, Didier
Montet, Ramesh C. Ray.
   p. cm.
 Includes bibliographical references and index.
 ISBN 978-1-57808-574-3 (hardcover)
1. Fishes--Diseases. 2. Fishes--Genetics. 3.Microbial
biotechnology. 4. Aquaculture. I. Montet, Didier. II. Ray,
Ramesh C.
 SB171.A72 2009
 639.3--dc22

                                                        2009003772

The views expressed in this book are those of the author(s) and the publisher does not assume responsibility for the authenticity of the findings/conclusions drawn by the author(s). Also no responsibility is assumed by the publishers for any damage to the property or persons as a result of operation or use of this publication and/or the information contained herein.

Preface

This is the second volume of the 2-volume series. Except for the first two chapters, this volume deals with post harvest aspects of aquaculture, pathogens associated with sea foods, bioremediation of fish waste management and advalorization of fish waste products. In Chapter 1, Pintado and his colleagues have broadly discussed the various strategies such as formation of bio-films, culture of microalgae and bacteriophages, prebiotics and probiotics, biofilters, etc and associated mechanisms for the control of bacterial infections in marine larval rearing. Gomez-Gill and his colleagues in Chapter 2 have dealt exclusively with the applications of probiotics in larval culture.

In Chapter 3, Leroi and Joffraud have discussed the various spoilage bacteria associated with fresh, preserved and processed fish and shrimp. Bari and his colleagues in Chapter 4 have focused on sea food diseases and illness, consumers' perception of seafood safety and quality and other health-related issues. Levin in Chapter 5 further elaborated the conventional and molecular techniques available for detection and enumeration of human infections and toxin producing bacteria that are native to marine environments and that are found associated with sea food.

In warm regions such as tropical countries, access to fresh fish can be a problem mainly in rural areas owing to the shortage of ice and lack of refrigeration. Fermentation is the most important way of preserving fish. Fermented fish is used both as flavouring and as a source of protein. Fish fermentation in the Southeast Asian region normally lasts for 3-9 months and the fish flesh may be liquefied or turned into a sauce or paste. Some of the products include *Nuoc-mam* of Vietnam and Cambodia, *Nam-pla* of Thailand, *Sushi* of Japan and *Patis* of the Philippines. The fermented whole fish products include *Plaa som, Som-fug, Hoi Dorng* from Thailand, *Jeotkals* from Korea, *Burong isda* from the Philippines, *Colombo cure* and *lona Ilishi* from India, Sri Lanka and Bangladesh.

In Chapter 6, Panda and her colleagues have given an overview of the different types of fish sauces, fish pastes and fermented fish products available globally with emphasis on the manufacturing processes, microbiology and proximate and nutrient compositions.

Chapter 7 exclusively describes the potentials of bio-prospecting microalgae, corals and other microorganisms as sources of valuable nutraceuticals and other compounds. Microalgae have industrial uses as raw material for carotenoids, vitamins, fatty acid supplements and as feed additives for poultry, livestock, fish and crustaceans. Likewise, corals provide numerous bioactive compounds including hormones, antibiotics and pharmaceuticals. The last three chapters in this volume deal with fish industry waste and by-products management. Worawattanamateekul and Ray in Chapter 8 have briefly discussed microbial remediation of fish and shrimp culture systems. In Chapter 9, Arvanitoyannis and his colleagues have described fish industry waste management with particular emphasis on the use of microorganisms. In the last chapter (10), Bhaskar and his colleagues have elaborately discussed microbial reclamation of fish waste especially with reference to fermentation techniques and application of microbial protease to recover value-added products like enzymes, carotenoids, poly-unsaturated fatty acids, etc. from fish industry wastes.

Didier Montet
Ramesh C. Ray

Contents

List of Contributors

Arvanitoyannis, Ioannis S.

Department of Agriculture Icthyology and Aquatic Environment, School of Agricultural Sciences, University of Thessaly, Fytokou Street, 38446 Nea Ionia Magnesias, Volos, Hellas, Greece

E-mail: parmenion@uth.gr

Balcázar, José Luis

Instituto de Investigacións Mariñas (CSIC), Eduardo Cabello n° 6, 36208 Vigo, Galicia, Spain

Present address: Institut Català de Recerca de l'Aigua. Parc Científic i Tecnològic de la Universitat de Girona, Emili Grahit 101, 17003, Girona, Catalonia, Spain

Bari, Md. Latiful

Centre for Advanced Research in Science, University of Dhaka, Dhaka-1000, Bangladesh, Tel.: 880-2-9661920-59 Ext 4721, Fax: 880-2-8615583

E-mail:latiful@univdhaka.edu

Bhaskar, N.

Department of Meat, Fish and Poultry Technology, Central Food Technological Research Institute, Mysore 570 020, India

E-mail: bhasg3@yahoo.co.in

El Sheikha, Aly F.

Minufiya University, Faculty of Agriculture, Department of Food Science and Technology, 32511 Shibin El Kom, Minufiya Government, Egypt, Tel.: 33 4 67 61 57 28, Fax: 33 4 67 61 44 44

E-mail: elsheikha_aly@yahoo.com

Gomez-Gil, Bruno

CIAD, A.C. Mazatlan Unit for Aquaculture and Environmental Management. AP 711 Mazatlan, Sinaloa, Mexico 82000

E-mail: bruno@ciad.mx

Isshiki, Kenji

Division of Marine Life Science, Research Faculty of Fisheries Science, Hokkaido University, 3-1-1, Minato-cho, Hakodate, Hokkaido 041-8611, Japan, Tel. & Fax: 81-138-40-5564
E-mail: isshiki@fish.hokudai.ac.jp

Joffraud Jean-Jacques

Ifremer, Laboratoire Science et Technologie de la Biomasse Marine, BP 21105, 44311, Nantes, France

Kassaveti, Aikaterini

Department of Agriculture Icthyology and Aquatic Environment School of Agricultural Sciences, University of Thessaly, Fytokou Street, 38446 Nea Ionia Magnesias, Volos, Hellas, Greece

Kawamoto, Shinichi

Food Hygiene Laboratory, National Food Research Institute, Kannondai-2-1-12, Tsukuba 305- 8642, Japan, Tel.: 81-29-838-8008, Fax: 81-29-838-7996
E-mail: taishi@affrc.go.jp

Leroi, Françoise

Ifremer, Laboratoire Science et Technologie de la Biomasse Marine, BP 21105, 44311, Nantes, France, Tel.: +33 2 40 37 41 72, Fax: 33 2 40 37 40 71
E-mail: fleroi@ifremer.fr

Levin Robert, E.

Department of Food Science, Massachusetts Agricultural Experiment Station, University of Massachusetts, Amherst, MA 01003, Fax: 413-545-1262, Tel.: 413-545-0187
E-mail: relevin@foodsci.umass.edu

Mahendrakar, N.S.

Department of Meat, Fish and Poultry Technology, Central Food Technological Research Institute, Mysore 570 020, India

Makridis, Pavlos

Institute of Aquaculture, Hellenic Center for Marine Research, P.O. Box 2214, Iraklion, Crete, Greece

Montet, Didier

Centre de Coopération Internationale en Recherche Agronomique pour le Développement, CIRAD, UMR Qualisud, TA 95B/16, 34398 Montpellier Cedex 5, France, Tel.: 33 4 67 61 57 28, Fax: 33 4 67 61 44 44
E-mail: didier.montet@cirad.fr

Panda, Smita H.

Agri-Bioresource Research Foundation, 81, District Centre, Chandra-sekharpur, Bhubaneswar 751 010, Orissa, India; Tel./Fax: 91-674-2351046

E-mail: agribiores6@rediffmail.com

Pintado, José

Instituto de Investigacións Mariñas (CSIC), Eduardo Cabello n° 6, 36208 Vigo, Galicia, Spain

E-mail: pintado@iim.csic.es

Planas, Miquel

Instituto de Investigacións Mariñas (CSIC), Eduardo Cabello n° 6, 36208 Vigo, Galicia, Spain

Prol, María J.

Instituto de Investigacións Mariñas (CSIC), Eduardo Cabello n° 6, 36208 Vigo, Galicia, Spain

Ray, Ramesh C.

Regional Centre, Central Tuber Crops Research Institute, Bhubaneswar 751 019, Orissa, India, Tel./Fax: 91-674-2470528

E-mail: rc_rayctcri@rediffmail.com

Roque, Ana

IRTA-Sant Carles de la Rápita, Crta. Poble Nou, Km 5.5, Sant Carles de la Rápita, Spain 43540

Sachindra, N.M.

Department of Meat, Fish and Poultry Technology, Central Food Technological Research Institute, Mysore 570 020, India

Suresh, P.V.

Department of Meat, Fish and Poultry Technology, Central Food Technological Research Institute, Mysore 570 020, India

Soto-Rodriguez, Sonia

CIAD, A.C. Mazatlan Unit for Aquaculture and Environmental Management. AP 711 Mazatlan, Sinaloa, Mexico 82000

Varzakas, Theodoros

School of Plant Sciences, TEI Kalamatas, Greece

Venugopal, V.

Former Scientific Officer, Food Technology Division, Bhabha Atomic Research Centre, Mumbai 400 076, India

E-mail: venugopalmenon@hotmail.com

Worawattanamateekul, Wanchai

Department of Fishery Products, Faculty of Fisheries, Kasetsart University, Bangkok 10900, Thailand, Tel.: 66-2-5790113 press 4088, Fax: 66-2-9428644 press 12

E-mail: ffiswcw@ku.ac.th

Yeasmin, Sabina

Department of Genetic Engneeering and Biotechnology, Dhaka University, Dhaka-1000, Bangladesh, Tel.: 880-2-9661920-59 Ext 7721, Fax: 880-2-8615583

E-mail: y_sabina01@yahoo.com

New Strategies for the Control of Bacterial Infections in Marine Fish Larval Rearing

José Pintado,[1,*] *María J. Prol,*[1] *José Luis Balcázar,*[1,2]
Miquel Planas[1] and *Pavlos Makridis*[3]

INTRODUCTION

One of the limiting steps of marine fish aquaculture expansion is the supply of juvenile fish. Intensive fish larvae production is highly susceptible to the proliferation of bacteria, which may cause poor growth or mass mortality of the larvae. In most cases, mortality cannot be attributed to specific pathogens, but to the proliferation of opportunistic bacteria (Olafsen, 1993, 2001). The rearing environment, with high larval densities and high load of organic matter (from faeces, dead larvae, debris, or from live feed) is highly susceptible to bacterial growth. The control of bacteria in rearing systems and live feed production is an important factor for the survival of fish larvae (Planas and Cunha, 1999; Skjermo and Vadstein, 1999; Dhert et al., 2001) and the increase in survival of larvae treated with antibiotics supports this fact (Gatesoupe, 1982, 1989; Pérez-Benavente and Gatesoupe, 1988).

In aquatic environments, the use of antibiotics induces the development of resistances (Sarter and Guichard, 2009, Volume 1 of this Series), which can be transferred to other bacteria, including pathogenic bacteria to fish or even humans (Cabello, 2006; Sarter et al., 2007). Therefore, the use of antibiotics should be kept to a minimum. Vaccination of juveniles has

---

[1]*Instituto de Investigacións Mariñas (CSIC), Eduardo Cabello nº 6, 36208 Vigo, Galicia, Spain*
[2]*Present address: Institut Català de Recerca de l'Aigua. Parc Científic i Tecnològic de la Universitat de Girona, Emili Grahit 101, 17003 Girona, Catalonia, Spain*
[3]*Institute of Aquaculture, Hellenic Center for Marine Research, P.O. Box 2214, Iraklion, Crete, Greece*
Corresponding author: E-mail: pintado@iim.csic.es

drastically reduced the use of chemotherapeutics in aquaculture, increasing the survival of farmed fish. However, fish larvae have an undeveloped immune system, relying on maternal antibodies and non-specific immune response, which prevent the use of vaccination (Vadstein, 1997).

Usual approaches to control bacterial growth in intensive rearing of marine fish larvae are based on preventive measures, leading to the development of a clean environment by water treatment processes (e.g., filtration, UV-irradiation, ozonization and disinfectants). Disinfection of fish eggs (Salvesen and Vadstein, 1995), and measures for control of bacterial microbiota in live feed such as, disinfection of rotifer eggs for the production of axenic cultures (Douillet, 1998; Dhert et al., 2001), treatment with hydrogen peroxide (Giménez et al., 2006), or ultraviolet radiation for partial decontamination (Munro et al., 1993, 1999) have been proposed, but would be difficult to implement at the industrial scale. Complete elimination of bacteria from the organisms and culture system is not possible, while disinfection implies in most cases a loss of a stable microbial population, with dominance of slow-growing bacteria (*K*-strategists) (Pianka, 1970; Andrews and Harris, 1986). Disturbance of these stable bacterial populations may promote, in a nutrient-rich environment (due to feed and excretions) as are larval culture systems, the rapid colonization by opportunistic bacteria (*r*-strategists), with high growth rates and potentially harmful to fish larvae (Skjermo and Vadstein, 1993).

Therefore, at present, efforts have been made towards new strategies for the control of bacterial microbiota in the rearing systems that, by avoiding the use of antibiotics and disinfectants, would lead to a more environmentally friendly and sustainable aquaculture. This approach would be in accordance with the Ecosystem Approach for the sustainable growth and expansion of aquaculture, promoted by FAO in The State of the World Fisheries and Aquaculture (FAO, 2007).

This chapter discusses these new strategies. Most of them have been proposed or developed at the laboratory scale and some of them in pilot-plant trials. Only a few have been applied at an industrial scale.

BACTERIA IN FISH LARVAE

Fish larvae at hatching are commonly colonized by very few bacteria, which may originate from the egg surface, or from the surrounding water. During the yolk-sac stage, marine fish larvae drink seawater to osmoregulate. They thus accumulate bacteria and microalgae in their gut (Reitan et al., 1998). The numbers of bacteria during this phase of rearing are low [<100 CFU (Colony Forming Units) per larva], whereas the biodiversity of microbiota can be quite high (Hansen and Olafsen, 1999).

As soon as fish larvae start to capture and ingest live prey, the numbers of bacteria in the larvae increase exponentially up to values as high as 10^5 per larva within a few days (Munro et al., 1994; Muroga et al., 1987). Members of α- and γ-proteobacteria are quite common in fish larvae (Nakase et al., 2007; Schulze et al., 2006). These bacteria are located mainly in the larval gut, where as has been proved disinfection of the surface of the larvae does not decrease the numbers of culturable bacteria to a noticeable degree. However, the biodiversity of microbiota associated with the gut is quite low, as only a few species dominate. *Vibrio* species are the main component of the gut microbiota in rearing conditions and are mainly derived from bacteria associated with live prey (Fjellheim et al., 2007; Skjermo and Vadstein, 1993; Verschuere et al., 1997).

After the mouth opening and the onset of exogenous feeding, it is relatively easy to manipulate the bacterial diversity in the gut by using bacteria bioencapsulated in the prey, rotifers or *Artemia metanauplii* (Makridis et al., 2000a). This phase is critical in many aspects for the successful rearing of marine fish larvae as pioneer species, i.e., bacterial populations that firstly colonize the gut, have a competitive advantage compared with bacteria that reach the gut at a later stage, when available adhesion sites have been reduced and environmental conditions are influenced by the bacterial populations already established (Hansen and Olafsen, 1999; Skjermo and Vadstein, 1999).

A main characteristic of the type of colonization in the gut of larvae is that species established during this period are mainly opportunistic, taking advantage of the input of nutrients in a "new" environment (larval gut). The establishment of excessive numbers of bacteria in weakened larvae may result in high mortalities, but mortalities in healthy larvae, not exposed to environmental stress and fed on high quality prey, are low and not dependent on the numbers of bacteria present in the gut (Makridis et al., 2008; Salvesen et al., 1999).

At the stage of weaning, the numbers of bacteria decrease, and relatively few bacteria are measured in the fish gut per unit of gut weight compared with earlier stages (Makridis et al., 2008). The development of the stomach and its acid barrier makes direct influence of species established in the intestinal microbiota by addition of probiotics in the feed more complicated.

NEW STRATEGIES FOR BACTERIAL INFECTIONS CONTROL IN MARINE FISH LARVAE

The development of new strategies is based mostly on the study and knowledge of microbial diversity and ecology in larviculture systems (Fig. 1.1). The strategies can be classified as based on:

- Acting on the microbial communities of the system, including or not the introduction of live microorganisms, and
- Acting on the host-microbe interactions, which include the use of immunostimulants.

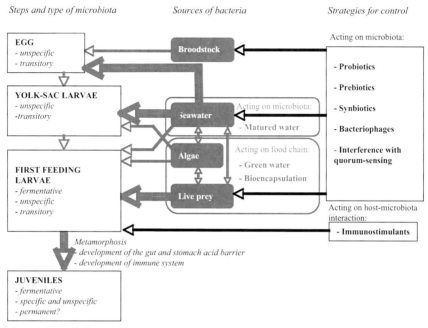

Fig. 1.1 Steps and interactions in bacterial colonization of fish larvae, sources of bacteria and strategies for bacterial infections control.

These new strategies can be applied to the different developmental stages in fish larvae, from egg to juvenile stage (Fig. 1.1). Each step of larval development, is characterized by a different type of associated microbiota, which is introduced to the system by different ways: *via* water (in eggs and yolk-sack larvae) and *via* live prey (feeding larvae). Some strategies imply one single manipulation (e.g., addition of probiotics to the water), whereas others (e.g., bioencapsulation) involve the manipulation of the food chain. All these facts should be considered when selecting the strategy to control infections.

Acting on the Microbial Communities of the System

Promoting Beneficial Bacteria in the System

Matured Water

A major change that occurs during transfer of seawater from the sea to a closed container is that water comes in contact with large solid surfaces in relation to its relatively small volume. These surfaces offer a new substrate for bacteria present in the water and thereby enhance the proliferation of opportunistic bacteria. Immediate use of this non-mature water has often led to mass mortalities during the rearing of marine fish larvae. The final aim of matured water is to reduce the percentage of opportunistic bacteria in the inflow water and thereby in the rearing system. A strategy has been developed where seawater was retained in a large container where material with high surface to volume ratio was placed under strong aeration for a long time period (optimally several weeks). Seawater is filtered through a membrane and the total numbers of bacteria are thus heavily reduced. The inflow rate of seawater and thereby the supply of nutrients is kept quite low inducing the proliferation of slow-growing, *K*-strategist bacterial species.

This approach involves two steps. At first, a biofilm is formed on the surface of the material, at the interface with water, where bacteria are established at a high density, thereby favouring *K*-strategists species. At the same time, a high bacterial load should be avoided in seawater in the rearing tanks. The second step of this strategy is the transfer of biofilm members from the biofilm to the water. A residence time of at least 24 h is considered necessary prior to the use of matured water in the rearing of marine fish larvae. A high residence time is accomplished by the use of large containers in combination with low inflow/outflow rates.

The basic principle involved in the use of matured water is that, as non-opportunistic and less harmful bacterial species are at an increased level in inflow water, they will prevent the proliferation of harmful opportunistic species. Slow-growing bacteria will thereby be present in high numbers during the first days after hatching and may have a competitive advantage over opportunistic bacteria, which proliferate at later stages of rearing.

Matured water has been successfully applied in several experiments with marine fish larvae, as in the case of incubation of Atlantic halibut yolk-sac larvae (Skjermo et al., 1997) and with first feeding of turbot larvae (Salvesen et al., 1999). It should be noted that in recirculation systems, there is an analogy with the matured water system, as there is material with high surface to volume ratio in the biofilter, which offers the necessary support for the establishment of a biofilm with bacterial populations at a high density. Stable bacterial conditions are normally observed in recirculation systems (Attramadal et al., 2005).

Green Water

Use of microalgae has often resulted in successful rearing of marine fish larvae, as determined by low mortalities, high growth and viable fry (Howell, 1979; Naas et al., 1991; Reitan et al., 1993). There are two different variants related to the application of microalgae in rearing of marine fish larvae. In the first case, microalgae are cultured in large tanks and when optimal conditions are reached these tanks are stocked with fish eggs or larvae (green water technique). In this case, production of microalgae takes place in the same tank as the rearing of the larvae. In a more common approach, production of microalgae takes place is separate culture tanks, the microalgae are harvested regularly and added to the rearing tanks of the larvae (pseudogreen water technique).

Use of microalgae has several implications for different aspects of rearing of the larvae. Marine fish larvae drink seawater to osmoregulate and ingest microalgae particles, which are accumulated in the gut. The phenomenon of "green gut" has been shown in larval rearing of many marine fish species (Cahu et al., 1998). Filling of the gut seems to have an effect on the stimulation of production of digestive enzymes in the larval gut (Hjelmeland et al., 1988).

In the case when fish larvae are stocked at a low density, average residence time of live food organisms can be up to several days. Microalgal cells are available for the live food organisms present in the tank and may prevent the incidence of starved live food organisms (Makridis and Olsen, 1999).

Addition of microalgae changes the physicochemical parameters such as light, colour of the tank water, the concentration of ammonia and other harmful compounds, and reduces light intensity in the water mass and the stress caused by direct light on the larvae.

Another aspect of addition of microalgae is that microalgae are normally produced in non-axenic cultures and high numbers of bacteria are present in the microalgae cultures (Salvesen et al., 2000). Microalgae are usually added together with their growth medium, so bacteria from microalgae cultures are added in the rearing tanks as well and may have an effect on the bacterial communities in the rearing tanks. Bacterial strains isolated from microalgae have shown *in vitro* antibacterial activity against fish pathogens (Makridis et al., 2006). The antibacterial effect of microalgae can be involved in a strategy to reduce the bacterial load in live feed organisms, where rotifers or *Artemia* are incubated in microalgae cultures (Makridis et al., 2006). It has been shown that an excessive bacterial content in live feed may overload the bacterial load in the rearing system, and that larvae avoid prey with high bacterial content (Pérez-Benavente and Gatesoupe, 1988).

Prebiotics

Prebiotics are generally non-digestible diet compounds which cause a beneficial effect on the host by selecting growth or activating metabolism of one or a limited number of health-promoting bacteria (Gibson and Roberfroid, 1995). Research on application of prebiotics to aquaculture is still limited (Burr et al., 2005; Gatesoupe, 2005; Gatlin et al., 2007). The capacity of non-digestible carbohydrates to support the growth of probiotic bacteria has been proved *in vitro* (Rurangwa et al., 2009), and in *in vivo* trials with juvenile tilapia fed on dietary yeast culture or short chain fructo-oligosaccharides, which modified the intestinal microbial communities (Zhou et al., 2009). In fish larviculture, prebiotics have a potential use in first feeding and during the weaning of larvae. Mahious et al. (2006), in assays with inulin and oligosaccharides in the weaning of turbot (*Psetta maxima*) larvae, demonstrated a positive effect of fructo-oligosaccharides on growth and a predominance of *Bacillus* sp. in bacterial isolates.

Adding Beneficial Bacteria

Probiotics

Probiotics, defined by FAO/WHO (2001) as "live microorganisms which, when administered in adequate amounts, confer a health benefit on the host", constitute a potential tool in the reduction of mortalities in the rearing of aquatic organisms (Gatesoupe, 1999; Verschuere et al., 2000; Gram and Ringø, 2005; Vine et al., 2006; Kesarcodi-Watson et al., 2008). In rearing systems, where host and microorganisms share the same aquatic environment, the definition of probiotic was broadened by Verschuere et al. (2000) to also include the microorganisms which could exert a beneficial effect not colonizing the host (intestinal track, gills or skins) but being present in the water. This broadens the concept, as microorganisms acting on water quality or as biocontrol agents in water are also included.

The use of probiotics in aquaculture has been reviewed by Austin and Brunt (2009), in the Volume 1 of the present edition. The different modes of action of probiotics include: production of inhibitory compounds (antibiotic, bacteriocins, organic acids, oxygen peroxide), competence for limiting nutrients (siderophores), enzymatic contribution to digestibility of feed, competence in adhesion to mucus, stimulation of immune system or improvement of water quality. As mentioned by Austin and Brunt (2009), there is still a lack of knowledge on the precise mode of action in most probiotics, and in some cases the beneficial effect requires the combination of several mechanisms, e.g., attachment and growth in mucus can be a previous requisite to exert antagonistic activity or immunostimulation on the host.

Most of the published work on probiotics in larviculture is applied research, with few evidences of the precise mode of action (Vine et al., 2006). Moreover, many probiotic mechanisms have been studied *in vitro*, which does not guarantee an *in vivo* effect (Gram et al., 2001).

Hatching or early developmental stages, before first feeding, is an adequate period for the use of probiotics. At these stages, bacteria colonizing the eggs or larvae may prevent the further colonization of opportunistic bacteria originating from the diet (Hansen and Olafsen, 1999; Olafsen, 2001; Carnevali et al., 2004). Another advantage is that, in contrast to adults, there is no acidic stomach barrier in most marine fish larvae, and potential probiotics would not need to resist acidity (Tanaka et al., 1996; Ronnestad et al., 2000).

Probiotics in marine larviculture have been recently reviewed by Vine et al. (2006). The study of probiotics in fish larvae, summarized in Table 1.1, has usually focussed on the use of commercially available or selected terrestrial lactic acid bacteria (Gatesoupe, 1991, 1994; Planas et al., 2004; Villamil et al., 2003; Carnevalli et al., 2004; Plante et al., 2007; Arndt and Wagner, 2007; Benetti et al., 2008a, b; Suzer et al., 2008; Abelli et al., 2009). Commercial yeasts have also been proposed as probiotics (Tovar-Ramirez et al., 2002, 2004).

A better strategy, that avoids the introduction of non-native microorganisms to the system, is to select probiotic candidates among strains isolated from healthy fish (Ottesen and Olafsen 2000; Olsson et al., 1992; Westerdahl et al., 1991; Meuer et al., 2006) or hatchery facilities (Huys et al., 2001; Hjelm et al., 2004a, b). The efficacy of autochthonous probiotics is likely to be highest in the host species the bacteria were isolated from (Verschuere et al., 2000).

Generally, selection is based on the antagonistic effect of the potential probiotic against pathogenic bacteria responsible for high mortalities in larvae, such as *Vibrio anguillarum* (Olsson et al., 1992; Planas et al., 2006), *Vibrio* sp. (Gatesoupe, 1994; Vine et al., 2004a, b) or *Aeromonas* sp. (Ringø and Vadstein, 1998; Vine et al., 2004a, b), but *in vitro* growth and adhesion to mucus, immunostimulation or enhancement of enzymatic activity have been also used as criteria for probiotic selection (Gatesoupe 1994; Suzer et al., 2008; Picchetti et al., 2009; Vine et al., 2004a, b). It is important to note that probiotics can be added directly to water or introduced to the system, previously bioencapsulated in live prey (Table 1.2). In some cases, the final effect may depend on the selected way of delivery.

Synbiotics

A synbiotic is defined as a combination of a probiotic with a prebiotic, usually a selective substrate, which ensures the growth and the permanence of the associated probiotic (Ziemer and Gibson, 1998). In aquaculture, the utility of synbiotics has been proposed by Vine et al. (2006) but at the present moment only *in vitro* trials have been conducted to assay the potential of non-digestible carbohydrates and other prebiotics on the growth of probiotics (Rurangwa et al., 2009).

Table 1.1 Probiotics and prebiotics evaluated to be used in marine fish larval rearing.

Probiotic strain	Isolation/source	Targeted fish species	Mechanism	Application	Reference
Bacteria					
Gram-positive					
Arthrobacter sp.	Commercial probiotic strain Renogen®, Novartis)	Haddock (*Melanogrammus aeglefinus*)	Modification of bacterial microbiota	Water	Plante et al., 2007
Carnobacterium sp.	Live food	Turbot (*Psetta maxina*)	Enhancement of dietary value of live food	Rotifers	Gatesoupe, 1994
Bacillus subtillis, + *B. licheniformis* + *B. megaterium*	Commercial probiotic mixture (EcoPro™, Eco Microbials)	Cobia (*Rachycentron canadum*)	Unknown	Rotifers and *Artemia* nauplii	Benetti et al., 2008a, b
Strain J84	Stomach of juvenile Senegalese sole (*Solea senagalensis*)	Senegalese sole (*Solea senagalensis*)	Antagonism	Water	Makridis et al., 2008
Lactobacillus spp	Commercial probiotic mixture	Cutthroat trout (*Onchorrynchus clarkia pleuriticus*)	Unknown	*Artemia* nauplii	Arndt and Wagner, 2007
Lactobacillus spp.	Commercial probiotic mixture (Biotexin and Protexin, Novartis)	Gilthead sea bream (*Sparus aurata*)	Enhancement of enzymatic activity	Water and live food (rotifers and *Artemia* nauplii)	Suzer et al., 2008
Lactobacillus delbrueckii	Gut of adult European seabass (*Dicentrarchus labrax*)	European seabass (*Dicentrarchus labrax*) Gilthead sea bream (*Sparus aurata*)	Stimulation of immune system Stimulation of immune system	Rotifers and *Artemia* nauplii Rotifers and *Artemia* nauplii	Picchietti et al., 2009 Abelli et al., 2009
Lactobacillus fructivorans + *Lactobacillus plantarum*	Human faeces	European seabass (*Dicentrarchus labrax*) and Gilthead sea bream (*Sparus aurata*)	Stimulation of immune system	Rotifers and *Artemia* nauplii	Abelli et al., 2009

Table 1.1 contd...

Table 1.1 contd...

Probiotic strain	Isolation/source	Targeted fish species	Mechanism	Application	Reference
Lactobacillus fructivorans	Gut of gilthead sea bream (*Sparus aurata*)	Gilthead sea bream (*Sparus aurata*)	Antagonism	Rotifers, *Artemia* nauplii and inert food	Carnevali et al., 2004
Lactobacillus helveticus + *Streptococcus thermophilus*	Commercial probiotic mixture (Adjulact 1000, Lallemand)	Turbot (*Psetta maxima*)	Enhancement of dietary value of live food	Rotifers	Gatesoupe, 1991, 2008
Lactobacillus plantarum	Commercial probiotic Bel Industries)	Turbot (*Psetta maxima*)	Enhancement of dietary value of live food	Rotifers	Gatesoupe, 1991,2008
Lactobacillus plantarum	Rotifers	Turbot (*Psetta maxima*)	Enhancement of dietary value of live food	Rotifers	Gatesoupe, 1994, 2008
Lactobacillus plantarum	Atlantic cod (*Gadus morhua*)	Halibut (*Hippoglossus hipoglossus*) eggs	Enhancement of non -specific defence	Water	Ottesen and Olafsen, 2000
Lactobacillus plantarum	Human faeces	Gilthead sea bream (*Sparus aurata*)	Antagonism	Rotifers, *Artemia* nauplii and inert food	Carnevali et al., 2004
Paracoccus sp.	Rotifers	Gilthead sea bream (*Sparus aurata*)	Antagonism	Water	Makridis et al., 2005
Pediococcus acidolactici	Commercial probiotic mixture (Bactocell)	Pollock (*Pollachius pollachius*)	Antagonism	*Artemia* nauplii	Gatesoupe, 2002
Gram-negative					
Aeromonas sp.	*Artemia nauplii*	Gilthead sea bream (*Sparus aurata*)	Antagonism	Water	Makridis et al., 2005
Cytophaga sp.	Rotifers	Gilthead sea bream (*Sparus aurata*)	Antagonism	Water	Makridis et al., 2005
Strain PB52	Turbot (*Psetta maxima*) larvae	Turbot (*Psetta maxima*)	Colonization of larval gut	Water and rotifers	Makridis et al., 2000a

Phaeobacter 27-4	Turbot (*Psetta maxima*) rearing tank walls	Turbot (*Psetta maxima*)	Antagonism	Rotifers	Planas et al., 2006
Roseobacter spp.	Turbot (*Psetta maxima*) rearing tank walls	Turbot (*Psetta maxima*)	Antagonism	Water	Hjelm et al., 2004b
Roseobacter sp.	Rotifers	Gilthead sea bream (*Sparus aurata*)	Antagonism	Water	Makridis et al., 2005
Ruegeria sp.	Rotifers	Gilthead sea bream (*Sparus aurata*)	Antagonism	Water	Makridis et al., 2005
Shewanella sp.	Rotifers	Gilthead sea bream (*Sparus aurata*)	Antagonism	Water	Makridis et al., 2005
Shewanella sp. (2J27)	Hindgut of juvenile Senegalese sole (*Solea senegalensis*)	Senegalese sole (*Solea senegalensis*)	Antagonism	Water	Makridis et al., 2008
Vibrio sp.	Turbot (*Psetta maxima*) larvae	Turbot (*Psetta maxima*)	Antagonism	Rotifers	Gatesoupe, 1997
Vibrio sp. 2J18	Midgut of juvenile Senegalese sole (*Solea senegalensis*)	Senegalese sole (*Solea senegalensis*)	Antagonism	Water	Makridis et al., 2008
Vibrio mediterranei	Gilthead sea bream (*Sparus aurata*) larvae	Turbot (*Psetta maxima*)	Antagonism	Water	Huys et al., 2001
Vibrio pelagius	Turbot (*Psetta maxima*) larvae	Turbot (*Psetta maxima*)	Antagonism	Water	Ringo and Vadstein, 1998
Vibrio salmonicida	Gut of Atlantic salmon (*Salmo salar*)	Halibut (*Hippoglossus hipoglossus*) eggs	Enhancement of non-specific defense	Water	Ottesen and Olafsen, 2000
Strain 4:44	Turbot (*Psetta maxima*) larvae	Turbot (*Psetta maxima*)	Colonization of larval gut	Water and rotifers	Makridis et al., 2000a
Yeasts					
Sacharomyces cerevisiae	Goteborg University Collection	European seabass (*Dicentrarchus labrax*)	Adhesion to the gut	Inert food	Tovar-Ramirez et al., 2002

Table 1.1 contd...

Table 1.1 contd...

Probiotic strain	Isolation/source	Targeted fish species	Mechanism	Application	Reference
	Commercial product	Nile tilapia (*Oreochromis niloticus*)	Colonization of larval gut	Inert food	Meurer et al., 2006
Debaryomyces hansenii	Fish gut	European seabass (*Dicentrarchus labrax*)	Enhancement of enzymatic activity	Inert food	Tovar- Ramirez et al., 2002

Prebiotic

Identity	Isolation/source	Used with	Mechanism	Application	Reference
Inulin, Oligosacharides (oligofructose, lactosucrose)	Commercial chicory inulin (Raftiline ST) Commercial chicory inulin hydrolisate (Raftilose P95) Commercial lactosucrose LS-55P.	Weaning turbot (*Psetta maxima*) larvae	Enhancement of *Bacillus* sp growth	Inert food	Mahious et al., 2006

Table 1.2 The immunostimulants used in marine fish larval rearing

Targeted fish species	Immunostimulant	Results	Reference
Atlantic cod (*Gadus morhua*)	Cod milt proteins	Better survival post challenge	Pedersen et al., 2004
Atlantic cod (*Gadus morhua*)	Lipopolysaccharide	Better survival post challenge	Magnadottir et al., 2006
Atlantic cod (*Gadus morhua*)	β-1,3/β-1,6 glucans	Better larval survival and growth	Skjermo et al., 2006
Common dentex (*Dentex dentex*)	β-1,3/β-1,6 glucans	Better larval survival and growth	Efthimiou, 1996
Chilean flounder (*Paralichthys adspersus*)	β-glucans and mannan -oligosaccharides	Better larval survival and growth	Piaget et al., 2007
Sea bream (*Sparus aurata*)	Lipopolysaccharide	Better survival post challenge	Hanif et al., 2005
Turbot (*Psetta maxima*)	Alginate	Better larval survival and growth	Conceicao et al., 2004

Biofilters

In aquaculture, the concept of probiotics has been extended to bacteria which can exert their beneficial effect on the host by modification of the microbial community associated with fish or water, improving the quality of the environment (Verschuere et al., 2000). Thus, strategies of water treatment with addition of beneficial bacteria would also be included.

Recirculating aquaculture systems (RAS) are the latest form of the fish farming production system, more adapted to fulfil environmental constraints (e.g., generating low emissions or making coastal location unnecessary). RAS are typically an indoor system that allows farmers to control environmental conditions all-year round, considered better than conventional flow-through systems from an environmental (Neori et al., 2007) and bio-security (Pruder, 2004) point of view. In the last few years, RAS systems have being investigated for cultivation of different species of fish larvae (Faulk and Holt, 2005; Martins et al., 2009). Water quality of aquaculture facilities equipped with RAS is maintained by a diversity of microbial community associated with biofilters (Gross et al., 2003; Sugita et al., 2005; Michaud et al., 2006). Recirculation of water tends to stabilize the microbial community and make it more persistent to perturbations (Attramadal et al., 2005). Biofilters colonized with non-opportunistic bacteria have been previously used for production of matured water, which increased survival of Atlantic halibut yolk sac larvae (Skjermo et al., 1997) and improved the growth of turbot larvae (Salvesen et al., 1999) by improving microbial water quality.

Some probiotic bacteria with antagonistic activity towards fish pathogens also have the ability to produce biofilms. Alpha proteobacteria belonging to the *Roseobacter clade* have shown to produce tropodithietic acid (Brinkhoff et al., 2004; Bruhn et al., 2005; Porsby et al., 2008), which inhibits different bacterial groups (Brinkhoff et al., 2004), including some fish pathogens (Ruíz-Ponte et al., 1999; Bruhn et al., 2005; Porsby et al., 2008). Production of this antibiotic is related to the attachment of the probiotic to surfaces and to biofilm formation (Bruhn et al., 2007; Porsby et al., 2008). Therefore, biofilters could be inoculated and colonized by biofilm-forming probiotic bacteria, improving the permanence of probiotics in the rearing system and protecting the larvae against opportunistic and potential pathogenic bacterial strains.

Inhibiting or Inactivating Detrimental Bacteria in the System

Bacteriophages

Bacteriophages are viruses that infect bacteria. It has been suggested that in nature bacteriophages are one of the main causes of mortalities in bacterial populations. The principle in using bacteriophages in aquaculture is their application for the control of bacterial diseases (Imbeault et al., 2006; Park et al., 2000). The advantage of such a strategy is avoiding the of use of antibiotics and other therapeutic agents. The drawback of this approach is that the ability of bacteriophages to reduce the numbers of bacteria is limited by its high specificity. Production of virus particles specific for the pathogenic bacterial strains in question at each time is therefore required. In addition, resistance to bacteriophages is quickly developed after a relatively short period of time due to short generation time of bacteria.

The following are the advantages of bacteriophage therapy against pathogenic bacteria:

  i) The high specificity results in protection of intestinal microbiota and
  ii) The ability of propagation.

Once susceptible bacteria are present in the fish, then few additions of bacteriophage may be sufficient for protection against pathogens (Nakai and Park, 2002).

In larviculture, there is a limited application of bacteriophages as mortalities are seldom attributed to specific pathogens. However, in a particular hatchery it can occur that problems with mortalities in larval rearing tanks are related to a specific pathogen. In such cases, a tailor-made bacteriophage therapy could be an alternative solution. Bacteriophages have shown encouraging results in the case of protection of shrimp larvae against *Vibrio harveyi* (Karunasagar et al., 2007).

Interference with Quorum Sensing

Quorum sensing is a mechanism by which bacteria coordinate the expression of certain genes in response to their population density by producing, releasing and detecting small signal molecules, inducing a physiological response which implies an ecological advantage (Defoirdt et al., 2004). Gram negative bacteria use acylated homoserine lactones (AHLs) and/or autoinducer 2 (AI-2) as signal molecules (Fuqua et al., 2001; Miller and Bassler, 2001; Whitehead et al., 2001), whereas gram positive bacteria use secreted peptides (Dunny and Leonard, 1997; Miller and Bassler, 2001).

Different bacterial phenotypes are controlled by a quorum sensing system, the most important being the expression of virulence factors and biofilm formation (De Kievit and Iglewski, 2000; Miller and Bassler, 2001; Whitehead et al., 2001). An AI-2 mediated system, is responsible for the virulence of the fish pathogen *Vibrio harveyi* (Gómez-Gil et al., 2004) toward gnotobiotic *Artemia franciscana* (Defoirdt et al., 2005), being also involved in the growth-retarding effect of this bacterium toward gnotobiotic *Brachionus plicatilis* (Tihn et al., 2007). Although these two models of gnotobiotic *Artemia* and gnotobiotic rotifers are difficult to compare directly, it seems that *V. harveyi* quorum sensing systems might operate in a host-dependent way (Deforidt et al., 2005; Tinh et al., 2007).

Disruption of quorum sensing was suggested as a new anti-infective strategy, as a consequence of its importance in virulence development in pathogenic bacteria (Finch et al., 1998). Several techniques have been developed for this purpose, such as inhibition of signal molecule biosynthesis, application of quorum sensing antagonists or agonistic analogues, chemical or enzymatic inactivation and biodegradation of quorum sensing molecules.

Disruption of quorum sensing might constitute an alternative approach in aquaculture for preventing or combating infections caused by pathogens that regulate virulence factor expression by a quorum sensing system (Defoirdt et al., 2004). Halogenated furanones produced by the marine red algae *Delisea pulchra* (Manefield et al., 1999) protected rotifers, *Artemia* and rainbow trout from the negative effect of vibrios when added in adequate concentrations (Rasch et al., 2004; Defoirdt et al., 2006; Tihn et al., 2007). Furanones show structural similarity with AHL molecules and thus the mechanism of action would be based on binding to LuxR proteins (Manefield et al., 1999) present in pathogens as *V. harveyi* (Henke and Bassler, 2004).

The introduction of potential probiotic bacteria able to degrade quorum sensing molecules might be useful as a biocontrol agent in aquaculture, but the exploitation of these kind of bacteria remains a challenge for future research (Tihn et al., 2008). It has been suggested that the positive effect of *Bacillus* spp. strains used as probiotics in aquaculture (Moriarty, 1998; Rengpipat et al., 2003) might partly be due to inactivation of quorum sensing molecules, apart from the production of antagonistic substances (Dong et al., 2000; Molina et al., 2003; Dong et al., 2004). A direct immunomodulatory activity was produced in several mammalian systems by a quorum sensing molecule from *Pseudomonas aeruginosa* (Hooi et al., 2004; Ritchie et al., 2005; Thomas et al., 2006).

The importance of quorum sensing *in situ* gene expression regulation is becoming more evident for pathogenic bacteria, but hardly for putative probiotic bacteria. Some problems could arise, such as the development of resistances (Zhu et al., 1998), or lack of specificity, as not all bacteria that are found to contain quorum sensing systems are pathogens. Therefore, we need to know more about the importance of quorum sensing molecules on their *in vivo* metabolism and their interaction with the non-pathogenic microbial community and with the host.

Acting on Host-Microbe Interactions

Immunostimulants

Fish are intimately in contact with a complex and dynamic microbial world, and a large fraction of these microbes cling to and colonize epithelial surfaces. In rare circumstances microbes cause disease, either directly, by damaging or traversing epithelial layers, or indirectly, by inducing tissue-damaging inflammatory responses. If microbial pathogens invade the host, innate and adaptive defence mechanisms are activated for preventing further spread of the infection (Gómez and Balcázar, 2008).

The fish immune system possesses two integral components: (i) the innate, natural or non-specific defence formed by a series of cellular and humoral components, and (ii) the adaptive, acquired or specific immune system characterized by the humoral immune response through the production of antibodies and by the cellular immune response mediated by T-lymphocytes.

However, it is believed that marine fish larvae do not have the ability to develop adaptive immunity during the early stages of development. In this regard, the innate immune system is probably the major defence against microbes (Vadstein, 1997). These parameters include phagocytes, various lectins, lytic enzymes, antibacterial peptides and proteinase inhibitors (Magnadottir et al., 2005). A variety of substances could, at least theoretically, activate these innate parameters. These substances or immunostimulants have recently been defined as "a naturally occurring compound that modulates the immune system by increasing the host's resistance against diseases that in most circumstances are caused by pathogens" (Bricknell and Dalmo, 2005). In general, these are compounds of bacterial, fungal or plant origin, especially composed of polysaccharides, which activate the pattern recognition receptors/proteins of the immune system resulting in a varied immune response.

Previous studies have demonstrated that administration of immunostimulants to fish is beneficial, providing improved protection against bacterial and, to a lesser extent, viral infections (Vadstein, 1997; Bricknell and Dalmo, 2005). Thus, the immunomodulation of larval fish has been proposed as a potential method for improving larval survival by increasing the innate responses until its adaptive immune response is sufficiently developed to generate an effective response to the pathogen.

Lipopolysaccharides (LPS) are one of the immunostimulants widely used in experimental fish larval aquaculture, as they can trigger various immune components. Sea bream (*Sparus aurata*) larvae when treated with LPS exhibited significantly high anti-protease activity, lysozyme activity and total globulin level, and also conferred protection against *Photobacterium damsela* (Hanif et al., 2005). Similarly, Magnadottir et al. (2006) reported enhanced survival by immunizing Atlantic cod (*Gadus morhua*) with LPS.

The range of potential immunostimulants for larval fish is quite limited and a short review can be found in Table 1.2. Although these studies have demonstrated some beneficial effects of immunostimulants in terms of survival and growth, the relative importance of the immune mechanisms involved remains unknown. It is also important to consider the specificity of immunostimulants, since the stimulation may be too intense, which may become detrimental or even lethal to the host. Moreover, the knowledge of the functions of different immunostimulators may be used to stimulate, more specifically, those parts of the immune system that may be more relevant in certain situations (Vadstein, 1997). The recent introduction of genomic and proteomic tools might be of special interest in this context.

Strategies Based on the Food Chain

Large scale use of probiotics in the rearing of marine fish larvae involves cultivation of bacteria in artificial medium, preparation of a stable form of these bacteria and development of a process for the transfer of probiotics to the larval gut either by direct addition to water in the rearing tanks or by bioencapsulation in live feed. The approach of addition of bacteria to the water is simpler, but implies that the added bacteria remain in the water for some time, accumulate in rotifers or *Artemia* and thereby get introduced in the larval gut. A far more efficient approach is the bioencapsulation of bacteria in live feed. This is the only possible route in the case of allochthonous bacteria of tellurian origin tested in larviculture, such as lactic acid bacteria and bacillus strains (Gatesoupe, 1994). These bacteria are not normally encountered in marine systems, but when introduced in the larvae may have positive effects by excreting antimicrobial compounds or by other ways of limiting the growth of harmful bacteria.

Bioencapsulation of bacteria is a relatively predictable process in the case of rotifers, but more variable in the case of *Artemia metanauplii*, as the size of bacteria may be decisive for the efficiency of the process (Makridis et al., 2000b). *Artemia metanauplii* have a filtering device which enables them to filter large bacterial cells or aggregates of bacteria, although the efficiency of this process is not so high. Bioencapsulation has a dual purpose as on one hand it provides probiotic bacteria to the larvae and on the other hand replaces the microbiota originally found in the live feed, acting as a control agent of opportunistic or pathogenic bacteria in live prey.

New Molecular Tools

The elucidation of the underlying mechanisms of action and the development of new strategies for microbial control requires the identification and characterization of the bacterial communities associated with the different parts of the rearing system. If probiotic or pathogenic bacterial strains are introduced during *in vivo* challenges, the specific detection and quantification of introduced strains is also required.

Less than 1% of total bacteria in natural seawater systems are culturable (Hansen and Olafsen, 1999; Kjelleberg et al., 1993) and thus the use of culture dependent methods leads to underestimate microbial diversity. Moreover, culture methods are time consuming. PCR (Polymerase Chain Reaction)-based techniques, such as Denaturing Gradient Gel Electrophoresis (DGGE) and real-time PCR, as well as non-PCR based techniques, as Fluorescence In Situ Hybridization (FISH), have gained importance in microbial ecology studies (Amann et al., 1990; Skovus et al., 2007, Nguyen et al., 2007) and have been applied to different fish and shellfish species (Griffiths et al., 2001; Sandaa et al., 2003; Jensen et al., 2004; Goarant and Merien, 2006; Labreuche et al., 2006; Schulze et al., 2006; Prol et al., 2009). These methods typically rely on detection of microbial DNA to infer information on the microbial species present in a sample and provide a unique perspective of the components of the microbiota; however, as with culture based methods, they also have limitations. These limitations range from technical problems, such as obtaining representative genomic DNA and suitable primers, to conceptual problems such as defining and using meaningful taxonomic units of diversity (species) (Forney et al., 2004).

DGGE is a reliable and rapid method to study variations of dominant bacteria and to characterize complex microbial populations (Muyzer et al., 1993). The general principle of DGGE is the separation of fragments of the individual rRNA genes based on differences in chemical stability or melting temperature of the target genes. Seasonal variations of bacterial communities associated with live preys (Rombaut et al., 2001; McIntosh et al., 2008) and different fish (Griffiths et al., 2001; Jensen et al., 2004) and shellfish (Bourne et al., 2004; Payne et al., 2006; Sandaa et al., 2003) species have been successfully characterized by using DGGE. Recently, this fingerprint technique has also been used for detection and characterization of *Vibrio* strains in cod larvae (Reid et al., 2009) and applied to the study of inmunostimulants in aquaculture (Liu et al., 2008).

Real-time PCR is a specific and sensitive method for quantification of bacteria (Klein, 2002). This technique is based on the use of specific primers and fluorescent molecules, which permit measuring the quantity of a specific amplicon at each PCR cycle. Two main chemistries have been developed: SYBR Green, which is based on the use of a double stranded DNA dye and TAQMAN, based on specific oligonucleotide probes which have a fluorescent dye and a quencher emitting fluorescence only when new copies of the target gene are created. In aquaculture, different real-time PCR protocols have been designed, focusing on detection and quantification of pathogenic vibrios in crustaceans (Goarant and Merien, 2006), molluscs (Labreuche et al., 2006) or fish (Prol et al., 2009). Recently, real-time PCR has demonstrated to be a reliable technique for specific detection and quantification of introduced pathogenic and probiotic strains in first feeding turbot larvae during a challenge trial (Prol et al., 2009).

FISH techniques, using rRNA target probes, have been developed for the *in situ* identification of single microbial cells (Amann et al., 1990). This method is based on the hybridization of synthetic oligonucleotide probes to specific regions within the bacterial ribosome and does not require cultivation. Due to its speed and sensitivity, this technique is considered a powerful tool for phylogenetic, ecological, diagnostic and environmental studies in microbiology (Bottari et al., 2006). In aquaculture, this technique has been used to determine bacterial composition in a Pacific white shrimp (*Litopenaeus vannamei*) hatchery (García and Olmos, 2007) and in red sea bream (*Pagrus major*) and marble goby (*Oxyeleotris marmoratus*) larvae (Nakase et al., 2007). Recently, a probiotic photosynthetic bacteria mixture used in aquaculture, was characterized by FISH, applying a group of species-specific probes (Qi et al., 2009).

In studies of host-microorganisms interactions, gene profiling can be carried out by microarray technology, which allows quantification of the transcriptional status of thousands of genes simultaneously (Douglas, 2006) and can help to elucidate global patterns of gene expression (Arcand et al., 2004). For the construction of a cDNA microarray, cDNA fragments or EST (Expressed Sequenced Tags) sequences for the fish species of interest must be prepared from cDNA library or PCR-amplified products. Transcriptome sequences of many fish species are now available in the National Centre for Biotechnology Information (NCBI) GenBank database, and these collected EST sequences can be used to synthesize oligonucleotides and can be spotted onto array chips to produce cDNA microarrays. Microarray analyses have been used to search for immune-related genes that are expressed following DNA vaccination of infection by fish pathogens (Byon et al., 2005; Kurobe et al., 2005) and to study environmental influences on gene expression of fish larvae (Williams et al., 2003), fry (Koskinen et al., 2004) and adults (Lam et al., 2006). Hossain et al. (2006) found, by using microarray technology, that the differences between virulence determinant functions of different pathogens can be reflected by specific host gene expression patterns, suggesting their adaptive survival strategies inside the host.

CONCLUSIONS AND PERSPECTIVES

The intensive rearing of marine fish larvae is highly susceptible to bacterial infections. Environmental and health concerns have resulted in a progressive reduction of the use of chemotherapeutics and to the search for alternatives, which should be based on considering bacteria as an active part of the rearing ecosystems. The selection of the appropriate strategy for microbial control would rely on a deep understanding of the microbial ecology in fish larvae, live food and rearing systems, of the origin of bacterial infections (specific pathogens or unspecific opportunistic bacteria) and of the host-bacteria interactions.

Tailored solutions should be developed for each case. In this sense, matured water, green water, biofilters or immunostimulants are more unspecific (general) tools, than prebiotics, probiotics or symbiotics. All of them are directed to prevent or to enhance particular bacterial groups and would be suitable when larval mortality is attributed to opportunistic bacteria. The interference with quorum sensing would be applied to prevent specifically bacterial infections due to pathogens which use quorum sensing regulated virulence expression. The use of bacteriophages, highly specific, would be a strategy to be used only when the aetiology of infections has been proved to correspond to a specific pathogen.

How the host reacts to the presence of a particular microbiota should be a subject of further research. Long-term effects also need to be studied. The development of new molecular tools will certainly improve our knowledge on microbial ecology and host-microbe interactions, and on the underlying *in vivo* mechanisms of action of the different strategies available but not sufficiently understood in many cases.

ABBREVIATIONS

AI 2	Autoinducer 2
AHLs	Acylated homoserine lactones
CFU	Colony forming units
DGGE	Denaturing gradient gel electrophoresis
EST	Expressed sequenced tags
FISH	Fluorescence in situ hybridization
LPS	Lipopolysaccharides
NCBI	National centre for biotechnology information
PCR	Polymerase chain reaction
RAS	Recirculating aquaculture systems

REFERENCES

Abelli, L., Randelli, E., Carnevali, O. and Picchietti, S. (2009). Stimulation of gut immune system by early administration of probiotic strains in *Dicentrarchus labraz* and *Sparus aurata*. Trends Comp. Endocrinol. Neurobiol. 1163: 340-342.

Amann, R.I., Binder, B.J., Olson, R.J., Chisholm, S.W., Deveroux, R. and Stahl, D.A. (1990). Combination of 16s rRNA-targeted oligonucleotide probes with flow cytometry for analyzing mixed microbial populations. Appl. Environ. Microbiol. 56: 1919-1925.

Andrews, J.H. and Harris, R.F. (1986). r- and K-selection in microbial ecology. Adv. Microb. Ecol. 9: 99-147.

Arcand, S.L., Mes-Masson, A.M., Provencher, D., Hudson, T.J. and Tonin, P.N. (2004). Gene expression microarray analysis and genome databases facilitate the characterization of a chromosome 22 derived homogeneously staining region. Mol. Cardinog. 41: 17-38.

Arndt, R.E. and Wagner, E. (2007). Enriched *Artemia* and probiotic diets improve survival of Colorado River cutthroat trout larvae and fry. North Am. J. Aquacult. 69: 190-196.

Attramadal, K., Xue, R., Salvesen, I., Olsen, Y. and Vadstein, O. (2005). Microbial environment in a flow-through and a recirculating system for intensive rearing of co larvae (*Gadus morhua* L.). Eur. Aquacult. Soc. Sp. Publ. 36: 11-14.

Austin, B. and Brunt, J.W. (2009). The use of probiotics in aquaculture. In: Aquaculture Microbiology and Biotechnology, Volume 1. D. Montet and R.C. Ray (eds). Science Publishers, Enfield, USA pp. 185-207

Benetti, D.D., Orhun, M.R., Sardenberg, B., O'Hanlon, B., Welch, A., Hoenig, R., Zink, I., Rivera, J.A., Denlinger, B., Bacoat, D., Palmer, K. and Cavalin, F. (2008a). Advances in hatchery and grow-out technology of cobia *Rachycentron canadum* (Linnaeus). Aquacult. Res. 39: 701–711

Benetti, D.D., Sardenberg, B., Welch, A., Hoenig, R., Orhun, M.R. and Zink, I. (2008b). Intensive larval husbandry and fingerling production of cobia *Rachycentron canadum*. Aquaculture 281: 22-27.

Bottari, B., Ercolini, D., Gatti, M. and Neviani, E. (2006). Application of FISH technology for microbiological analysis: current state and prospects. Appl. Microbiol. Biotechnol. 73: 485-494.

Bourne, D.G., Young, N., Webster, N., Payne, M., Salmon, M., Demel, S. and Hall, M. (2004). Microbial community dynamics in a larval aquaculture system of the tropical rock lobster, *Panulirus ornatus*. Aquaculture 242: 31-51.

Bricknell, I. and Dalmo, R.A. (2005). The use of immunostimulants in fish larval aquaculture. Fish Shellfish Immunol. 19: 457-472.

Brinkhoff, T., Bach, G., Heidorn, T., Liang, L., Schlingloff, A. and Simon, M. (2004). Antibiotic production by a *Roseobacter clade*-affiliated species from the German Wadden Sea and its antagonistic effects on indigenous isolates. Appl. Environ. Microbiol. 70: 2560-2565.

Bruhn, J.B., Nielsen, K.F., Hjelm, M., Hansen, M., Bresciani, J., Schulz, S. and Gram, L. (2005). Ecology, inhibitory activity and morphogenesis of a marine antagonistic bacterium belonging to the *Roseobacter clade*. Appl. Environ. Microbiol. 71: 7263-7270.

Bruhn, J.B., Gram, L. and Belas, R. (2007). Production of antibacterial compounds and biofilm formation by *Roseobacter* species are influenced by culture conditions. Appl. Environ. Microbiol. 73: 442-450.

Burr, G., Gatlin, D. and Ricke, S. (2005). Microbial ecology of the gastrointestinal tract of fish and the potential application of prebiotics and probiotics in finfish aquaculture. J. World Aquacult. Soc. 36: 425-436.

Byon, J.Y, Ohira, T., Hirono, I. and Aoki, T. (2005). Use of a cDNA microarray to study immunity against viral hemorrhagic septicemia (VHS) in Japanese flounder (*Paralichthys olivaceus*) following DNA vaccination. Fish. Shellfish. Immunol. 18: 135-147.

Cabello, F.C. (2006). Heavy use of prophylactic antibiotics in aquaculture: a growing problem for human and animal health and for the environment. Environ. Microbiol. 8: 1137-1144.

Cahu, C.L., Zambonino Infante, J.L., Peres, A., Quazuguel, P. and Le Gall, M.M. (1998). Algal addition in sea bass (*Dicentrarchus labrax*) larvae rearing: Effect on digestive enzymes. Aquaculture 161: 479-489.

Carnevali, O., Zamponi, M.C., Sulpizio, R., Rollo, A., Nardi, M., Orpianesi, C., Silvi, S., Caggiano, M., Polzonetti, A.M. and Cresci, A. (2004). Administration of probiotic strain to improve sea bream wellness during development. Aquacult. Int. 12: 377-386.

Conceicao, L.E.C., Skjermo, J., Skjak-Bræk, G. and Verreth, J.A.J. (2004). Effect of an immunostimulating alginate on protein turnover of turbot (*Scophthalmus maxumus* L.) larvae. Fish Physiol. Biochem. 24: 207-212.

De Kievit, T.R. and Iglewski, B.H. (2000). Bacterial quorum sensing in pathogenic relationships. Infect. Immun. 68: 4839-4849.

Defoirdt, T., Boon, N, Bossier, P. and Verstraete, W. (2004). Disruption of bacterial quorum sensing: an unexplored strategy to fight infections in aquaculture. Aquaculture 240: 69-88.

Defoirdt, T., Bossier, P., Sorgeloos, P. and Verstraete, W. (2005). The impact of mutations in the quorum sensing systems of *Aeromonas hydrophila, Vibrio anguillarum* and *Vibrio harveyi* on their virulence towards gnotobiotically cultured *Artemia franciscana*. Environ. Microbiol. 7: 1239-1247.

Defoirdt, T., Crab, R., Wood, T.K., Sorgeloos, P., Verstraete, W. and Bossier, P. (2006). Quorum sensing –disrupting brominated furanones protect the gnotobiotic brine shrimp *Artemia franciscana* from pathogenic *Vibrio harveyi, Vibrio campbellii,* and *Vibrio parahaemolyticus* isolates. Appl. Environ. Microbiol. 72: 6419-6423.

Dhert, P., Rombaut, G., Suantika, G. and Sorgeloos, P. (2001). Advancement of rotifer culture and manipulation techniques in Europe. Aquaculture 200: 129-146.

Dong, Y.H., Xu, J.L., Li, X.Z. and Zhang, L.H. (2000). AiiA, an enzyme that inactivates the acylhomoserine lactone quorum-sensing signal and attenuates the virulence of *Erwinia carotovora*. Proc. Natl. Acad. Sci. USA 97: 3526-3531.

Dong, Y.H., Zahng, X. F., Xu, J.L. and Zhang, L.H. (2004). Insecticidal *Bacillus thuringiensis* silences *Erwinia carotovora* virulence by a new form of microbial antagonism, signal interference. Appl. Environ. Microbiol. 70: 954-960.

Douglas, S.E. (2006). Microarray studies of gene expression in fish. Omics 10: 474-489.

Douillet, P.A. (1998). Disinfection of rotifer cysts leading to bacteria free populations. J. Exp. Mar. Biol. Ecol. 224: 183-192.

Dunny, G.M.and Leonard, B.A.B. (1997). Cell-Cell communication in Gram-positive bacteria. Annu. Rev. Microbiol. 51: 527-564.

Efthimiou, S. (1996). Dietary intake of β-1,3/1,6 glucans in juvenile dentex (*Dentex dentex*), Sparidae: effects on growth performance, mortalities and non-specific defense mechanisms. J. Appl. Ichthyol. 12: 1-7.

FAO/WHO (2001). Evaluation of health and nutritional properties of powder milk and live lactic acid bacteria. Food and Agriculture Organization of the United Nations and World Health Organization Expert Consultation Report. FAO, Rome, Italy.

FAO (2007). The State Of World Fisheries and Aquaculture 2006. FAO, Rome, Italy.

Faulk, C.K. and Holt, G.J. (2005). Advances in rearing cobia *Rachycentron canadum* larvae in recirculating aquaculture systems: Live prey enrichment and greenwater culture. Aquaculture 249: 231-243.

Finch, R.G., Prithcard, D.I., Bycroft, B.W., Williams, P. and Stewart, G.S.A.B. (1998). Quorum sensing: a novel target for anti-infective therapy. J. Antimicrob. Chemother. 42: 569-571.

Fjellheim, A.J., Playfoot, K.J., Skjermo, J. and Vadstein, O. (2007). *Vibrionaceae* dominates the microflora antagonistic towards *Listonella anguillarum* in the intestine of cultured Atlantic cod (*Gadus morhua* L.) larvae. Aquaculture 269: 98-106.

Forney, L.J., Zhou, X. and Brown, C.J. (2004). Molecular microbial ecology: Land of the one-eyed king. Curr.Opin. Microbiol. 7: 210-220.

Fuqua, C., Parsek, M.R. and Greenberg, E.P. (2001). Regulation of gene expression by cell-to-cell communication: acyl-homoserine lactone quorum sensing. Annu. Rev. Genet. 35: 439-468.

García, A.T. and Olmos, J.S. (2007). Quantification by fluorescent in situ hybridization of bacteria associated with *Litopenaeus vannamei* larvae in Mexican shrimp hatchery. Aquaculture 262: 211-218.

Gatesoupe, F.J. (1982). Nutritional and antibacterial treatments of live food organisms: the influence on survival, growth rate and weaning success of turbot (*Scopththalmus maximus*). Ann. Zootech. 31: 353-368.

Gatesoupe, F.J. (1989). Further advances in the nutritional and antibacterial treatments of rotifers as food for turbot larvae, *Scopththalmus maximus* L. In: Aquaculture—A Biotechnology in Progress. N. De Pauw, N.E. Jaspers, H. Achefors, and N. Wilkins (eds). European Aquaculture Society, Bredene, Belgium pp. 721-730

Gatesoupe, F.J. (1991). The effect of three strains of lactic bacteria on the production rate of rotifers, *Brachionus plicatilis,* and their dietary value for larval turbot, *Scophthalmus maximus*. Aquaculture 96: 335-342.

Gatesoupe, F.J. (1994). Lactic acid bacteria increase the resistance of turbot larvae, *Scophthalmus maximus*, against pathogenic *Vibrio*. Aquat. Living Resour. 7: 277-282.

Gatesoupe, F.J. (1997). Siderophore production and probiotic effect of *Vibrio* sp associated with turbot larvae, *Scophthalmus maximus*. Aquatic Living Resour. 10: 239-246.

Gatesoupe, F.J. (1999). The use of probiotics in aquaculture. Aquaculture 180: 147-165.

Gatesoupe, F.J. (2002). Probiotic and formaldehyde treatments of *Artemia nauplii* as food for larval Pollack, *Pollachius pollachius*. Aquaculture 212: 347-360.

Gatesoupe, F.J. (2005). Probiotics and prebiotics for fish culture, at the parting of the ways. Aqua Feeds: Formulation & Beyond 2: 3-5.

Gatesoupe, F.J. (2008). Updating the importance of lactic acid bacteria in fish farming: natural occurrence and probiotic treatments. J. Mol. Microbiol. Biotech. 14: 107-114.

Gatlin, D.M., Barrows, F.T., Brown, P., Dabrowski, K., Gaylord, T.G., Hardy, R.W., Herman, E., Hu, G., Krogdahl, Å., Nelson, R., Overturf, K., Rust, M., Sealey, W., Skonberg, D., Souza, E.J., Stone, D., Wilson, R. and Wurtele, E. (2007). Expanding the utilization of sustainable plant products in aquafeeds: a review. Aquacult. Res. 38: 551-579.

Gibson, G.R. and Roberfroid, M.B. (1995). Dietary modulation of the human colonic microbiotica: introducing the concept of prebiotics. J. Nutr. 125: 1401-1412.

Giménez, G., Padrós, F., Roque, A., Estévez, A. and Furones, D. (2006). Bacterial load reduction of live prey for fish larval feeding using Ox-Aquaculturer. Aquacult. Res. 37: 1130-1139.

Goarant, C. and Merien, F. (2006). Quantification of *Vibrio penaeicida*, the etiological agent of syndrome 93 in new Caledonian shrimp, by real-time PCR using SYBR Green I chemistry. J. Microbiol. Methods 67: 27-35.

Gómez, G.D. and Balcázar, J.L. (2008). A review on the interactions between gut microbiota and innate immunity of fish. FEMS Immunol. Med. Microbiol. 52: 145-154.

Gómez-Gil, B., Soto-Rodríguez, S., García-Gasca, A., Roque, A., Vázquez-Juárez, R. and Thompson, F.L. (2004). Molecular identification of *Vibrio harveyi*-related isolates associated with diseased aquatic organisms. Microbiol. SGM 150: 1769-1777.

Gram, L., Lovold, T., Nielsen, J., Melchiorsen, J. and Spanggaard, B. (2001). *In vitro* antagonism of the probiont *Pseudomonas fluorescens* strain AH2 against *Aeromonas salmonicida* does not confer protection of salmon against furunculosis. Aquaculture 199: 1-11.

Gram, L. and Ringø, E. (2005). Prospects of fish probiotics. In: Microbial Ecology in Growing Animals, W.Holzapfel and P. Naughton (eds). Elsevier, Edinburgh, UK 55: 379-417.

Griffiths, S., Melville, K., Cook, M. and Vincent, S. (2001). Profiling of bacterial species associated with haddock larviculture by PCR amplification of 16S rDNA and denaturing gradient gel electrophoresis. J. Aquat. Anim. Health. 13: 355-363.

Gross, A., Nemirovsky, A., Zilberg, D., Khaimov, A., Brenner, A., Snir, E., Ronen, Z. and Nejidat, A. (2003). Soil nitrifying enrichments as biofilter starters in intensive recirculating saline water aquaculture. Aquaculture 223: 51-62.

Hanif, A., Bakopoulos, V., Leonardos, I. and Dimitriadis, G.J. (2005). The effect of sea bream (*Sparus aurata*) broodstock and larval vaccination on the susceptibility by *Photobacterium damsela* subsp. *piscicida* and on the humoral immune parameters. Fish Shellfish Immunol. 19: 345-361.

Hansen, G.H. and Olafsen, J.A. (1999). Bacterial interactions in early life stages of marine cold water fish. Microb. Ecol. 38: 1-26.

Henke, J.M. and Bassler, B.L. (2004). Three parallel quorum sensing systems regulated gene expression in *Vibrio harveyi*. J. Bacteriol. 186: 6902-6914.

Hjelm, M., Bergh, Ø., Riaza, A., Nielsen, J., Melchiorsen, J., Jensen, S., Duncan, H., Ahrens, P., Birkbeck, H. and Gram, L. (2004a). Selection and identification of autochthonous potential probiotic bacteria from turbot larvae (*Scolphthalmus maximus*) rearing units. Syst. Appl. Microbiol. 27: 360-371.

Hjelm, M., Riaza, A., Formoso, F., Melchiorsen, J. and Gram, L. (2004b). Seasonal incidence of autochtonous antagonistic bacteria, *Roseobacter* spp. and *Vibrionaceae*, in a turbot larvae (*Scophthalmus maximus*) rearing system. Appl. Environ. Microbiol. 70: 7288-7298.

Hjelmeland, K., Pedersen, B.H. and Nilssen, E.M. (1988). Trypsin content in intestines of herring larvae, *Clupea harengus*, ingesting inert polysterene spheres of live crustasea prey. Mar. Biol. 98: 331-335.

Hooi, D.S.W., Bycroft, B.W., Chhabra, S.R., Williams, P. and Pritchard, D.I. (2004). Differential immune modulatory activity of *Pseudomonas aeruginosa* quorum-sensing signal molecules. Infect. Immun. 72: 6463-6470.

Hossain, H., Tchatalbachev, S. and Chakraborty, T. (2006). Host gene expression profiling in pathogen-host interactions. Curr. Opin. Immunol. 18: 422-429.

Howell, B.R. (1979). Experiments on the rearing of larval turbot, *Scophthalmus maximus* L. Aquaculture 18: 215-225.

Huys, L., Dhert, P., Robles, R., Ollevier, F., Sorgeloos, P. and Swings, J. (2001). Search for beneficial bacteria strains for turbot (*Scophthalmus maximus* L.) larviculture. Aquaculture 193: 25-37.

Imbeault, S., Parent, J., Blais, J.F., Lagace, M. and Uhland, C. (2006). Use of bacteriophages to control antibiotic-resistant *Aeromonas salmonicida* polulations. Rev. Sci. Eau/J. Water Sci. 19: 275-282.

Jensen, S., Øvreas, L., Bergh, Ø. and Torsvik, V. (2004). Phylogenetic analysis of the bacterial communities associated with larvae of Atlantic halibut propose succession from a uniform normal flora. Syst. Appl. Microbiol. 27: 728-736.

Karunasagar, I., Shivu, M.M., Girisha, S.K. and Krohne, G. (2007). Biocontrol of pathogens in shrimp hatcheries using bacteriophages. Aquaculture 268: 288-292.

Kesarcodi-Watson, A., Kaspar, H., Lategan, M.J. and Gibson, L. (2008). Probiotics in aquaculture: The need, principles and mechanisms of action and screening processes. Aquaculture 274: 1-14.

Kjelleberg, S., Albertson, N., Flärdh, K., Holmquist, L., Jouper-Jaan, A., Marouga, R., Ostling, J., Svenblad, B. and Wichart, D. (1993). How do non-differentiating bacteria adapt to starvation?. Antonie van Leeuwenhoek 63: 333-341.

Klein, D. (2002). Quantification using real-time PCR technology: applications and limitations. Trends Mol. Med. 8: 257-260.

Koskinen, H., Pehkonen, P., Vehniainen, E., Krasnov, A., Rexroad, C., Afanasyev, S., Molsa, H. and Oikari, A. (2004). Response of rainbow trout transcirptome to model chemical contaminants. Biochem. Biophys. Res. Commun. 320: 745-753.

Kurobe, T., Yasuike, M., Kimura, T., Hirono, I. and Aoki, T. (2005). Expression profiling of immune-related genes from Japanese flounder *Paralichthys olivaceus* kidney cells using cDNA microarrays. Dev. Comp. Immunol. 29: 515-523.

Labreuche, Y., Lambert, C., Soudant, P., Boulo, V., Huvet, A. and Nicolas, J.L. (2006). Cellular and molecular hemocyte responses of the Pacific oyster *Crassostrea gigas*, following bacterial infection with *Vibrio aestuarianus* strain 01/32. Microbes and Infection 8: 2715-2724.

Lam, S.H., Winata, C.L., Tong, Y., Korzh, S., Lim, W.S., Korzh, V. Spitsbergen, J., Mathavan, S., Miller, L.D, Liu, E.T. and Gong, Z. (2006). Transcriptome kinetics of arsenic-induced adaptive response in zebrafish liver. Physiol. Genomics 27: 351-361.

Liu, Y., Zhou, Z., Yao, B., Shi, P., He, S., Hølvold, L.B. and Ringø, E. (2008). Effect of intraperitoneal injection of immunostimulatory substances on allochthonous gut microbiota of Atlantic salmon (*Salmo salar* L.) determined using denaturing gradient gel electrophoresis. Aquacult. Res. 39: 635-646.

Magnadottir, B., Lange, S., Gudmundsdottir, S., Bøgwald, J. and Dalmo, R.A. (2005). Ontogeny of humoral immune parameters in fish. Fish Shellfish Immunol. 19: 429-439.

Magnadottir, B., Gudmundsdottir, B.K., Lange, S., Steinarsson, A., Oddgeirsson, M., Bowden, T., Bricknell, I., Dalmo, R.A. and Gudmundsdottir, S. (2006). Immunostimulation of larvae and juveniles of cod, *Gadus morhua* L.J. Fish Dis. 29: 147-155.

Mahious, A.S., Gatesoupe, F.J., Hervi, M., Metailler, R. and Ollevier, F. (2006). Effect of dietary inulin and oligosaccharides as prebiotics for weaning turbot, *Psetta maxima* (Linnaeus, C. 1758). Aquacult. Int. 14: 219-229.

Makridis, P. and Olsen, Y. (1999). Protein depletion of the rotifer *Brachionus plicatilis* during starvation. Aquaculture 174: 343-353.

Makridis, P., Fjellheim, A.J., Skjermo, J. and Vadstein, O. (2000a). Colonization of the gut in first feeding turbot by bacterial strains added to the water or bioencapsulated in rotifers. Aquacult. Int. 8: 367-380.

Makridis, P., Fjellheim, J.A., Skjermo, J. and Vadstein, O. (2000b). Control of the bacterial flora of *Brachionus plicatilis* and *Artemia franciscana* by incubation in bacterial suspensions. Aquaculture 185: 207-218.

Makridis, P., Martins, S., Vercauteren, T., Van Driessche, K., Decamp, O. and Dinis M.T. (2005). Evaluation of candidate probiotic strains for gilthead sea bream larvae (*Sparus aurata*) using an in vivo approach. Lett. Appl. Microbiol. 40: 274–277

Makridis, P., Alves Costa, R. and Dinis, M.T. (2006). Microbial conditions and antimicrobial activity in cultures of two microalgae species. *Tetraselmis chuii* and *Chlorella minutissima*, and effect on bacterial load of enriched *Artemia* metanauplii. Aquaculture 255: 76-81.

Makridis, P., Martins, S., Reis, J. and Dinis, M.T. (2008). Use of probiotic bacteria in the rearing of Senegalese sole (*Solea senegalensis*) larvae. Aquacult. Res. 39: 627-634.

Manefield, M., deNys, R., Kumar, N., Read, R., Givskov, M., Steinberg, P. and Kjelleberg, S. (1999). Evidence that halogenated furanones from *Delisea pulchra* inhibit acylated homoserine lactone (AHL)-mediated gene expression by displacing the AHL signal from its receptor protein. Microbiology 145: 283-291.

Martins, C.I.M., Pistrin, M.G., Ende, S.S.W., Eding, E.H. and Verreth, J.A.J. (2009). The accumulation of substances in Recirculating Aquaculture Systems (RAS) affects embryonic and larval development in common carp *Cyprinus carpio*. Aquaculture 291: 65-73.

McIntosh, D., Ji, B., Forward, B.S., Puvanendran, V., Boyce, D. and Ritchie, R. (2008). Culture-independent characterization of the bacterial populations associated with cod (*Gadus morhua* L.) and live feed at an experimental hatchery facility using denaturing gradient gel electrophoresis. Aquaculture 275: 42-50.

Meurer, F., Hayashi, C., da Costa, M.M., Mauerwerk, V.L. and Freccia, A. (2006). *Saccharomyces cereivisiae* as probiotic for Nile tilapia during the sexual reversión phase under a sanitary challenge. Brazi. J. Anim. Sci. 35: 1881-1886.

Michaud, L., Blancheton, J.P., Bruni, V. and Piedrahita, R. (2006). Effect of particulate organic carbon on heterotrophic bacterial populations and nitrification efficiency in biolofical filters. Aquac. Eng. 34: 224-233.

Miller, M.B. and Bassler, B.L. (2001). Quorum sensing in bacteria. Annu. Rev. Microbiol. 55: 165-199.

Molina, L., Constantinescu, F. Michel, L., Reimmann, C., Duffy, B. and Défago, G. (2003). Degradation of pathogen quorum-sensing molecules by soil bacteria: a preventive and curative biological control mechanism. FEMS Microbiol. Ecol. 45: 71-81.

Moriarty, D.J.W. (1998). Control of luminous *Vibrio* species in penaeid aquaculture ponds. Aquaculture 164: 351-358.

Munro, P.D., Barbour, A. and Birkbeck, T.H. (1993). Bacterial flora of rotifers *Brachionus plicatilis*; evidence for a major location on the external surface and methods for reducing the rotifer bacterial load. In: Fish Farming Technology. H Reinertsen, L.A. Dahle, L. Jørgensen, and K. Tvinnereim (eds). A.A. Balkema, Rotterdam, The Netherlands, pp. 93-100.

Munro, P.D., Barbour, A. and Birkbeck, T.H. (1994). Comparison of the gut bacterial flora of start-feeding larval turbot reared under different conditions. J. Appl. Bacteriol. 77: 560-566.

Munro, P.D., Henderson, R.J., Barbour, A. and Birkbeck, T.H. (1999). Partial decontamination of rotifers with ultraviolet radiation: the effect of changes in the bacterial load and flora of rotifers on mortalities in start-feeding larval turbot. Aquaculture 170: 229-244.

Muroga, K., Higashi, M. and Keitoku, H. (1987). The isolation of intestinal microflora of farmed red seabream (*Pagrus major*) and black seabream (*Acanthopagrus schlegeli*) at larval and juvenile stages. Aquaculture 65: 79-88.

Muyzer, G., de Waal, E.C. and Uitterlinden, A.G. (1993). Profiling of complex microbial populations by denaturing gradient gel electrophoresis analysis of polymerase chain reaction-amplified genes coding for 16S rRNA. Appl. Environ. Microbiol. 59: 695-700.

Naas, K.E., Naess, T. and Harboe, T. (1991). Enhanced first feeding of halibut larvae (*Hippoglossus hippoglossus* L.) in green water. In: Larvi '91. Fish and Cruastacean Larviculture Symposium. P. Lavens, P. Sorgeloos, E. Jaspers and F. Ollevier (eds). European Aquaculture Society, Special Publication No. 24, Gent, Belgium.

Nakai, T. and Park, S.C. (2002). Bacteriophage therapy of infectious diseases in aquaculture. Res. Microbiol. 153: 13-18.

Nakase, G., Nakagawa, Y., Miyashita, S., Nasu, T., Senoo, S., Matsubara, H. and Eguchi, M. (2007). Association between bacterial community structures and mortality of fish larvae in intensive rearing systems. Fish. Sci. 73: 784-791.

Neori A., Krom M.D. and van Rijn J. (2007). Biogeochemical processes in intensive zero-effluent marine fish culture with recirculating aerobic and anaerobic biofilters. J. Exp. Marine Biol. Ecol. 349: 235-247.

Le Nguyen, D.D., Ha Ngoc, H., Dijoux, D., Montet, D. and Loiseau, G. (2007). Determination of fish origin by using 16S rDNA fingerprinting of bacterial communities by PCR- DGGE: application to Pangasius fish from Viet Nam. Food Control 19: 454-460.

Olafsen, J.A. (1993). The microbial ecology of fish aquaculture. In: Salmon Aquaculture. K.Heen, R.L. Monahan. and F. Utter (eds). Fishing New Books, Oxford, UK, pp. 166-175.

Olafsen J.A. (2001). Interaction between fish larvae and bacteria in marine aquaculuture. Aquaculture 200: 223-257.

Olsson, J.C., Westerdahl, A., Conway, P. and Kjelleberg, S. (1992). Intestinal colonization potential of turbot (*Scophthalmus maximus*) and dab (*Limanda limanda*)-associated bacteria with inhibitory effects against *Vibrio anguillarum*. Appl. Environ. Microbiol. 58: 551-556.

Ottesen, O.H. and Olafsen, J.A. (2000). Effects on survival and mucous cell proliferation of Atlantic halibut, *Hippoglossus hippoglossus* L., larvae following microflora manipulation. Aquaculture 187: 225-238.

Park, S.C., Shimamura, I., Fukunaga, M., Mori, K.I. and Nakai, T. (2000). Isolation of bacteriophages specific to a fish pathogen, *Pseudomonas plecoglossicida*, as a candidate for disease control. Appl. Environ. Microbiol. 66: 1416-1422.

Payne, M.S., Hall, M.R., Bannister, R., Sly, L. and Bourne, D.G. (2006). Microbial diversity within the water column of a larval rearing system for the ornate rock lobster (*Panulirus ornatus*). Aquaculture 258: 80-90.

Pedersen, G.M., Gildberg, A. and Olsen, R.L. (2004). Effects of including cationic proteins from cod milt in the feed to Atlantic cod (*Gadus morhua*) fry during a challenge trial with *Vibrio anguillarum*. Aquaculture 233: 31-43.

Pérez-Benavente, G. and Gatesoupe, F.J. (1988). Bacteria associated with cultured rotifers and *Artemia* are detrimental to larval turbot, *Scophthalmus maximus* L. Aquacult. Eng. 7: 289-293.

Piaget, N., Vega, A., Silva, A. and Toledo, P. (2007). Effect of the application of β-glucans and mannan-oligosaccharides (βG MOS) in an intensive larval rearing system of *Paralichthys adspersus* (Paralichthydae). Invest. Mar. 35: 35-43.

Pianka, E.R. (1970). On *r*- and *K*-selection. Am. Nat. 104: 592-597.

Picchietti, S., Fausto, A.M., Randelli, E., Carnevali, O., Taddei, A.R., Bounocore, F., Scapigliati, G. and Abelli, L. (2009). Early treatment with *Lactobacillus delbrueckii* strain induces an increase in intestinal T-cells and granulocytes and modulates immune-related genes of larval *Dicentrarchus labrax* (l.). Fish Shellfish Immunol. 26: 368-376.

Planas, M. and Cunha, I. (1999). Larviculture of marine fish: Problems and perspectives. Aquaculture 177: 171-190.

Planas, M., Vázquez, J.A., Marques, J., Pérez Lomba, R., González, M.P. and Murado, M.A. (2004). Enhancement of rotifer (*Brachionus plicatilis*) growth by using terrestrial lactic acid bacteria. Aquaculture 240: 313-329.

Planas, M., Pérez-Lorenzo, M., Hjelm, M., Gram, L., Fiksdal, I.U. and Bergh, Ø, Pintado, J. (2006). Probiotic effect in vivo of *Roseobacter* strain 27-4 against *Vibrio* (*Listonella*) *anguillarum* infections in turbot (*Scophthalmus maximus* L.) larvae. Aquaculture 255: 323-333.

Plante, S, Pernet, F., Hache, R., Ritchie, R., Ji, B.J. and McIntosh, D. (2007). Ontogenic variations in lipid class and fatty acid composition of haddock larvae *Melanogrammus aeglefinus* in relation to changes in diet and microbial environment. Aquaculture 263: 107-121.

Porsby, C.H., Nielsen, K.F. and Gram, L. (2008). *Phaeobacter* and *Ruegeria* species of the *Roseobacter clade* colinize separate niches in a Danish turbot (*Scophthalmus maximus*)-rearing farm and antagonize *Vibrio anguillarum* under different growth conditions. Appl. Environ. Microbiol. 74: 7356-7364.

Prol, M.J., Bruhn, J.B., Pintado, J. and Gram, L. (2009). Real-time PCR detection and quantification of fish probiotic *Phaeobacter* strain 27-4 and fish pathogenic *Vibrio* in microalgae, rotifer, *Artemia* and first feeding turbot (*Psetta maxima*) larvae. J. Appl. Microbiol. 106: 1292-1303.

Pruder, G.D. (2004). Biosecurity: application in aquaculture. Aquacultural Engineering 32: 3-10.

Qi, Z., Zhang, X., Boon, N. and Bossier, P. (2009). Probiotics in aquaculture of China—Current state, problems and prospect. Aquaculture 290: 15-21.

Rasch, M., Buch, C., Austin, B., Slierendrecht W.J., Ekmann, K.S., Larsen, J.L., Johansen, C., Riedel, K., Eberl, L., Givskov, M. and Gram, L. (2004). An inhibitor of bacterial quorum sensing reduces mortalities caused by vibriosis in rainbow trout (*Onchorrynchus mykiss*, Walbaum). Syst. Appl. Microbiol. 27: 350-359.

Reid, H.I., Treasurer, J.W., Adam, B. and Birkbeck, T.H. (2009). Analysis of bacterial populations in the gut of developing cod larvae and identification of *Vibrio logei*, *Vibrio anguillarum* and *V. splendidus* as pathogens of cod larvae. Aquaculture 288: 36-43.

Reitan, K.I., Rainuzzo, J.R., Oie, G. and Olsen, Y. (1993). Nutritional effects of algal addition in first-feeding of turbot (*Scophthalmus maximus* L.) larvae. Aquaculture 118: 257-275.

Reitan, K.I., Natvik, C.M. and Vadstein, O. (1998). Drinking rate, uptake of bacteria and microalgae in turbot larvae. J. Fish Biol. 53: 1145-1154.

Rengpipat, S., Tunyanun, A., Fast, A.W., Piyatiratitivorakul, S. and Menasveta, P. (2003). Enhanced growth and resistance to *Vibrio* challenge in pond-reared black tiger shrimp *Penaeus monodon* fed a *Bacillus probiotic*. Dis. Aquat. Org. 55: 169-173.

Ringø, E. and Vadstein, O. (1998). Colonization of *Vibrio Pelagius* and *Aeromonas caviae* in early developing turbot, *Scophthalmus maximus* (L.) larvae. J. Appl. Microbiol. 84: 227-233.

Rithchie, A.J., Jansson, A., Stallberg, J., Nilsson, P., Lysght, P. and Cooley M.A. (2005). The *Pseudomonas aeruginosa* quorum-sensing molecule N-3-(oxododecanoyl)-L-homoserine lactone inhibits T-cell differentiation and cytokine production by a mechanism involving an early step in T-cell activation. Infect. Immun. 73: 1648-1655.

Rombaut, G., Suantika, G., Boon, N., Maertens, S., Dhert, P., Top, E., Sorgeloos, P. and Verstraete, W. (2001). Monitoring of the evolving diversity of the microbial community present in rotifer cultures. Aquaculture 198: 237-252.

Rønnestad, I., Conceicão, L.E.C., Aragão, C. and Dinis, M.T. (2000). Free amino acids are absorbed faster and assimilated more efficiently than protein in postlarval Senegal sole (*Solea senegalensis*). J. Nutr. 130: 2809-2812.

Rurangwa, E., Laranja, J.L., Van Houdt, R., Delaedt, Y., Geraylou, Z., Van de Wiele, T., Loo, J., Van Craeyveld, V., Courtin, CM; Delcour, J.A. and Ollevier, F. (2009). Selected nondigestible carbohydrates and prebiotics support the growth of probiotic fish bacteria mono-cultures in vitro. J. Appl. Microbiol. 106: 932-940.

Ruíz-Ponte, C., Samain, J.F., Sánchez, J.L. and Nicolas, J.L. (1999). The benefit of *Roseobacter* species on the survival of scallop larvae. Mar. Biol. 1: 52-59.

Salvesen, I. and Vadstein, O. (1995). Surface disinfection of eggs from marine fish: evaluation of four chemicals. Aquacult. Int. 3: 1-18

Salvesen, I., Skjermo, J. and Vadstein, O. (1999). Growth of turbot (*Scophthalmus maximus* L.) during first feeding in relation to the proportion of r/K-strategists in the bacterial community of the rearing water. Aquaculture 175: 337-350.

Salvesen, I., Reitan, K.I., Skjermo, J. and Oie, G. (2000). Microbial environments in marine larviculture: Impacts of algal growth rates on the bacterial load in six microalgae. Aquacult. Int. 8: 275-287.

Sandaa, R.A., Magnesen, T., Torkildsen, L. and Bergh, Ø. (2003). Characterisation of the bacterial community associated with early stages of great scallop (*Pecten maximus*), using denaturing gradient gel electrophoresis (DGGE). Syst. Appl. Microbiol. 26: 302-311.

Sarter, S. and Guichard, B. (2009). Bacterial antibiotic resistance in aquaculture. In: Aquaculture Microbiology and Biotechnology, Volume 1. D. Montet and R.C. Ray (eds). Science Publishers, Enfield, USA pp. 133-157.

Sarter, S., Hoang Nam, K. N., Le Thanh, H., Lazard, J. and Montet, D. (2007). Antibiotic resistance in Gram-negative bacteria isolated from farmed catfish. Food Control 18: 1391-1396.

Schulze, A.D., Alabi, A.O., Tattersall-Sheldrake, A.R. and Miller, K.M. (2006). Bacterial diversity in a marine hatchery: Balance between pathogenic and potentially probiotic bacterial strains. Aquaculture 256: 50-73.

Skjermo, J. and Vadstein, O. (1993). Characterization of the bacterial flora of mass cultivated *Brachionus plicatilis*. Hydrobiologia 255/256: 185-191.

Skjermo, J. and Vadstein, O. (1999). Techniques for microbial control in the intensive rearing of marine larvae. Aquaculture 177: 333-343.

Skjermo, J., Salvesen, I., Oie, G., Olsen, Y. and Vadstein, O. (1997). Microbially matured water: A technique for selection of a non-opportunistic bacterial flora in water that may improve performance of marine larvae. Aquacult. Int. 5: 13-28.

Skjermo, J., Størseth, T.R., Hansen, K., Handa, A. and Øie, G. (2006). Evaluation of β-(1→3, 1→6)-glucans and High-M alginate used as immunostimulatory dietary supplement during first feeding and weaning of Atlantic cod (*Gadus morhua* L.). Aquaculture 261: 1088-1101.

Skovus, T.L., Holmström, C., Kjelleberg, S. and Dahllö, I. (2007). Molecular investigation of the distribution, abundance and diversity of the genus *Pseudoalteromonas* in marine samples. FEMS Microbiol. Ecol. 61: 348-361.

Sugita, H., Nakamura, H. and Shimada, T. (2005). Microbial communities associated with filter materials in recirculating aquaculture systems of freshwater fish. Aquaculture 243: 403-409.

Suzer, C, Coban, D. Kamaci, H.O., Saka, S., Firat, K., Otgucuolu, O. and Kucuksari, H., (2008). *Lactobacillus* spp bacteria as probiotics in gilthead sea bream (*Sparus aurata*, L.) larvae: Effects on growth performance and digestive enzyme activities. Aquaculture 280: 140-145.

Tanaka, M., Kawai, S., Seikai, T. and Burke, J.S. (1996). Development of the digestive organ system in Japanese flounder in relation to metamorphosis and settlement. Mar. Fresh. Behav. Physiol. 28: 19-31.

Thomas, G.L., Bohner, C.M., Williams, H.E., Walsh, C.M., Ladlow, M., Welch, M., Bryant, C.E. and Spring, D.R. (2006). Immunomodulatory effects of *Pseudomonas aeruginosa* quorum sensing small molecule probes on mammalian macrophages. Mol. Biosyst. 2: 132-137.

Tihn, N.T.N., Linh, N.D., Wood, T.K., dierckens, K., Sorgeloos, P. and Bossier, P. (2007). Interference with the quorum sensing systems in a *Vibrio harveyi* strain alters the growth rate of gnotobiotically cultured rotifer *Brachionus plicatilis*. J. Appl. Microbiol. 103: 194-203.

Tihn, N.T.N., Dierckens, K., Sorgeloos, P. and Bossier, P. (2008). A review of the functionality of probiotics in larviculture food chain. Mar. Biotechnol. 10: 1-12.

Tovar- Ramirez, D., Zambonini, J., Cahu, C., Gatesoupe, F.J., Vázquez-Juarez, R. and Lesel, R. (2002). Effect of live yeast incorporation in compound diet on digestive enzyme activity in sea bass (*Dicentrarchus labraz*) larvae. Aquaculture 204: 113-123.

Tovar-Ramírez, D., Infante, J.Z., Cahu, C., Gatesoupe, F.J. and Vázquez-Juárez, R. (2004). Influence of dietary live yeast on European sea bass (*Dicentrarchus labrax*) larval development. Aquaculture 234: 415-427.

Vadstein, O. (1997). The use of immunostimulation in marine larviculture: possibilities and challenges. Aquaculture 155: 401-417.

Verschuere, L., Dhont, J., Sorgeloos, P. and Verstraete, W. (1997). Monitoring Biolog patterns and r/K-strategists in the intensive culture of *Artemia* juveniles. J. Appl. Microbiol. 83: 603-612.

Verschuere, L., Rombaut, G., Sorgeloos, P. and Verstraete, W. (2000). Probiotic bacteria as biologic control agents in aquaculture. Microbiol. Mol. Microbiol. Mol. Biol. Rev. 64: 655-671.

Villamil, L., Tafalla, C., Figueras, A. and Novoa, B. (2003). Evaluation of immunomodulatory effects of lactic acid bacteria in turbot (*Scophthalmus maximus*). Clin. Diagn. Lab. Immunol. 9: 1318-1323.

Vine, N.G., Leukes, W.D., Kaiser, H., Daya, S., Baxter, J. and Hecht, T. (2004a). Competition for attachment of aquaculture candidate probiotic and pathogenic bacteria on fish intestinal mucus. J. Fish Dis. 27: 319-326.

Vine, N.G., Leukes, W.D. and Kaiser, H. (2004b). *In vitro* growth characteristics of five candidate aquaculture probiotics and two fish pathogens grown in fish intestinal mucus. FEMS Microbiol. Lett. 231: 145-152.

Vine, N.G., Leukes, W.D. and Kaiser, H. (2006). Probiotics in marine larviculture. FEMS Microbiol. Rev. 30: 404-427.

Westerdahl, A., Olsson, J.C., Kjelleberg, S. and Conway, P.L. (1991). Isolation and characterization of turbot (*Scophtalmus maximus*)-associated bacteria with inhibitory effects against *Vibrio anguillarum*. Appl. Environ. Microbiol. 57: 2223-2228.

Whitehead, N.A., Barnard, A.M.L., Slater, H., Simpson, N.J.L. and Salmond, G.P.C. (2001). Quorum-sensing in gram negative bacteria. FEMS Microbiol. Rev. 25: 365-404.

Williams, T.D., Gensberg, K., Minchin, S.D. and Chipman, J.K. (2003). A DNA expression array to detect toxic stress responses in European flounder (*Paralichthys fesus*). Aquat. Toxicol. 65: 141-157.

Zhou, Z., Liu, Y., He, S., Shi, P, Gao, X., Yao, B. and Ringø, E. (2009). Effects of dietary potassium diformate (KDF) on growth performance, feed conversion and intestinal bacterial community of hybrid tilapia (*Oreochromis niloticus* ♀× *O. aureus* ♂). Aquaculture 291: 89-94.

Zhu, J., Beaber, J.W., Moré, M.I., Fuqua, C., Eberhard, A. and Winans, S.C. (1998). Analogs of the autoinducer 3-oxooctanoyl-homoserine lactone strongly inhibit activity of the TraR protein of *Agrobacterium*. Mol. Microbiol. 27: 289-297.

Ziemer, C. J. and Gibson, G.R. (1998). An overview of probiotcs, prebiotics and synbiotics in the functional food concept: Perspectives and future strategies. Int. Dairy J. 8: 473-479.

2

Probiotics in the Larval Culture of Aquatic Organisms

Bruno Gomez-Gil,[1],* Ana Roque[2] **and** *Sonia Soto-Rodriguez[1]*

INTRODUCTION

Diseases have been one of the major problems that affect any aquaculture enterprise and methods to control or eradicate them have been used widely with varying results. The most common methods of control are better management practices, vaccines, and the use of chemical compounds that eradicate or limit the growth of pathogens.

The control of bacterial pathogens has largely been done with the use of chemotherapeutants, particularly with antibiotics and antimicrobials in general. Antimicrobial use in larviculture of aquatic animals is especially intense and are usually applied directly into the rearing water. The use of antibiotics has been proven to affect the performance and survival of larvae (Soto-Rodriguez et al., 2006; Williams et al., 1992) and some authors have mentioned the risk of bacterial resistance (Sarter et al., 2007; Sarter and Guichard, 2009).

Resistance to these chemicals has been found in bacterial strains isolated from cultured aquatic organisms and facilities (Hernandez-Serrano, 2005; Sarter et al., 2007) although a direct link between use of antibiotics in aquaculture and acquisition and development of resistance have seldom been proven (WHO, 2006). Nevertheless, academia and the industry are looking for new ways to control infectious diseases. Since many microorganisms are able to out-compete, kill or inhibit the growth of others, it seems reasonable to presume that some beneficial bacteria are also capable of exerting these tasks against pathogenic bacteria inside the

[1]*CIAD, A.C. Mazatlan Unit for Aquaculture and Environmental Management. AP 711 Mazatlan, Sinaloa, Mexico 82000*
[2]*IRTA-Sant Carles de la Rápita, Crta. Poble Nou, Km 5.5, Sant Carles de la Rápita, Spain 43540*
*Corresponding author: E-mail: bruno@ciad.mx

gut of reared animals. Microorganisms that are capable of protecting the host against harmful pathogens are called probionts. While a large amount of probiotic research has been done with farm animals and humans, there is comparatively less research on aquatic animals (Austin and Brunt, 2009) and even less in the larval culture of these animals. Much of the work done in aquaculture provides little or no evidence on the role of probiotic microorganisms inside the host, but there is scientific evidence that also proves that some bacteria are capable of limiting the action of pathogens. In Chapter 1, Pintado and his colleagues have discussed several new strategies including the use of prebiotics and probiotics in marine fish larval rearing. This chapter deals exclusively on the use of probiotics in the larval culture of aquatic animals.

DEFINITIONS

Probiotics

The definition of probiotics has evolved mainly by the study of farm animals or humans, terrestrial animals, but not necessarily applicable to aquatic animals. Fuller (1989) proposed the definition of a probiont as "a live microbial feed which beneficially affects the host animal by improving its intestinal balance". Because there is a strong interaction between the organism and its aquatic environment, an environment where opportunistic pathogens can and do grow, the definition has to also encompass this environment. A general definition of probiotic that takes into account the environment has been proposed by Verschuere and collaborators (2000)—"a live microbial adjunct which has a beneficial effect on the host by modifying the host-associated or ambient microbial community, by ensuring improved use of the feed or enhancing its nutritional value, by enhancing the host response towards disease, or by improving the quality of its environment". This definition incorporates new properties that might not be associated with a probiont, as advised by Gomez-Gil et al. (2000), a probiont should not be confused with a feed additive, such as a growth promoter, nor a bioremediation agent. The principal goal of a probiont is to eliminate a pathogenic treat, diminish it, or to enhance the immunological response of the host to it. Therefore, the definition of Verschuere et al. (2000) can be simplified to "a live microbial adjunct, which has a beneficial effect on the host by modifying the host-associated or ambient microbial community and so enhances the host response towards disease". The microbe should not only be alive, but it should be kept alive (Gatesoupe, 1999). If the microbes are dead, then they can be considered as feed additives due to their passive action. It is desired that they actively enhance the host's response towards disease by means of colonization, production of metabolites, etc.

Prebiotics

This term was introduced to define "a non-digestible food ingredient that beneficially affects the host by selectively stimulating the growth and/or activity of one or a limited number of bacteria in the colon" (Gibson and Roberfroid, 1995) and later redefined as "a selectively fermented ingredient that allows specific changes, both in the composition and/or activity in the gastrointestinal microflora that confers benefits upon the well-being and health" (Robertfroid, 2007). A question could be if microorganisms, at a certain point, could also be considered as prebiotics? Not according to the definition provided because at some point, the host could digest the microorganisms, and that they are also not fermented ingredients. It is more suitable for feed ingredients that specifically enhance the activity of certain microorganisms, such as non-digestible carbohydrates (starch), some peptides, proteins, and lipids (Gibson and Roberfroid, 1995). In synthesis, a probiotic is a live microorganism and a prebiotic is an ingredient, and both confer benefits upon the well-being and health of the host.

Selection of Probiotics
Sources

Sources of potential probiotic strains have been either microbial culture collections, or from the environment of the cultured organism (Gram and Hjelm, 2002; Maeda et al., 1997). From the environment, common sources are successful rearing cycles (Planas et al., 2006), culture water (Abidi, 2003; Balcazar and Rojas-Luna, 2007; Li et al., 2006), and uncontaminated seawater (Abidi, 2003; Garriques and Arevalo, 1995; Gomez-Gil et al., 2002). Sometimes, they have also been isolated from the gut of healthy organisms, especially in the case of fishes (Ringo et al., 2007; Spanggaard et al., 2001), or from other organs, such as the hepatopancreas of wild shrimps (Gullian et al., 2004). The theory behind selecting these sources is that bacteria colonizing these habitats under these circumstances are less likely to be pathogenic. That, of course, does not guarantee that they are capable of becoming a good probiont. It might be advisable to look for potential probiotic bacteria right in the "battlefield", that is, where they are constantly tested for fitness and survival against their "natural enemies", harmful bacteria that might be more capable of colonizing a certain host. These "battlefields" could be infected organisms, polluted culture water, etc. Strains coming from these "dangerous" sources should be tested for their pathogenicity. It is very rare that in a diseased aquatic organism, especially in invertebrates, only pathogenic bacteria are found. Most often, these pathogens dominate, more so in the latter stages of the infection, but other bacteria are also to be found and these could be the ones of interest.

Antagonistic Bacteria

The most common method to select potential probiotic bacteria has been by testing the inhibition potential (*via* production of antagonistic compounds) of candidate strains against known pathogens, mostly by growing both on an artificial medium (Gram and Hjelm, 2002; Gullian et al., 2004; Hai et al., 2007; Spanggaard et al., 2001) and seldom in an *in vivo* challenge (Makridis et al., 2005). As mentioned by Gram and Hjelm (2002), "a word of caution should be included since no studies have documented, (i) that inhibitory substances produced *in vitro* are actually effective *in vivo* or, (ii) that *in vitro* antagonism is a predictor of *in vivo* effect". Considering this, we have also to acknowledge that identifying antagonistic strains can be a hard process. Of 1018 isolates tested for their antagonistic properties in an agar diffusion assay, only 45 (4.4%) showed inhibitory properties against *Listonella anguillarum* (although commonly used, the synonym *Vibrio anguillarum* is not a valid name), a common fish pathogen (Spanggaard et al., 2001). And even so, only six of the 45 strains tested, showed any improvement in infected trout; that is 0.59% of the total strains tested. A similar result was obtained for strains isolated from a Chilean scallop (*Argopecten purpuratus*) culture system; only 2.2% of the 506 strains analyzed showed any inhibitory effect on a *V. anguillarum*-related pathogen (Riquelme et al., 1997). Prado and colleagues (2009) also showed inhibitory action in 52 of 523 strains tested (9.9%) to three *Vibrio* pathogens, but only 11 (2.1%) inhibited all three. One of these (PP-154) identified as *Phaeobacter* sp., was able to show antagonistic activity with more than 25 *Vibrio* strains as well as other pathogens, such as *Aeromonas hydrophila, A. salmonicida, Pseudomonas anguilliseptica, Tenacibaculum maritimum,* and *Streptococcus parauberis.*

Methods to Measure Antagonism

Many methods have been devised or adapted to evaluate the antagonism between potential probiotic bacterial strains and pathogenic strains. The most common, perhaps because of its simplicity is to culture both strains in a solid medium crossing each other (cross-streaking) and observing if an inhibition halo in the pathogenic strains is observed at the crossing point. This method has the disadvantage that it is very difficult to observe an inhibition.

Another common method is the evaluation of bacteriocin-like inhibitory substances (BLIS) (Gibson et al., 1998) or its modifications (Hai et al., 2007). This method has the same principle of the cross-streaking method but the tested strain is streaked, incubated and the bacterial growth removed mechanically and any remaining viable cells killed with chloroform vapors. The pathogenic strain is then streaked at a right angle and incubated. Inhibition is determined if the inhibition halo is twice the width of the probiotic strain growth (Hai et al., 2007).

Several other methods have been devised but with variable results, such as well-diffusion, disc-diffusion, PCR (Nitisinprasert et al., 2006), spot (Prado et al., 2009) and co-culture (Hai et al., 2007).

Methods to Test Probionts

According to the definitions proposed earlier, to be able to prove that an organism is indeed a probiont, it has to protect the host against the attack of a pathogen. There are basically two ways to test this, one is to administer the potential probiont to the host and wait until an outbreak starts and measure significant survival enhancements in those organisms exposed to the probiont against untreated organisms. The other method is to artificially challenge treated organisms with pathogens and observe the positive effect of the probiont (Ringo and Gatesoupe, 1998). Survival enhancements do not have to be the only parameter evaluated; less diseased larvae, weight gain, higher molting rates, etc. are also indications that the treated larvae coped with the infection better, although they did not die. But a requirement is that an infection by a pathogen was established, and this infection significantly reduced the performance of the larvae. If this cannot be proven, the microorganism can only be classified as a "potential probiont".

Bacteria-free organisms, called "gnotobiotic", are important as model organisms to understand the true action of a pathogen and/or probiont, because the results will not be masked by the normal bacterioflora present in the gut. Only a few studies have been done with gnotobiotic marine animals; germ-free sea bass (*Dicentrarchus labrax*) larvae have recently been produced (Dierckens et al., 2009) and tested with potential pathogens. These organisms will be of capital importance to understand host-microbe interactions.

PROBIOTICS IN LARVAL CULTURE: GRAM-POSITIVE BACTERIA

In the present review, only data will be presented where it has been proven that the tested probionts do somehow protect the host against the action of pathogens, or at least the immune system was improved, although this last aspect is difficult to evaluate due to the small size of the larvae.

Lactic Acid Bacteria (LAB, order Lactobacillales)

Several species of LAB have been used as probiotics, mainly because many of them are part of the normal microflora of the intestine of fishes (Ringo and Gatesoupe, 1998). *Carnobacterium divergens* (previously identified as

Lactobacillus plantarum) have been shown to protect 9-d-old larvae of turbot (*Scophthalmus maximus*) when inoculated at high densities (10^7 CFU/mL) against a pathogenic *Vibrio splendidus* (Vibrio P), with survival differences of as much as 79% (Gatesoupe, 1994). *C. divergens* was also tested against a pathogenic *V. pelagius* in turbot larvae but the results were inconclusive as larvae mortalities were, at most, only delayed (Ringo, 1999).

Rainbow trout fry (1-2 g) and fingerlings were used to evaluate the probiotic strain K1 of *Carnobacterium* sp., a strain capable of inhibiting several bacterial pathogens (Jöborn et al., 1997; Robertson et al., 2000). This strain could still be isolated from the intestines of fish fry after 10 d after the end of the probiotic feeding period. In this study, only juveniles (20 g) were challenged with two pathogens (*Aeromonas salmonicida* and *Yersinia ruckeri*) with significant survival enhancements in those organisms fed with the probiont. If this probiont can also protect larvae or fry is still a matter of speculation (Gildberg and Mikkelsen, 1998).

Although no infection challenges have been carried out with Gilt-head sea bream (*Sparus aurata*) larvae to test potential probiotics, experiments have been done to evaluate the stress response and stimulation of the gut immune system of larvae. Larvae exposed to *Lactobacillus fructivorans* and *L. plantarum* colonized the gut differently after 35, 66, and 90 d post hatching (Carnevali et al., 2004) and affected the bacterial genera decreasing potential pathogenic groups such as Enterobacteriaceae and *Staphylococcus* (Rollo et al., 2006). Stress response of the larvae was significantly influenced by the administration of the potential probionts; larvae exposed to pH stress survived better as compared to untreated controls (Rollo et al., 2006). The gut immune system of *S. aurata* and *Dicentrarchus labrax* was stimulated by the addition of *L. delbrueckii* subsp. *delbrueckii* and *L. fructivorans* combined with *L. plantarum* orally via live carriers (*Artemia* sp. nauplii and rotifers) (Abelli et al. 2009).

Atlantic cod (*Gadus morhua*) fry have also been challenged with a pathogenic strain of *L. anguillarum* (strain LFI 1243) by supplementing their feed with an unspecified strain of *Carnobacterium divergens* (Gildberg and Mikkelsen, 1998; Gildberg et al., 1997). Significant reductions in mortalities in treated fry were observed 12 d after infection but cumulative mortalities level out after 4 wk. The probiotic bacterium was supplied to the fry at a density of 2×10^9 CFU/g feed. In this study, fry were also offered a diet with intact muscle protein (no bacteria added), and those fish had the lowest cumulative mortalities (Gildberg et al., 1997). Mixed cultures *in vitro* of both bacterial strains did not show an obvious inhibition of the pathogen (Gildberg et al., 1997). Previously, this strain (identified as *L. plantarum*) was fed to 5 d old larvae and after 4 d the bacterial flora was dominated by it as compared to the controls (Strom and Ringo, 1993). These results make it difficult to evaluate the true action of the probiont

C. divergens; it was capable of colonizing the larvae (Strom and Ringo, 1993), could inhibit pathogenic *L. anguillarum* (Gildberg and Mikkelsen, 1998), but gives a temporary and marginal protection to the fry.

Lactobacillus acidophilus (unspecified strain) was used in a diet to test the protection it confers to the tiger shrimp (*Penaeus monodon*) postlarvae challenged with a pathogenic strain of *Vibrio harveyi* (Immanuel et al., 2007). Postlarvae fed with the probiont or with a combination of lactobacilli and yeast (*Saccharomyces creviciae*), survived significantly better at the end of 10 d; 94.3 and 82.3% respectively versus 26.5% of untreated postlarvae. The *V. harveyi* load in the hepatopancreas and muscle tissues was also significantly reduced by orders of magnitude (Immanuel et al., 2007). A strain of *Enterococcus faecium* (MC13) was also tested in *P. monodon* postlarvae challenged with *V. harveyi* and *V. parahaemolyticus* (Swain et al., 2009). MC13 significantly reduced the mortality of the postlarvae, 84% survival against 60% when challenged with a strain of *V. harveyi*, however little protection was conferred against *V. parahaemolyticus*.

The probable mechanisms of action of LAB could be the inhibition of pathogens (vibrios) by the production of lactic and acetic acids and not by bacteriocins which cannot access to the plasmatic membrane of Gram negative bacteria (Vazquez et al., 2005).

Bacillaceae

Bacillus sp. spores (strain IP 5832) improved the survival of larvae significantly when included with rotifers by reducing the density of an opportunistic bacterial strain (Gatesoupe, 1991). The taxonomic status of this potential pathogens is not clear as it is referred as *Vibrio* sp. by Gatesoupe (1991) citing a previous article where it is identified as *Aeromonas* sp. (Gatesoupe, 1990). With the tiger shrimp (*Penaeus monodon*) larvae, another *Bacillus* spp. (strain S11) used mixed with the feed as fresh cells or lyophilized and challenged with *V. harveyi* D331 (Rengpipat et al., 1998). After 10 d, larvae survival was significantly higher than the untreated larvae with any of the different feeds prepared with the probiont. When the larvae were dissected and their microbial loads analyzed, those larvae that received the probiotic feed, *Bacillus* sp. S11 was the principal bacteria found at densities of up to 10^8 CFU/g of gut (Rengpipat et al., 1998). When the experiment was repeated with a different culture system (recirculation), the probiotic effect measured as survival and growth was significant only at certain days post challenge (Rengpipat et al., 2000). The immune response of probiotic treated larvae was evaluated and a significant enhancement of percent phagocytosis, phagocytic index, total hemocytes, and phenoloxidase was observed (Rengpipat et al., 2000). They speculate that the mode of action of S11 was competitive exclusion and also provided both cellular and humoral immune defense responses (Rengpipat et al., 1998; Rengpipat et al., 2000).

Streptococcaceae

Streptococcus phocae (PI80) was isolated from the intestine of shrimp (*Penaeus indicus*) and conferred protection to *Penaeus monodon* postlarvae (PL-55, 55 d after they molted from the last mysis stage) when challenged with a *V. harveyi* strain (Swain et al., 2009). The survival attained with this strain was 92% against 60% in the control (*V. harveyi,* only). However, this strain did not protect the postlarvae against a strain of *V. parahaemolyticus.* The authors speculated that the mode of action is the production of bacteriocins and other substances capable of making a pore in the cell membrane of pathogens, which leads to the efflux of K + ions, although no evidence is presented.

PROBIOTIC IN LARVAL CULTURE: GRAM-NEGATIVE BACTERIA

Vibrios (Vibrionaceae)

Turbot larvae often experience mass mortalities apparently caused by different pathogenic vibrios and other Gram-negative bacteria (Hjelm et al., 2004). To control these pathogens, several Gram-negative and positive bacteria have been tested. A strain of *Vibrio pelagius* introduced in the rearing water improved the survival of larvae compared to fish exposed to *Aeromonas caviae*, a species associated with high larval mortalities, and with the control group (Ringo and Vadstein, 1998). This result might point out that *V. pelagius* served more as a feed additive than a true probiont. A very desirable and maybe indispensable requirement for a probiotic strain to function as such is its ability to colonize the gut of the host. A strain of *V. mediterranei* (Q40) was observed as a first colonizer of the gut of turbot larvae and thus be able to prevent the colonization of potential pathogens (Huys et al., 2001). Larvae treated this way also showed improved survival, but since no infection challenges were done, it is difficult to say that this strain is indeed a probiont. "Vibrio E" (phenotypically close to *V. alginolyticus*) was challenged against the pathogenic strain "Vibrio P" described previously (Gatesoupe, 1997). In this case, a higher survival percentage was observed when a higher density of the probiont was inoculated, but in any case, significantly lower than uninfected larvae. Vibrio E is a siderophore producing strain that can grow in iron-chelating media and it was proposed that both strains competed for iron, and since the pathogen Vibrio P cannot produce siderophores, it had a competitive disadvantage against the probiont. The same effect was obtained when the bacterial siderophore deferoxamine was added, thus reinforcing the competition for iron hypothesis (Gatesoupe, 1997).

A strain of *V. alginolyticus* (Ili) was inoculated to white shrimp (*Penaeus vannamei*) larvae (zoea II stage) and later they were challenged with White Spot Syndrome Virus (WSSV) (Rodriguez et al., 2007). Survival was enhanced during the first 52 h post challenge, but final survival rate was no different than the untreated larvae. β-glucans were also offered alone and in combination with probiotics; the combination treatment provided a significant survival of the postlarvae after 292 h. Treated (probiotics + β-glucans) larvae were reared until juveniles, challenged with WSSV and some immunological parameters evaluated. It was found, that this treatment modifies the immune response of juvenile shrimps and influences WSSV prevalence and shrimp survival in ponds (Rodríguez et al., 2007).

Pseudomonadaceae and Aeromonadaceae

Pseudomonas sp. previously identified as *Vibrio* sp. (strain 11) significantly protected Chilean scallop (*Argopecten purpuratus*) larvae when they were immersed for 1 h prior to being challenged with a *L. anguillarum*-related (VAR) pathogen at a density of 5×10^3 cells/mL (Riquelme et al., 1997). This strain, isolated from microalgae, showed an inhibitory activity against the pathogen with the double-layer method (Dopazo et al., 1988); the pathogenicity of the VAR strain is based on digestive tract invasion and production of exotoxins (Riquelme et al., 1995). The larvae required a period of about 6 h to significantly incorporate this strain when exposed at a density of 10^6 cells/mL (Riquelme et al., 2000). Re-inoculation of the strain has to be done every 2-3 d to ensure its presence in the bacterial microbiota of the larvae (Riquelme et al., 2001).

Aeromonas media (strain A199), which displayed bacteriocin-like inhibitory substances (BLIS) against several pathogens of aquatic organisms, also protected Pacific oyster (*Crassostrea gigas*) larvae when challenged with *V. tubiashii* (Gibson et al., 1998). The probiotic strain was added at a final density of 10^4 CFU/mL and 1 h later, the pathogen at 10^2, 10^3, and 10^5 CFU/mL. After 72-96 h, regardless of the density inoculated of the pathogen, the larval mortalities were no different than those of the control larvae (no pathogen added), but were significantly different than those where no probiont was used (only pathogen). It has been correctly pointed out that the association between BLIS activity and probiotic activity is circumstantial (Gibson et al., 1998) and further research is needed.

Rhodobacterales

Roseobacter sp. (strain 27-4) was inoculated to rotifers and then delivered to the larvae challenged with a pathogenic *Vibrio anguillarum* (strain 90-11-287 serotype O1), mortalities were reduced significantly (between 10 and 30%) after 10 d (Planas et al., 2006). The strain was detected in the gastrointestinal lumen of the larvae but apparently could not colonize it as it disappeared or diminished after inoculation stopped (Planas et al., 2006). *Roseobacter* could not inhibit the growth of the pathogens in co-culture assays (Hjelm et al., 2004), nor was it able to reduce *V. anguillarum* counts in the larvae or better colonize the larvae, so it is speculated that *Roseobacter* is able to perform the antagonistic effect only at specific sites in the larvae (Planas et al., 2006). This strain of *Roseobacter* was isolated from a healthy turbot larvae rearing system in Spain and showed a strong inhibitory action against *V. anguillarum, V. splendidus,* and *Pseudoalteromonas* sp. by the well diffusion agar assay (Hjelm et al., 2004). Another strain of *Roseobacter*, BS107 isolated from scallop (*Pecten maximus*) larval cultures in Brest, France, was inoculated directly and its cell extracts to *P. maximus* larvae and challenged with *L. anguillarum* A496 (Ruiz-Ponte et al., 1999). Significant reduction in the mortalities of larvae was recorded only with the cell extracts of the potential probiont. Since the live BS107 strain could not protect the larvae against action of the pathogen, it still cannot be considered a true probiont. The cell extract therefore acted as antibacterial substances that controlled the pathogen.

Other bacteria have been tested (Decamp, 2008; Gatesoupe, 1989; Gatesoupe, 1991), but in all of them, the experimental designs did not include (or it was not clear) a pathogen and thus it can be questioned if they acted as probionts, feed additives or adjuvant, or immunostimulants.

UNKNOWN BACTERIA

Kesarcodi-Watson et al. (2009) showed that 40 of 69 unidentified bacterial colonies isolated from healthy aquaculture facilities were able to protect Greenshell™ mussel larvae (*Perna canaliculus*) when challenged with two pathogenic vibrios. Larval survival with these probionts improved between 21.3 and 87.4% . It was shown that a pre-exposure time of 20 h to the probiont prior to the pathogen inoculation positively influenced the outcome of the assay. The results of this study suggest that a competitive exclusion mechanism plays a significant role in the protective action of these probionts if they are allowed to colonize the larvae before a pathogen enters the system.

Table 2.1 Proven probiotic bacteria for the larval culture of marine organisms

Probiont	Host larvae	Pathogen	Results	Reference
Gram-positive probionts				
Bacillus sp. IP 5832	Turbot (*Scophtalmus maximus*)	*Vibrio* sp. or *Aeromonas* sp.	Higher survival	Gatesoupe (1991)
Bacillus sp. S11	Tiger shrimp (*Penaeus monodon*)	*Vibrio harveyi* D331	Enhanced growth and survival.	Rengpipat et al. (1998, 2000)
Carnobacterium divergens	Turbot (*S. maximus*)	*V. splendidus* (strain *Vibrio* P)	Higher survival	Gatesoupe (1994)
Carnobacterium divergens		*Listonella anguillarum* LFI 1243	Higher survival after 12 days, none after 4 weeks	Gildberg and Mikkelsen (1998)
Enterococcus faecium (MC13)	Tiger Shrimp (*P. monodon*) postlarvae	*V. harveyi*	Higher survival	Swain et al., 2009
Streptococcus phocae (PI80)	Tiger Shrimp (*P. monodon*) postlarvae	*V. harveyi*	Higher survival	Swain et al., 2009
Gram-negative probionts				
Vibrio E (possibly *V. alginolyticus*)	Turbot (*S. maximus*)	*V. splendidus* (strain *Vibrio* P)	Higher survival	Gatesoupe (1997)
V. alginolyticus Ili	White shrimp (*Penaeus vannamei*)	White Spot Syndrome Virus	Higher survival after 52 h	Rodriguez et al. (2007)
Pseudomonas sp. 11	Chilean scallop (*Argopecten purpuratus*)	*L. anguillarum*-related	Higher survival	Riquelme et al. (1997)
Aeromonas media A199	Pacific oyster (*Crassostrea gigas*)	*Vibrio tubiashii*	Higher survival after 72–96 h	Gibson et al. (1998)
Roseobacter sp. 27-4	Turbot (*S. maximus*)	*L. anguillarum* 90-11-287	Higher survival	Planas et al. (2006)
Different bacterial colonies	Greenshell™ mussel (*Perna canaliculus*)	*Vibrio splendidus* and *V. coralliilyticus/neptunius*-like (*Vibrio* sp. D01)	Higher survival, 20 h pre-exposure	Kesarcodi-Watson et al. (2009)

SUMMARY AND CONCLUSION

In general, Lactic Acid Bacteria have been proven to be the most promising group of probiotic bacteria. This might be, in part, because the intestinal microbiota of normal fishes can be composed mainly by LAB and therefore are able to colonize the exposed surfaces of epithelial cells (Ringo and Gatesoupe, 1998) by adhering to the intestinal mucus (Jöborn et al., 1997). These LAB have demonstrated to inhibit the colonization of the fish intestinal mucus by pathogenic bacteria (Balcazar et al., 2008) and able to produce inhibitory substances (Vazquez et al., 2005) that affect mainly Gram-negative pathogens (Ringo and Gatesoupe, 1998). In shrimps, the gut microbiota of healthy organisms is also composed by Gram-positives, more than 75% of the taxa found by molecular methods in *Litopenaeus vannamei* are members of the Firmicutes; a group that includes Bacillales and LAB (Li et al., 2007). There are relatively few commercial products available as probiotics that are useful in hatcheries, and many of these are composed of LAB or *Bacillus* spp.

Future perspectives to consider for LAB research include evaluating the permanence of the probiont within the host, mainly the ability to colonize the host's intestinal tract and stay there for enough time to prevent pathogen colonization. Also, the ability to be effective under different external and internal conditions has to be estimated. Usually, research has focused on evaluating the efficacy of the probiont under controlled laboratory conditions, but seldom at commercial facilities where environmental parameters could fluctuate greatly. Another area worth studying is devising strain combinations that are capable of enhancing the probiotic effect as compared to single strain formulations.

ABBREVIATIONS

BLIS   Bacteriocin-like inhibitory substances
LAB   Lactic acid bacteria
WSSV   White Spot Syndrome Virus

REFERENCES

Abelli, L., Randelli, E., Carnevali, O., Picchietti, S. (2009). Stimulation of gut immune system by early administration of probiotic strains in *Dicentrarchus labrax* and *Sparus aurata*. In: Trends in Comparative Endrocrinology and Neurobiology. H. Vaudry, E.W. Roubos, G. Coast and M. Vallarino (eds). Ann. New York Acad.Sci., New York, USA pp. 340-342.

Abidi, R. (2003). Use of probiotics in larval rearing of new candidate species. Aquaculture Asia 8:15-16.

Austin, B. and Brunt, J.W. (2009). The use of probiotics in aquaculture. In: Aquaculture Microbiology and Biotechnology. Volume 1. D. Montet and R.C. Ray (eds). Science Publishers, Enfield, New Hampshire, USA pp. 185-207.

Balcazar, J.L. and Rojas-Luna, T. (2007). Inhibitory activity of probiotic *Bacillus subtilis* UTM 126 against *Vibrio* species confers protection against vibriosis in juvenile shrimp (*Litopenaeus vannamei*). Curr. Microbiol. 55: 409-412.

Balcazar, J.L., Vendrell, D., de Blas, I., Ruiz-Zarzuela, I., Muzquiz, J.L. and Girones, O. (2008). Characterization of probiotic properties of lactic acid bacteria isolated from intestinal microbiota of fish. Aquaculture 278: 188-191.

Carnevali, O., Zamponi, M.C., Sulpizio, R., Rollo, A., Nardi, M., Orpianesi, C., Silvi, S., Caggiano, M., Polzonetti, A.M. and Cresci, A. (2004). Administration of probiotic strain to improve sea bream wellness during development. Aquacult. Int. 12: 377-386.

Decamp, O. (2008). Probiotics for shrimp larviculture: review of field data from Asia and Latin America. Aquacult. Res. 39: 334-338.

Dierckens, K., Rekechi, A., Laureau, S., Sorgeloos, P., Boon, N., Van den Broeck, W., Bossier, P. (2009). Development of a bacterial challenge test for gnotobiotic sea bass (*Dicentrarchus labrax* L.). Environ. Microbiol. 11: 526-533.

Dopazo, C.P., Lemos, M.L., Lodeiros, C., Bolinches, J., Barja, J.L. and Toranzo, A.E. (1988). Inhibitory activity of antibiotic-producing marine bacteria against fish pathogens. J. Appl. Bacteriol. 65: 97-101.

Fuller, R. (1989). Probiotics in man and animals. J. Appl. Bacteriol. 66: 365-378.

Garriques, D. and Arevalo, G. (1995). An evaluation of the production and use of a live bacterial isolate to manipulate the microbial flora in the commercial production of *Penaeus vannamei* postlarvae in Ecuador. In: Swimming Through Troubled Waters. C.L. Browdy and J.S. Hopkins (eds). World Aquaculture Society, Baton Rouge, Lousiana, USA pp. 53-59.

Gatesoupe, F.J. (1989). Further advances in the nutritional and antibacterial treatments of rotifers as food for turbot larvae, *Scophthalmus maximus* L. In: Aquaculture—A Biotechnology in Progress. N. De Pauw, E. Jaspers, H. Ackefors and N. Wilkins (eds). European Aquaculture Society, Bredene, Belgium pp. 721-730.

Gatesoupe, F.J. (1990). The continuous feeding of turbot larvae, *Scophthalmus maximus*, and control of the bacterial environment of rotifers. Aquaculture 89: 139-148.

Gatesoupe, F.J. (1991). The effect of three strains of lactic bacteria on the production rate of rotifers, *Brachionus plicatilis*, and their dietary value for larval turbot, *Scophthalmus maximus*. Aquaculture 96: 335-342.

Gatesoupe, F.J. (1994). Lactic acid bacteria increase the resistance of turbot larvae, *Scophthalmus maximus*, against pathogenic vibrio. Aquat. Living Resour. 7: 277-282.

Gatesoupe, F.J. (1997). Siderophore production and probiotic effect of *Vibrio* sp. associated with turbot larvae, *Scophthalmus maximus*. Aquat. Living Resour. 10: 239-246.

Gatesoupe, F.J. (1999). The use of probiotics in aquaculture. Aquaculture 180: 147-165.

Gibson, G.R. and Roberfroid, M.B. (1995). Dietary modulation of the human colonic microbiota: Introducing the concept of prebiotics. J. Nutr. 125: 1401-1412.

Gibson, L.F., Woodworth, J. and George, A.M. (1998). Probiotic activity of *Aeromonas media* on the Pacific oyster, *Crassostrea gigas*, when challenged with *Vibrio tubiashii*. Aquaculture 169:1 11-120.

Gildberg, A. and Mikkelsen, H. (1998). Effects of supplementing the feed to Atlantic cod (*Gadus morhua*) fry with lactic acid bacteria and immuno-stimulating peptides during a challenge trial with *Vibrio anguillarum*. Aquaculture 167: 103-113.

Gildberg, A., Mikkelsen, H., Sandaker, E. and Ringo, E. (1997). Probiotic effect of lactic acid bacteria in the feed on growth and survival of fry of Atlantic cod (*Gadus morhua*). Hydrobiologia 352: 279-285.

Gomez-Gil, B., Roque, A. and Turnbull, J.F. (2000). The use and selection of probiotic bacteria for use in the culture of larval aquatic organisms. Aquaculture 191: 259-270.

Gomez-Gil, B., Roque, A. and Velasco, G. (2002). Culture of the bacterial strain C7b, a potential probiotic bacterium, with the microalgae *Chaetoceros muelleri* Aquaculture 211: 43-48.

Gram, L. and Hjelm, M. (2002). The selection and potential use of aquatic bacteria as fish probiotics. ICES Council Meeting Documents, ICES, Copenhagen, Denmark.

Gullian, M., Thompson, F. and Rodriguez, J. (2004). Selection of probiotic bacteria and study of their immunostimulatory effect in *Penaeus vannamei*. Aquaculture 233: 1-14.

Hai, N.V., Fotedar, R. and Buller, N. (2007). Selection of probiotics by various inhibition test methods for use in the culture of western king prawns, *Penaeus latisulcatus* (Kishinouye). Aquaculture 272: 231-239.

Hernandez-Serrano, P. (2005). Responsible use of antibiotics in aquaculture, FAO Fisheries Technical Paper No. 469, Rome. Italy, 97p.

Hjelm, M., Bergh, O., Riaza, A., Nielsen, J., Melchiorsen, J., Jensen, S., Duncan, H., Ahrens, P., Birkbeck, H. and Gram, L. (2004). Selection and identification of autochthonous potential probiotic bacteria from turbot larvae (*Scophthalmus maximus*) rearing units. System Appl. Microbiol. 27: 360-371.

Huys, L., Dhert, P., Robles, R., Ollevier, F., Sorgeloos, P. and Swings, J. (2001). Search for beneficial bacterial strains for turbot (*Scophthalmus maximus* L.) larviculture. Aquaculture 193: 25-37.

Immanuel, G., Citarasu, T., Sivaram, V., Michael Babu, M. and Palavesam, A. (2007). Delivery of HUFA, probionts and biomedicine through bioencapsulated Artemia as a means to enhance the growth and survival and reduce the pathogenicity in shrimp *Penaeus monodon* postlarvae. Aquacult. Int. 15: 13-152.

Jöborn, A., Olsson, J.C., Westerdahl, A., Conway, P.L. and Kjelleberg, S. (1997). Colonization in the fish intestinal tract and production of inhibitory substances in intestinal mucus and faecal extracts by *Carnobacterium* sp. strain K1. J. Fish Dis. 20: 383-392.

Kesarcodi-Watson, A., Kaspar, H., Lategan, M.J. and Gibson, L. (2009). Screening for probiotics of Greenshell (TM) mussel larvae, *Perna canaliculus*, using a larval challenge bioassay. Aquaculture 296: 159-164.

Li, J., Tan, B., Mai, K., Ai, Q., Zhang, W., Xu, W., Liufu, Z. and Ma, H. (2006). Comparative study between probiotic bacterium *Arthrobacter* XE-7 and chloramphenicol on protection of *Penaeus chinensis* post-larvae from pathogenic vibrios. Aquaculture 253: 140-147.

Li, K., Zheng, T.L., Yun, T. and Yuan, J.J. (2007). Bacterial community structure in intestine of the white shrimp, *Litopenaeus vannamei*. Wei Sheng Wu Xue Bao 47: 649-653.

Maeda, M., Nogami, K., Kanematsu, M. and Hirayama, K. (1997). The concept of biological control methods in aquaculture. Hydrobiologia 358: 285-290.

Makridis, P., Martins, S., Vercauteren, T., Van Driessche, K., Decamp, O. and Dinis, M.T. (2005). Evaluation of candidate probiotic strains for gilthead sea bream larvae (*Sparus aurata*) using an in vivo approach. Lett. Appl. Microbiol. 40: 274-277.

Nitisinprasert, S., Pungsunggworn, N., Wanchaitanawong, P., Loiseau, G., and Montet, D. (2006). *In Vitro* adhesion assay of lactic acid bacteria, *Escherichia coli* and *Salmonella* sp. by microbiological and PCR methods. Songklanakarin J. Sci. Technol. 28: 99-106.

Planas, M., Perez-Lorenzo, M., Hjelm, M., Gram, L., Uglenes Fiksdal, I., Bergh, O. and Pintado, J. (2006). Probiotic effect in vivo of *Roseobacter* strain 27-4 against *Vibrio* (*Listonella*) *anguillarum* infections in turbot (*Scophthalmus maximus* L.) larvae. Aquaculture 255: 323-333.

Prado, S., Montes, J., Romalde, J.L. and Barja, J.L. (2009). Inhibitory activity of *Phaeobacter* strains against aquaculture pathogenic bacteria. Int. Microbiol. 12: 107-114.

Rengpipat, S., Phianphak, W., Piyatiratitivorakul, S. and Menasveta, P. (1998). Effects of a probiotic bacterium on black tiger shrimp *Penaeus monodon* survival and growth. Aquaculture 167: 301-313.

Rengpipat, S., Rukpratanporn, S., Piyatiratitivorakul, S. and Menasaveta, P. (2000). Immunity enhancement in black tiger shrimp (*Penaeus monodon*) by a probiont bacterium (*Bacillus* S11). Aquaculture 191: 271-288.

Ringo, E. (1999). Does *Carnobacterium divergens* isolated from Atlantic salmon, *Salmo salar* L., colonize the gut of early developing turbot, *Scophthalmus maximus* L., larvae? Aquacult. Res. 30: 229-232.

Ringo, E. and Gatesoupe, F.J. (1998). Lactic acid bacteria in fish: a review. Aquaculture 160: 177-203.

Ringo, E. and Vadstein, O. (1998). Colonization of *Vibrio pelagius* and *Aeromonas caviae* in early developing turbot (*Scophthalmus maximus* L.) larvae. J. Appl. Microbiol. 84: 227-233.

Ringo, E., Salinas, I., Olsen, R.E., Nyhaug, A., Myklebust, R. and Mayhew, T.M. (2007). Histological changes in intestine of Atlantic salmon (*Salmo salar* L.) following *in vitro* exposure to pathogenic and probiotic bacterial strains. Cell Tissue Res. 328: 109-116.

Riquelme, C., Hayashida, G., Toranzo, A.E., Vilches, J. and Chavez, P. (1995). Pathogenicity studies on a *Vibrio anguillarum*-related (VAR) strain causing an epizootic in *Argopecten purpuratus* larvae cultured in Chile. Dis. Aquat. Org. 22: 135-141.

Riquelme, C., Araya, R., Vergara, N., Rojas, A., Guaita, M. and Candia, M. (1997). Potential probiotic strains in the culture of the Chilean scallop *Argopecten purpuratus* (Lamarck, 1819). Aquaculture 154: 17-26.

Riquelme, C., Araya, R. and Escribano, R. (2000). Selective incorporation of bacteria by *Argopecten purpuratus* larvae: implications for the use of probiotics in culturing systems of the Chilean scallop. Aquaculture 181: 25-36.

Riquelme, C., Jorquera, M.A., Rojas, A.I., Avendaño, R.E. and Reyes, N. (2001). Addition of inhibitor-producing bacteria to mass cultures of *Argopecten purpuratus* larvae (Lamarck, 1819). Aquaculture 192: 111-119.

Robertfroid, M. (2007). Prebiotics: the concept revisited. J. Nutr. 137: 830-837.

Robertson, P.A.W., O'Dowd, C., Burrells, C., Williams, P. and Austin, B. (2000). Use of *Carnobacterium* sp. as a probiotic for Atlantic salmon (*Salmo salar* L.) and rainbow trout (*Oncorhynchus mykiss*, Walbaum). Aquaculture 185: 235-243.

Rodriguez, J., Espinosa, Y., Echeverria, F., Cardenas, G., Roman, R. and Stern, S. (2007). Exposure to probiotics and [beta]-1,3/1,6-glucans in larviculture modifies the immune response of *Penaeus vannamei* juveniles and both the survival to White Spot Syndrome Virus challenge and pond culture. Aquaculture 273: 405-415.

Rollo, A., Sulpizio, R., Nardi, M., Silvi, S., Orpianesi, C., Caggiano, M., Cresci, A. and Carnevali, O. (2006). Live microbial feed supplement in aquaculture for improvement of stress tolerance. Fish Physiol. Biochem. 32: 167-177.

Ruiz-Ponte, C., Samain, J.F., Sanchez, J.L. and Nicolas, J.L. (1999). The benefit of a *Roseobacter* sp. on the survival of scallop larvae. Mar. Biotechnol. 52-59.

Sarter, S. and Guichard, B. (2009). Bacterial antibiotic resistance in aquaculture. In: Aquaculture Microbiology and Biotechnology, Volume 1. D. Montet and R.C. Ray (eds). Science Publishers, Enfield, New Hampshire, USA pp. 133-157.

Sarter, S., Hoang Nam Kha, N., Le Thanh, H., Lazard J., and Montet D. (2007). Antibiotic resistance in Gram-negative bacteria isolated from farmed catfish. Food Control 18: 1391-1396.

Soto-Rodriguez, S., Armenta, M. and Gomez-Gil, B. (2006). Effects of enrofloxacin and florfenicol on survival and bacterial population in an experimental infection with luminescent *Vibrio campbellii* in shrimp larvae of *Litopenaeus vannamei* Aquaculture 255: 48-54.

Spanggaard, B., Huber, I., Nielsen, J., Sick, E.B., Pipper, C.B., Martinussen, T., Slierendrecht, W.J. and Gram, L. (2001). The probiotic potential against vibriosis of the indigenous microflora of rainbow trout. Environ. Microbiol. 3: 755-765.

Strom, E. and Ringo, E. (1993). Changes in the bacterial composition of early developing cod, *Gadus morhua* (L.) larvae following inoculation of *Lactobacillus plantarum* into the water. In: Physiology and Biochemical Aspects of Fish Development. B.T. Walther and H.J. Fyhn (eds). University of Bergen, Bergen, Sweden, pp. 226-228.

Swain, S.M., Singh, C. and Arul, V. (2009). Inhibitory activity of probiotics *Streptococcus phocae* PI80 and *Enterococcus faecium* MC13 against Vibriosis in shrimp *Penaeus monodon*. World J. Microbiol. Biotechnol. 25: 697-703.

WHO (2006). Report of a Joint FAO/OIE/WHO Expert Consultation on Antimicrobial Use in Aquaculture and Antimicrobial Resistance: Seoul, Republic of Korea, 13-16 June 2006. World Health Organization, WHO Press, Geneva, Switzerland.

Vazquez, J.A., Gonzalez, M.P. and Murado, M.A. (2005). Effects of lactic acid bacteria cultures on pathogenic microbiota from fish. Aquaculture 245: 149-161.

Verschuere, L., Rombaut, G., Sorgeloos, P. and Verstraete, W. (2000). Probiotic bacteria as biological control agents in aquaculture. Microbiol. Mol. Biol. Rev. 64: 655-671.

Williams, R.R., Bell, T.A. and Lightner, D.V. (1992). Shrimp antimicrobial testing: II. Toxicity testing and safety determinations for twelve antimicrobials with *Penaeus* shrimp larvae. J. Aquat. Anim. Health 4: 262-270.

3

Microbial Degradation of Seafood

*Françoise Leroi** and *Jean-Jacques Joffraud*

INTRODUCTION

The flesh of living fish is sterile. However, the skin, mucus, gills and the gastro-intestinal tract contain significant microflora. The composition of the bacterial community is determined to a large extent by the bacteria in the aqueous medium surrounding the fish when it is still in its larval stage (Hansen and Olafsen, 1999) and thus it varies according to a large number of hydrological parameters. The bacterial contents commonly observed vary from 10^2 to 10^5 bacteria/cm^2 in the skin, 10^3 to 10^7 bacteria/cm^2 in the gills and 10^3 to 10^5 bacteria/g in faeces (Abgrall, 1988). At the time of fish death, there is first a loss of freshness due to autolytic enzyme activity. Then microorganisms present in the fish can contaminate the flesh by moving through the muscle fibres or by spreading during the processing stages (gutting, head cutting, filleting, etc.) and their activity leads to spoilage, characterised by off-odours, taste and degradation of texture.

This chapter is devoted to the bacterial degradation of the flesh of fish and shrimp, the most consumed shellfish in the world. Molluscs are not dealt with here because the process of bacterial degradation does not usually begin before consumption and molluscs are most frequently eaten alive or immediately after being cooked.

THE MICROFLORA OF LIVING FISH AND SHELLFISH

It is generally agreed that the flora of fish found in temperate waters consist of Gram-negative psychrotolerant bacteria, whose growth is possible at 0°C but optimal around 25°C. Among these, the majority belongs to the subclass γ of proteobacteria: *Pseudomonas*, *Shewanella*, *Acinetobacter*,

Ifremer, Laboratoire Science et Technologie de la Biomasse Marine, BP 21105, 44311, Nantes, France
*Corresponding author: Tel.: +33 2 40 37 41 72; Fax: 33 2 40 37 40 71; E-mail: fleroi@ifremer.fr

Aeromonas, Vibrio, Moraxella, Psychrobacter, Photobacterium, etc. and to a lesser extent the CFB *(Cytophaga-Flavobacter-Bacteroides)* group (Huber et al., 2004; Wilson et al., 2008). Nevertheless, Gram-positive bacteria such as *Micrococcus, Bacillus, Lactobacillus, Clostridium* or Coryneforms, may also be present in variable proportions (Shewan, 1971, 1977; Hobbs, 1983; Mudarris and Austin, 1988; Gram and Huss, 1996a; Gennari et al., 1999; Wilson et al., 2008). Some genera like *Vibrio, Photobacterium* and *Shewanella*, require the presence of salt to multiply and are thus typically found in seawater while *Aeromonas* is more common in freshwater even though it is often isolated from marine products (Hanninen et al., 1997). In tropical fish, the flora has the same composition overall (Al Harbi and Uddin, 2005; Emborg et al., 2005), but often with a greater proportion of Gram-positive bacteria (*Micrococcus, Bacillus,* Coryneforms) and enterobacteria (Devaraju and Setty, 1985; Liston, 1992; Huss, 1999).

The indigenous microflora of the gastro-intestinal tract of fish have been studied much more than those of the skin or the mucus due to their importance in digestion, nutrition and growth and in disease control in aquaculture (Ringo et al., 1995; Spanggaard et al., 2000). Although this environment is partially anaerobic, most researchers have observed a predominance of aerobic bacteria which is also present in the surrounding water and capable of surviving and multiplying in the particular medium of the gastro-intestinal tract (Cahill, 1990). This predominance of aerobic bacteria could be due to the collecting techniques, which are not always suitable for strict anaerobes (Burr et al., 2005). Nevertheless, Huber et al. (2004) have shown by molecular methods that the aerobic flora of rainbow trout intestine usually represents 50-90% of the total flora. Gram-negative bacteria dominate the intestinal flora. In general, *Aeromonas, Pseudomonas* and members of the *Flavobacterium/Cytophaga* group are most often found in the intestine of freshwater fish while *Vibrio, Acinetobacter,* and Enterobacteriaceae are more common in marine fish (Ringo et al., 1995; Ringo and Birkbeck, 1999). These are fermentative bacteria that develop rapidly in the gastro-intestinal tract due to the low pH, the lack of oxygen and the abundance of nutrients. Staphylococci have also been found to be the dominant flora in the intestine of the Arctic char (Ringo and Olsen, 1999). Although not predominant, lactic acid bacteria (LAB) *(Lactobacillus, Carnobacterium, Streptococcus, Leuconostoc, Lactococcus, Vagococcus)* have often been isolated from the gastro-intestinal tract of fish (Ringo and Gatesoupe, 1998).

Both the number and diversity of the microflora are probably widely underestimated because the majority of studies carried out so far have used classic microbiological methods involving growth on agar media. The cultivability of bacteria has been estimated to be sometimes less than 2% of the intestinal flora of rainbow trout (Huber et al., 2004), or even less than 0.01% of the skin flora (Bernadsky and Rosenberg, 1992). Pond et al. (2006) have identified a strict anaerobe (*Clostridium gasigenes*) in the intestinal flora of rainbow trout. Similarly, Kim et al. (2007) have shown the presence of *Clostridium* in the intestinal mucus. Moreover, molecular methods have enabled a new species belonging to the genus *Mycoplasma* to be detected for the first time in fish. It was found in abundance in the intestine of wild and farmed salmon (Holben et al., 2002).

The worldwide shrimp market is mainly composed of the Nordic shrimp (*Pandalus borealis*), which is only fished, and the tropical shrimp (*Penaeus* sp.), which can be fished or farmed and whose production has expanded rapidly in recent years. The deep-water tropical shrimp (*Parapenaeus longirostris*) is also found in Europe, particularly in the Spanish and Portuguese markets.

As in fish, the bacterial flora of shrimps depends on several factors including the species considered, the geographic location and environment, the temperature and salinity of the water, etc. However, overall, the same species of microorganisms are found in shrimps and in fish from a given geographical zone. In fresh tropical shrimps, the initial bacterial flora consists mainly of *Pseudomonas*, *Vibrio*, *Acinetobacter*, *Moraxella*, *Flavobacterium* and a high proportion of *Aeromonas* (Vanderzant et al., 1973; Jayaweera and Subasinghe, 1988; Jeyasekaran et al., 2006). In India, Gopal et al. (2005) have detected significant amounts of different species of *Vibrio,* including *V. parahaemolyticus*. Benner et al. (2004), working on Nicaraguan shrimps, reported a predominance of Coryneforms and *Moraxella* followed by lower levels of *Bacillus*, *Lactobacillus*, *Micrococcus*, *Proteus*, *Shewanella, Acinetobacter* and *Pseudomonas*. These results confirm those previously obtained by Matches (1982). Chinivasagam et al. (1996) have shown the influence of the fishing zone on the nature of the initial flora: mostly Gram-positive bacteria on shrimps fished at low depths and *Pseudomonas* on those caught in deep water. The nature of this initial flora has an effect on the shelf life and the organoleptic properties of shrimps linked to spoilage.

MICROBIAL CONTAMINATION OF FLESH AND ITS EVOLUTION DURING PRESERVATION

At fish death, the immune system collapses and bacteria can contaminate the flesh by moving through the muscle fibres. However, bacteria are found in much greater quantities on the skin than in the tissues and it is more likely that the spoilage of the whole fish is mostly due to bacterial enzymes that have spread through the tissues. On the other hand, the various stages of processing (evisceration, head removal, filleting, and trimming) contribute to spreading the bacteria naturally present in fish throughout the muscular tissue, thus accelerating spoilage.

Regarding the bacterial growth, first, there is a lag phase, generally short for fish found in temperate waters but whose length varies depending on the composition and state of the fish, the storage temperature and the species of bacteria. Next, the microorganisms begin a period of exponential growth and can reach level of 10^{6-8} CFU (Colony Forming Units)/g. Clearly, the rate of multiplication depends on the factors mentioned previously. Generally, tropical fish kept in ice have a longer lag phase (1 to 2 wk) than that of fish found in temperate waters and growth during the exponential phase is slower (Gram et al., 1990; Gram, 1995), probably because the microorganisms cannot adapt well to the low storage temperatures (Devaraju and Setty, 1985).

THE CONCEPT OF SPECIFIC SPOILAGE FLORA

Sensory spoilage is not always linked to the total number of microorganisms. In most seafood products organoleptic rejection occurs well after the total flora has reached its maximum. Although the flora of fish is varied, only certain microorganisms, called "specific spoilage microorganisms" are responsible for the production of unpleasant odours and flavours.

In general, in unprocessed fish, the chemical modifications that lead to sensory rejection are due to only one bacterial species (Hozbor et al., 2006). In the case of processed products, however, the mechanisms are often more complex as several bacterial groups can interact and contribute to the spoilage of the product (Stohr et al., 2001; Joffraud et al., 2006). The specific spoilage flora and associated chemical molecules vary according to the species of fish, the storage temperature, the type of packaging or processing and even the fishing season.

Identifying the microorganisms specifically responsible for spoilage is not easy. The common method consists in analysis of the products during storage (selective microbiological enumeration, sensory and chemical analyses) and establishment of correlations between different parameters. However, the enumerations are not always selective enough and it is often better to collect colonies from the total flora at the time of sensory spoilage and then check their capacity to produce bad odours in a pure culture. Tests done directly in the fish matrix always give more relevant results than using a laboratory medium or fish flesh extract (Truelstrup Hansen, 1995; Stohr et al., 2001) but are often more difficult to carry out as the fish must be sterilised without modifying its composition. The collections of fillets in aseptic conditions (Herbert et al., 1971) or low intensity ionisation (Joffraud et al., 1998; Jorgensen et al., 2000b) are the solutions most frequently used. Once the microorganisms potentially involved in spoilage have been identified, it is important to check their colonisation kinetics on naturally contaminated products and during "real" storage conditions (temperature, packaging). Potential spoilage flora in a model medium may not have the capacity to develop enough to reach levels that could cause perceptible sensory deterioration (Jorgensen et al., 1988). Furthermore, interactions with endogenous flora could modify the spoiling characteristics of microorganisms (Gram and Huss, 1996a).

BACTERIAL METABOLISM

Fish is a matrix that particularly favours microbial development. Despite a low percentage of carbohydrates (0.2 to 1.5% depending on the species), fish flesh is rich in non-protein, low molecular weight nitrogenous molecules that are rapidly metabolised by bacteria. These compounds include free amino acids, creatine, nucleotides, urea and trimethylamine oxide (TMAO). The high *post-mortem* pH (>6) and the low acidification during preservation, combined with the small quantity of carbohydrates present, enable the rapid growth of pH-sensitive psychrotrophic spoilage bacteria like *Shewanella putrefaciens*. Lastly, fatty fish are rich in polyunsaturated fatty acids that can be rapidly oxidised by either chemical chain reactions or lipolysis resulting from autolytic or bacterial enzyme activity.

In aerobiosis, the carbohydrates (ribose and lactate) can be metabolised into CO_2 and H_2O. In anaerobiosis, and in the presence of an electron acceptor such as TMAO some microorganisms are capable of anaerobic respiration that leads to the production of acetic acid. However, spoilage of fresh fish is rarely linked to the production of organic acids due to the low concentration of carbohydrates in the flesh. TMAO, whose concentration varies according to the species, plays an extremely important role in spoilage as certain microorganisms such as *Shewanella, Photobacterium* and *Aeromonas* reduced it to trymethyl amine (TMA), a pungent molecule responsible for the strong amine odour typical of rotten fish. TMAO is occasionally found in freshwater fish (Anthoni et al., 1990) but is generally associated with marine fish (Seibel and Walsh, 2002).

Urea found in large quantities in selachians can be metabolised into ammonia which has a strong, unpleasant odour. Deamination of amino acids also leads to the production of ammonia.

The breakdown of sulphurous amino acids found naturally in fish leads to the production of H_2S (hydrogen sulphide) (from cysteine), methylmercaptan and dimethyl-disulphide (from methionine). These molecules all play a part in the spoilage process (Shewan, 1974; Lee and Simard, 1984). Certain microorganisms such as *Shewanella, Photobacterium* and lactic acid bacteria (LAB) are capable of this production, but to an extent that varies depending on the strain.

The decarboxylation of amino acids leads to the formation of biogenic amines, which are often linked to spoilage (Veciana-Nogues et al., 1997) even though they have no particular odour in the product (Jorgensen et al., 2000a). Tyrosine is a precursor of tyramine and cadaverine. Arginine can be degraded into putrescine *via* the agmatine pathway in the presence of arginine decarboxylase, as is the case with *Photobacterium* (Jorgensen et al., 2000b). Putrescine can also be produced by the decarboxylation of ornithine, for example in the case of certain enterobacteria such as *Hafnia alvei* and *Serratia liquefaciens* (Grimont and Grimont, 1992; Sakzaki and Tamura, 1992). Histamine is a biogenic amine formed by the degradation of histidine. It can provoke allergic-like reactions of varying intensity (redness of the skin, swelling, headaches) and fish with high levels of histidine are strictly regulated (Scombridae and Clupeidae, CE regulation NO. 2073/2005). In fish, the principal producers are mesophilic enterobacteria such as *Morganella morganii, Hafnia alvei* and *Klebsiella pneumoniae* (Kim et al., 2002; Kim et al., 2004), but recently psychrotolerant microorganisms such as *Photobacterium phosphoreum* and *Morganella psychrotolerans* have been clearly incriminated in cases of histamine poisoning (Kim et al., 2002; Dalgaard et al., 2006).

SPOILAGE OF FRESH FISH AND SHELLFISH

Fish

Fish spoilage occurs at varying speeds depending on a large number of parameters. Different fish species display a variety of surface characteristics in terms of resistance of their skin texture and the composition of the mucus, which may contain antibodies and bacteriolytic enzymes. Geographical zone, in particular water temperature, also plays an important role as fish from tropical waters keep much longer in ice than those from temperate or cold waters. Rough handling of the fish can damage its integrity and accelerate spoilage. The storage conditions (temperature and packaging) also play a vital part in fish preservation.

When temperate-water sea fish (for example cod) are stored at chilled temperatures, bacteria such as *Shewanella putrefaciens* are the specific spoilage bacteria. Jorgensen et al. (1988) have shown that the number of bacterial cells is inversely correlated to the remaining shelf life, and consider that at approximately 10^8 CFU/g the sensory deterioration of the product is no longer acceptable. The spoilage is characterised by the production of TMA, H_2S and other sulphurous compounds such as methylsulphide and di-methylsulphide. *Sh. putrefaciens* consists of a heterogeneous group whose taxonomy has greatly changed over the last few years. Vogel et al. (2005) have shown that the large majority of microorganisms that produce H_2S isolated from Baltic Sea fish chilled for several days should now be identified not as *Sh. putrefaciens*, but as *Sh. baltica*. Furthermore, other minor strains have been identified as new species: *Sh. hafniensis*, *Sh. morhuae*, *Sh. glacialipiscicola* and *Sh. algidipiscicola* (Satomi et al., 2006; Satomi et al., 2007). Tropical sea or freshwater fish are spoiled by *Pseudomonas*. *Sh. putrefaciens* has also been isolated from these products but does not appear to play an important part in spoilage. This could be due to the inability of this microorganism to develop in the presence of a large number of *Pseudomonas* (Gram et al., 1990; Gram and Melchiorsen, 1996b). The spoilage caused by this microorganism can be differentiated from *Sh. putrefaciens* by the absence of TMA and sulphurous compounds and the appearance of fruity and rotten odours caused by aldehydes, ketones and esters. When fish is stored at room temperature, *Aeromonas* is more likely to spoil freshwater fish stored in aerobiosis, but it has also been shown that *Sh. putrefaciens* can be involved.

Effects of Storage Temperature on Spoilage and Shelf life

Storage temperature is the most influential factor for shelf life. Ratkowsky et al. (1982) have described models that express the relation between relative spoilage speed and storage temperature. The relative spoilage speed at a temperature T is defined as the ratio of the shelf life at 0°C to the shelf life at T°C. For example, if the shelf life of a cod is 14 d at 0°C and 6 d at 5°, the relative speed of spoilage will be equal to $14/6 = 2.3$, i.e., the spoilage will be 2.3 times faster at 5°C than at 0°C. Shelf life at 0°C differs depending on the species of fish and the method of preservation, but the effect of temperature on the relative spoilage speed R is constant and the following formula has been established: $\sqrt{R} = 1 + 0.1 * T°C$. This "square-root" model enables the shelf life of a product to be calculated at different temperatures if its shelf life is known at a certain temperature.

The bacteria responsible for the spoilage of fresh fish are different depending on the storage temperature. *Sh. putrefaciens*, *Sh. baltica* *Pseudomonas* sp., *Aeromonas* sp. and *Phosphobacterium phosphoreum* are the principal spoilage bacteria found between 0 and 5°C, and their quantity varies depending on the storage atmosphere. At 15-30°C, Enterobacteriaceae, Vibrionaceae and Gram-positive bacteria are responsible for spoilage. The "square-root" model described previously does not take into account these changes in flora. Nevertheless, the estimations of relative spoilage speed are satisfactory for whole fresh fish, vacuum-packed fish or modified atmosphere packaged (MAP) fish (Gibson and Ogden, 1986; Dalgaard and Huss, 1997). However, for tropical fish, the relative spoilage speeds at 20-30°C are more than double those estimated by the model. A separate model for tropical fish has therefore been developed for temperatures between 0 and 30°C by Dalgaard and Huss (1997). It establishes a linear correlation between the Neperian logarithm of the relative spoilage speed and the storage temperature (Ln $R = 0.12 * T°C$).

Super-chilling consists in storing products at temperatures between 0 and $-4°C$. The "square-root" model gives satisfactory results for products stored in this way (Dalgaard and Huss, 1997). This technique can greatly extend the shelf life of the product but often lower the product quality (water retention, texture, etc.).

The Effects of Storage Atmosphere on Shelf life: Anaerobiosis and Carbon dioxide (CO_2)

Vacuum-packaging or MAP with varying amounts of CO_2 (25-100%) are both widely used to preserve food. Numerous studies have been carried out on seafood products with very different results. Most frequently, high concentrations of CO_2 lead to a 30 to 60% longer shelf life. The efficiency is closely linked to the temperature, which must be as low as possible so that the gas can dissolve in the product (Sivertsvik et al., 2002). However, unlike in meat products, these techniques do not significantly lengthen fish shelf life. In fact, the number of *Pseudomonas* decreases due to the lack of oxygen in vacuum-packed temperate-water sea fish. *Sh. putrefaciens* can respire in anaerobiosis due to TMAO, and can develop up to 10^6-10^8 CFU/g, thus increasing the production of TMA (Gram et al., 1987; Jorgensen et al., 1988; Dalgaard et al., 1993). Nevertheless, it is probable that, below 10^8 CFU/g, *Sh. putrefaciens* is not the only microorganism responsible for fish spoilage as these authors have also identified the presence of *Ph. phosphoreum* on vacuum-packed cod. This microorganism, which had until recently escaped detection by microbiologists produces 10 to 100 times more TMA per cell than *Sh. putrefaciens*. Consequently, these two microorganisms can provoke the spoilage of vacuum-packed temperate-water sea fish and it is probably the initial quantity of each type that determines which one will dominate the flora. Furthermore, the growth rate of *Ph. phosphoreum* increases in anaerobiosis, which explains why this microorganism plays such an important part in the spoilage of packaged products such as cod (Dalgaard et al., 1993).

In the presence of CO_2, the growth of *Sh. putrefaciens* and *Pseudomonas* is greatly inhibited whereas *Ph. phosphoreum* is relatively resistant (Dalgaard et al., 1993; Dalgaard, 1995). It reduces TMAO to TMA without producing H_2S, which explains why MAP products are characterised by high levels of TMA without the odour of H_2S typical of spoiled fish stored in aerobiosis. *Ph. phosphoreum* can be eliminated if the raw material is first frozen, which increases the shelf life of thawed cod stored in a modified atmosphere at 2°C from 12 to 20 d (Guldager et al., 1998).

Numerous other seafood products have a shelf life similar to that of cod. As *Ph. phosphoreum* is widespread in the marine environment, it seems likely that this microorganism, or others resistant to CO_2, is responsible for the spoilage of packed seafood products (Van Spreekens, 1974; Dalgaard et al., 1993). Indeed, Emborg et al. (2002) have shown that *Ph. phosphoreum* is the specific spoilage bacteria of MAP salmon. Hovda et al. (2007) have shown that Atlantic halibut *(Hippoglossus hippoglossus)* packed in a CO_2/O_2 (50/50) enriched atmosphere has a microflora dominated by *Ph. phosphoreum* and, due to the presence of oxygen, *Pseudomonas* sp. as well as *Brochothrix thermosphancta* on samples at the end of their shelf life. *Ph. phosphoreum* has also been found on pollock (Rudi et al., 2004) and tuna from the Indian Ocean *(Thunnus albacares)* (Emborg et al., 2005).

Ph. phosphoreum constitutes a heterogenic group whose taxonomy has evolved over the last few years. Some strains isolated from seafood products have characteristics that are not homogenous, notably in luminous and non-luminous strains. For this reason, some of the strains isolated from cod fillets in a modified atmosphere have been reclassified as *Ph. iliopiscarium* (Ast and Dunlap, 2005).

The efficiency of MAP has also been demonstrated for warm-water fish. Drosinos et al. (1997) and Drosinos and Nychas (1996) have identified *Br. thermosphacta* as the dominant species on sea bream packed in a CO_2-enriched atmosphere (40%) at the end of their shelf life. This microorganism could play a part in spoilage. Emborg et al. (2005) have demonstrated that, for tropical tuna steaks, a CO_2/O_2 (40/60) modified atmosphere is better than vacuum-packing for inhibiting the growth and histamine production of the psychrotrophic, *Morganella*.

The combination of super-chilling and modified atmosphere enabled salmon fillets to retain a high level of quality after 24 d of storage due to the almost total inhibition of psychrotrophic bacterial growth (Sivertsvik et al., 2003). The shelf life is 3 to 4 times longer than that of a refrigerated product under air.

Shellfish

Amongst shellfish, Nordic and tropical shrimps are by far the most consumed products worldwide and this section will be wholly devoted to them.

Tropical shrimps (*Penaeus* sp.) are generally frozen shortly after catch and sold frozen before being processed (usually cooked) on land. Nordic shrimps *(Pandalus borealis)* are generally cooked after catch, sometimes peeled, and then frozen.

The few studies concerning this product have shown that shrimps are very sensitive to spoilage. This sensitivity is due to a relatively high pH (> 7) and the fact that the intestinal tract containing enzymes and bacterial flora is not removed immediately after catch. In addition, the muscle contains large quantities of non-protein, water-soluble molecules such as amino acids. These amino acids are used directly by the bacteria for their growth. Shrimps generally contain large amounts of arginine and glycine and small amounts of cysteine and methionine (Chinivasagam et al., 1998).

Shelf life is variable and depends on the nature of the initial flora and the spoilage bacteria, often unknown, that may develop during storage. It can be relatively short: 5 to 8 d for ice-packed *Penaeus merguiensis* (Gonçalves et al., 2003); 6 d at 0°C for *Parapenaeus longirostris* (Mendes et al., 2005). Other authors have found longer shelf lives for tropical shrimps: 13 to 16 d when ice-packed (Cann, 1974; Jayaweera and Subasinghe, 1988; Shamshad et al., 1990); 10 to 17 d when ice-packed or over 20 d when stored in liquid ice for different species of the genus *Penaeus* found in Australian waters (Chinivasagam et al., 1996).

The major cause of early shrimp spoilage is melanosis, which is the formation of black spots on the cephalothorax. This process is biochemical and is not due to bacterial activity, but rather an enzyme complex called polyphenol oxidase. The benzoquinones thus produced interact with amines, amino acids and oxygen to form the melanin responsible for this black colouration.

Other organoleptic changes are generated by the action of bacteria. Chinivasagam et al. (1996) have shown that when tropical Australian shrimps are ice-packed in an oxygen-rich environment, the growth of *Pseudomonas fragi* increases, whereas using liquid ice where oxygen is limited leads to the development of *Sh. putrefaciens*. The sensory characteristics of contaminated products are very different depending on the predominance of one or the other of these bacteria. To confirm these observations, Chinivasagam et al. (1998) have demonstrated experimentally that strains of *Ps. fragi* inoculated into shrimp juice produce odours of fruit and onion, whereas strains of *Sh. putrefaciens* give off a sulphurous odour. Moreover, these two bacteria display different profiles of volatile compounds.

Packaging fresh shrimps *(Parapenaeus longirostris)* in a modified atmosphere combining 40-45% CO_2 with 30 or 5% O_2, respectively, delays microbial growth and the production of total volatile nitrogen bases (TVBN) and TMA compared to air-packed or iced stored shrimp, especially at the end of storage (López-Caballero et al., 2002). Amine production and the low level of H_2S-producing microorganisms and enterobacteria suggest that *Ph. phosphoreum* could be involved in the spoilage of shrimps both in air and MAP. The production of biogenic amines (tyramine, putrescine, cadaverine, agmatine) during storage is higher for atmospheres with modified O_2 concentrations than for air. Gonçalves et al. (2003) have shown that, for the same species of shrimp and using the same gas compositions, shelf life can be prolonged at least by 2 d (9 d instead of 4 to 7 d with the ice storage). This type of packaging can also extend the shelf life of cooked whole shrimps *(Pandalus borealis)* by 200% compared to air-packed products (Sivertsvik et al., 1997).

SPOILAGE FLORA OF LIGHTLY PRESERVED PRODUCTS

Semi-preserved seafood products have undergone a very light preservation process, such as salting, drying, smoking or marinating, leading to a final pH greater than 5.0 and a salt content below 6% in the aqueous phase. Examples include smoked fish, carpaccio of salmon or trout, gravelax, marinated anchovy fillets, cooked products like peeled shrimps, preserved in brine or in a modified atmosphere etc. In terms of microbiology, these products are very fragile. Because the raw material follows a long preservation process with a good deal of handling, there is a significant risk of recontamination. Moreover, no step of the process involves the total elimination of microorganisms and the final product is often preserved for several weeks at low temperature before being eaten. This allows some psychrotrophic bacteria, which can spoil organoleptic qualities or present a risk to human health, to multiply.

Two examples representative of this product category, cold-smoked salmon and cooked peeled shrimps, have been discussed in this chapter due to their economic importance in the European market.

Cold-smoked Salmon

In France, there is a long tradition of salting, drying and cold-smoking seafood products, mainly salmon and herring. Nowadays, in the industrialised countries, the aim of smoking is not so much to ensure a long shelf life for the product but to give it a particular colour and taste. With more than 20,000 tonnes produced each year, France has always been the world leader in the production and consumption of smoked salmon. It was originally a luxury product, eaten only on special occasions and made by traditional methods. In recent decades, however, smoked salmon has become an everyday food, available in supermarkets all year round and produced on an industrial scale.

The production of smoked salmon involves different steps of filleting, salting, drying and smoking at low temperatures (< 25°C) but each plant has developed its own technology: salting by dry salt or by injection, smoking with different wood essences and different smoke generators etc (Duffes, 1999; Sérot et al., 2004; Knockaert, 2005; Cardinal et al., 2006). The final salt concentration is usually between 2.5 and 3.5% in the flesh while the smoke content, estimated by the quantity of total phenols, is between 2 and 20 µg/g. These products are generally sold sliced and vacuum-packed. The producer is responsible for the use-by date, which is usually 3 to 6 wk at 4°C in Europe. However, it is not uncommon to notice deterioration in taste from the end of the second week of preservation (Leroi et al., 2001; Cardinal et al., 2004).

It has been clearly shown that this organoleptic spoilage is mainly due to the microbial activity, with autolytic or chemical reactions of lipid oxidation playing a lesser role (Joffraud et al., 1998). For a very long time, however, the mechanisms of microbial spoilage of smoked fish were poorly understood. Although lactic flora have always been found to be dominant in this product (Magnusson and Traustadottir, 1982; Hildebrandt and Erol, 1988; Civera et al., 1995), their role in spoilage was not clear because no correlation between their number, nor that of the total flora, and the sensory deterioration could be identified (Rakow, 1977; Cann et al., 1984; Hildebrandt and Erol, 1988; Dodds et al., 1992; Huss et al., 1995; Truelstrup Hansen, 1995; Gram and Huss, 1996a). More recent studies have specified the composition of the flora of these products and their link with spoilage.

On leaving the factory, product contamination can range from 10^2 to 10^6 bacteria/g, the initial level varies mainly according to the factory and its hygiene standards and independently of the origin and state (fresh or frozen) of the raw material (Truelstrup Hansen et al., 1998; Leroi et al., 2001). Despite the presence of inhibitory factors like salt, smoke, preservation at low temperature and vacuum-packaging, bacterial growth in these products can be quite rapid. It is not unusual to observe total flora concentrations of 10^6 CFU/g from the first week of storage, which can regularly reach 10^{7-9} CFU/g before the use-by date (Cardinal et al., 2004; Dondero et al., 2004; Espe et al., 2004).

The initial flora of smoked salmon is often dominated by Gram-negative bacteria typical of fresh fish flora, such as *Shewanella, Photobacterium*, and *Aeromonas*, later identified as *Serratia* (Leroi et al., 1998). Nevertheless, along with *Photobacterium, Brochothrix, Yersinia* and *Carnobacterium* were also found (Olofsson et al., 2007). During vacuum-packed preservation at low temperatures, the diversity of genera decreases and Gram-positive bacteria, especially LAB, very often become predominant. Variations between factories (Leroi et al., 2001) and even between batches from the same factory have often been observed (Truelstrup Hansen et al., 1998), but the lactic flora always seems to dominate, regardless of the geographical provenance of the products analysed (processed in Europe). According to the authors, 50 to 90% of the colonies taken from the total flora of samples at the end of preservation were LAB (Leroi et al., 1998; Truelstrup Hansen et al., 1998). The technological parameters of salting and smoking can influence the final proportion of this group of bacteria (Leroi and Joffraud, 2000) so it should be possible to control the level of lactic flora by varying these parameters (Tomé et al., 2007).

In smoked salmon, the major species are *Carnobacterium maltaromaticum* and *Lactobacillus curvatus* or *sakei*. Other species such as *Carnobacterium divergens, Lactobacillus farciminis, Lb. alimentarius, Lb. plantarum, Lb. homohiochii, Lb. delbrueckii, Lb. casei, Lb. coryneformys, Leuconostoc mesenteroides, Enterococcus faecalis* and *Weisella kandleri* are more rarely isolated (Leroi et al., 1998; Truelstrup Hansen et al., 1998; Jorgensen et al., 2000b; Gonzales-Fandos et al., 2004; Rachman et al., 2004). Despite this predominance, quite often the number of enterobacteria, strains of *Ph. phosphoreum* and *Br. thermosphacta* are also fairly high (Truelstrup Hansen et al., 1998; Jorgensen et al., 2000a; Jorgensen et al., 2000b; Rachman et al., 2004). In contrast, yeasts are rarely present, except sometimes when the product contains a high concentration of salt or phenol (Leroi and Joffraud, 2000). Nevertheless, they do not reach high enough population levels to contribute to spoilage. To illustrate the complexity of the flora of smoked salmon, Gonzales-Fandos et al. (2004) have shown that out of 96 isolates coming from 30 batches of smoked salmon in Spain and preserved for 3 wk at 2°C, 49% were LAB, 20% enterobacteria, 16% micrococci (mostly coagulase-negative staphylococci), 5% *Br. thermosphacta*, 5% Gram-negative *(Moraxella, Acinetobacter* and *Pseudomonas)*, 2.5% mobile Aeromonadaceae and 2.5% *Bacillus.*

Recently, temporal temperature or denaturating gradient gel electrophoresis (TTGE, DGGE) molecular methods have been applied to smoked salmon. In some cases, these techniques have given similar results to the culture method (predominance of *Lactobacilli, Ph. phosphoreum* and *Ph. iliopiscarium* at the end of preservation) (Olofsson et al., 2007). In contrast, Cambon-Bonavita et al. (2001) found clones corresponding to Gram-negative bacteria *(Vibrio, Photobacterium, Enterobacteriaceae, Alteromonas)* and assumed that this technique is biased because it does not allow a good amplification of the DNA of Gram-positive bacteria. Nevertheless, in both studies, clones no doubt corresponding to new species of *Photobacterium* and *Vibrio* were identified by the culture-independent technique, underlining how these tools can complement classic culture methods.

The role of LAB in the sensory deterioration of smoked salmon is still not very clear. Several authors have shown that there is no correlation between the total lactic flora and sensory spoilage. Paludan.Müller et al. (1998), however, were able to prolong the shelf life of smoked salmon by inhibiting LAB with nisine, suggesting a possible spoilage effect by this bacterial group. By inoculating cubes of smoked salmon sterilised by ionisation, Stohr et al. (2001) clearly demonstrated that certain species of LAB spoilt a lot (i.e., *Lb. sakei)* while others had no effect at all (i.e., *Lb. alimentarius)*. However, the potential for spoilage seems to vary according to the strain tested. *Lb. sakei* generally produces sulphurous and acidic odours (Truelstrup Hansen et al., 1995; Nilsson et al., 1999; Stohr

et al., 2001), associated with the production of H₂S, acetic acid, ethyl and
n-propyl-acetate (Joffraud et al., 2001), but some strains of *Lb. sakei* do
not spoil the organoleptic qualities of this product (Weiss and Hammes,
2006). Similarly, *Lb. alimentarius*, which does not spoil smoked salmon, has
been identified as the bacteria responsible for the sensory deterioration
of marinated herrings (Lyhs et al., 2001). The role of Carnobacteria is still
under study (Laursen et al., 2005; Leisner et al., 2007). Many studies have
shown that they are probably not responsible for spoilage in smoked
salmon. Inoculation by different strains of *Carnobacterium maltaromaticum*
and *Cb. divergens* results in no or few changes in organoleptic qualities
(Leroi et al., 1996; Paludan.Müller et al., 1998; Duffes, 1999; Nilsson
et al., 1999). When the Carnobacteria reach a sufficient level, odours of
butter and plastic may be detected, associated with the production of 2,3-
butanedione (diacetyl) and 2,3-pentanedione (Joffraud et al., 2001; Stohr et
al., 2001), but these odours/flavours do not lead the product to be rejected
by a specialist jury (Brillet et al., 2005).

Among the other bacteria frequently found in smoked salmon, Stohr
et al. (2001) showed that *Serratia liquefaciens* spoiled a lot, releasing
odours of amines, cheese, acid or rubber, associated with the molecules
TMA, dimethyldisulphur, 2,3-butanediol and 2-pentanol (Joffraud et
al., 2001). However, *Se. liquefaciens* was considered to spoil less than *Lb.
sakei* as the unpleasant odours were perceived much later (Joffraud et al.,
2006). *Brochothrix thermosphancta* also leads to the sensory rejection of the
product due to the odours of blue cheese and plastic, well correlated with
2-heptanone and 2-hexanone. Nevertheless, it is quite rare for these bacteria
to reach sufficiently high levels in naturally contaminated products to be
the sole explanation for sensory rejection. Although strongly spoiling in
fresh fish packed in a modified atmosphere, *Ph. phosphoreum* seems to
play a more moderate role in the deterioration of smoked salmon. Weak
odours of "acid", "amine" and "feet" result in the product being judged
as moderately spoilt (Joffraud et al., 2006). Moreover, there is a great
variability according to the strain (Leroi et al., 1998; Stohr et al., 2001).
Jorgensen et al. (2000b) give much greater weight to the spoiling action
of this species. They have shown a good correlation between the sensory
quality of smoked salmon and the production of tyramine and histamine
(Jorgensen et al., 2000a), two potential chemical indicators for spoilage.
Shewanella putrefaciens, the most common spoilage bacteria in fresh fish,
and *Vibrio* sp. have never been implicated in the spoilage phenomena in
smoked salmon, even when inoculated at high concentrations.

Although the bacteria responsible for sensory deterioration are now
quite well identified, spoilage remains a complex phenomenon because the
interactions between all these bacterial groups change their metabolism.
For example, when *Lb. sakei* was co-inoculated with *Se. liquefaciens*, spoilage

was clearly delayed. Conversely, when *Cb. maltaromaticum* and *Vibrio* sp. were both present (two non-spoiling bacteria in pure culture) spoilage was increased significantly (Joffraud et al., 2006). Brillet et al. (2005) noted that *Cb. divergens* did not produce TVBN in pure culture in sterile smoked salmon but some TVBN was detected when the bacterium was added to commercial samples containing a natural endogenous flora. Some phenomena of metabolic interaction between LAB and enterobacteria have been explained by Jorgensen et al. (2000b). For example, a high production of putrescine (>200 µg/g) could not be due to the simple degradation of ornithine, whose concentration does not exceed 10 µg/g in salmon flesh. Arginine deaminase-positive LAB, like *Cb. divergens* and *Lb. sakei*, metabolise arginine, found in greater quantity in salmon, into ornithine, which is then converted to putrescine by ornithine decarboxylase-positive enterobacteria (i.e., *Se. liquefaciens* and *Hafnia alvei).*

All these results show that it is over-optimistic to predict the quality of lightly preserved seafood products using only one microbiological or biochemical parameter. Not only is the microflora very complex but also the organoleptic properties of spoilage are greatly varied, with odours that can be described as amine, sulphur, acidic, sour or "cabbage". Some authors succeeded in improving the organoleptic quality of smoked salmon by modifying the technological parameters, like salt, smoke or the addition of bioprotective bacteria, but have not demonstrated the inhibition of a particular group of bacteria (Leroi et al., 1996; Leroi et al., 2000). The situation is not the same in fresh fish, preserved in air or in a modified atmosphere or vacuum packed, where one or two well identified species dominate the flora and for which a predictive model of sensory spoilage has been suggested (Dalgaard et al., 2002). However, in the case of smoked salmon, a multiple approach enabled some authors to correlate quality to several microbiological and biochemical measurements. For example, Leroi et al. (2001) have put forward a predictive model for the remaining shelf life of smoked salmon, based on TVBN content and enumeration of the flora on Mann-Rogosa-Sharp (MRS) medium (selective for lactobacilli) at pH 5.5. If the TVBN content is less than 35 mg-N/100 g of flesh, the product is not spoilt. If it rises above this level, the number of bacteria on MRS medium must be considered. A product with 50 mg-N/100 g of TVBN will only be spoilt if the number of colonies on MRS is higher than 10^5 CFU/g. Another model, based on pH and the quantity of histamine and tyramine, has been developed by Jorgensen et al. (2000a).

Cooked Shrimps Packed in a Modified Atmosphere

France is the second largest importer of shrimps in Europe, and this product corresponds to the highest value of all seafood imported. Tropical shrimps account for 80% of the shrimps imported. Refrigerated cooked ready-to-eat shrimps represent the largest shrimp consumption in France. They may or may not be peeled, are often packed under a CO_2-enriched protective atmosphere, and can also pass through brine containing salt and different organic acids (citric, ascorbic, benzoic, etc.). Thanks to the increasing consumption, the shrimp market has been steadily expanding over the last 10 yr.

In the recent literature, knowledge of this product's microbiology has focused on the Nordic shrimp *(Pandalus borealis)*. The flora of fresh shrimps is greatly reduced during the cooking stage of processing (70-80°C throughout) and the spoilage flora is seemingly the result of a recontamination by Gram-positive bacteria before or during packaging. The microorganisms are then selected according to the different characteristics of the product and the storage conditions. More specifically, the shelf life and development of the microflora depend on the initial contamination, the characteristics of the product (i.e., brined or not) and the storage conditions (modified atmosphere, temperature). The combined effect of these different parameters could create a barrier effect against microbial development.

Brined Shrimps

From and Huss (1990) carried out one of the first studies on brined Nordic shrimps. The spoilage microflora was dominated at 5°C by yeasts and LAB *(Streptococcus* sp., *Lactobacillus* sp.). Another study by Einarsson and Lauzon (1995) on brined Nordic shrimps demonstrated the importance of benzoate and sorbate in preventing the development of flora at 4.5°C. When the flora is fully developed, it is dominated by coryneform bacteria and *Moraxella* sp.

Dalgaard and Jorgensen (2000) studied Nordic and tropical shrimps *(Penaeus* sp.) that had been brined, drained, packed in a modified atmosphere and stored at different temperatures between 0 and 25°C. They showed that the effect of temperature on shelf life was even greater than with other seafood products. For example, batches at 8°C had a shelf life 15 to 33 times longer than those stored at 25°C. Certain variations in the composition of the product, no matter how small, could have a large effect on this shelf life. Thus, shelf life at 5°C was over 3 mon for tropical shrimps containing 2.3% salt in an aqueous solution and over 6 mon for Nordic shrimps with 3.3% salt. The initial pH of these products was between 5.7 and 5.9. TVBN levels were relatively low when the products were spoilt (20 mg/100g at 15 and 25°C and 10 mg/100g at the end of the shelf life for tropical shrimps stored at lower temperatures). Most of the spoilage flora could not be identified, but LAB made up an important part of the total microflora, whatever was the storage temperature. Amongst these bacteria, several strains isolated from products stored at 15 and 25°C had been identified as *Enterococcus faecalis*. This microorganism was probably responsible for spoilage at these temperatures, but did not appear to develop much, if at all, at temperatures between 0 and 8°C. The authors also observed the presence of *Lactobacillus curvatus* on tropical shrimps stored at 0°C (Einarsson and Lauzon, 1995).

In continuing their studies on the flora isolated from these products, Dalgaard et al. (2003) observed an evolution of the microflora from bacilli to cocci between 5 and 8°C for Nordic shrimps and between 8 and 15°C for tropical shrimps. At 25°C, the flora was dominated by cocci. Most of the strains isolated at different temperatures had been identified as *Carnobacterium (Cb. divergens)* and *Enterococcus (En. faecalis)*. *En. faecalis* was present on products stored at 15°C and higher whereas *Cb. divergens* and *Lb. curvatus* were found on products stored at 8°C and lower. Three of the isolated strains correspond to an unknown *Carnobacterium* species. At high concentrations, *En. faecalis* and *En. durans* could represent a health risk for consumers. In this study, tyramine might represent a chemical indicator of spoilage, but at these levels it was low enough not to present a risk. Likewise, at low storage temperatures, the authors recommend that the product should contain at least 3% salt in an aqueous solution so as to avoid spoilage problems and health risks.

Mejlholm et al. (2008) studied Nordic shrimps that have been either brined or brined and drained and then packed in a modified atmosphere. They focused on the effect of hygiene (during industrial or manual processing), the composition of brine and the storage conditions (atmosphere and temperature) on the overall evolution of the microflora and the shelf life. Different groups of organic acids were tested: benzoic, citric and ascorbic or acetic, citric and lactic, including the effect of diacetate. The pH of brined shrimps was between 5.6 and 5.8. The shelf life depended on the nature of the organic acids, their concentration, the temperature and also the initial contamination, which had a significant effect. In fact, industrial products were more contaminated than products processed manually, which led to shorter shelf lives and a more diverse spoilage flora. This was especially true for brined and drained MAP shrimps, which had a shelf life of over 75 d at 7°C for manually processed products, but only 28-35 d for industrial products. Similarly, shrimps in brine composed of acetic, citric and lactic acid had shelf lives of 69-84 d and 42-49 d for manual and industrial products, respectively, at the same temperature. The modified atmosphere prolonged the shelf life of shrimps from 53-60 d for brined shrimps to over 75 d for the same shrimps that have been drained and packed in a modified atmosphere, although these results were not confirmed by subsequent tests.

Brine composition influenced the nature of the flora that would develop on the products. At the time of spoilage, industrially-produced shrimps in brine composed of benzoic, citric and ascorbic acid were host to *En. faecalis*-like, *Leuconostoc pseudomesenteroides*, coagulase-negative *Staphylococcus* and yeasts, whereas manually-processed shrimps also had *Ps. fluorescens*, *En. malodoratus*, *Cb. maltaromaticum* and *Lb. sakei*. Adding diacetate to the brine inhibited the growth of *Ps. fluorescens,* coagulase-negative *Staphylococcus* sp. and *Enterococcus* sp. When the brine contained acetic, citric and lactic acid, the microflora was composed exclusively of yeasts and LAB, such as *Lb. sakei,* which were responsible for unpleasant sour and buttery odours. When the yeasts dominated the flora, 10^6 CFU/g was enough to spoil the product.

Whatever the composition of the brine at the start of the process, the spoilage microflora of brined, drained and MAP shrimps was dominated by *Lb. sakei*. Spoilage of the industrial product led to sour flavours and a release of gas in the packaging, probably produced by *Lb. sakei*.

The spoilage of some brined shrimp samples with a reduced microflora was due to oxidation rather than the microbial activity.

The microbial ecosystem of commercial, cooked, peeled, brined and drained tropical shrimps (*Penaeus vannamei*) stored at 5°C and 15°C has been studied using a polyphasic approach combining culture-dependent and culture-independent methods including PCR-TTGE (Jaffrès et al., 2009). When samples were spoilt, bacterial strains were isolated and identified by phenotypic and molecular tests. LAB constituted the major group with principally *Carnobacterium* (*Cb. divergens, Cb. maltaromaticum* and *Cb. jeotgali*), *Enterococcus* (*En. faecalis and En. faecium*) and *Vagococcus* including the novel species *Vagococcus penaei* (Jaffrès et al., 2010). The other groups corresponded to *Brochothrix thermosphancta* and *Enterobacteriaceae* (*Serratia liquefaciens*). From PCR-TTGE profiles some of DNA fragments were assigned to those of standard strains (*Br. thermosphacta, Se. liquefaciens, En. faecalis, Cb. maltaromaticum* and *Cb. divergens*) and additional informations were provided by fragment cloning (*Psychrobacter* sp. and *Citrobacter* sp.).

Shrimps without Brine

Cooked, peeled, frozen or unfrozen Nordic shrimps without brine and packed in a modified atmosphere (50% CO_2) had an initial pH of 7.7, a salt content of 1.9% in aqueous solution and a TVBN of 10mg/100mg (Mejlholm et al., 2005). If the product was frozen for 4 mon preceding refrigerated storage then the initial concentration of microorganisms was reduced, but the rate of growth, the maximum level of flora and the shelf life remained the same. The latter was 26 d at 2°C, 16 d at 5°C and 10 d at 8°C. A low production of TVBN, lower than 20mg/100g, and no production of TMA, organic acids or biogenic amines were observed. From 116 isolates, Mejlholm et al. (2005) identified 60% *Cb. maltaromaticum*,

27% *Br. thermosphacta* and 13% *Psychrobacter* sp. When inoculated together, *Cb. maltaromaticum* and *Br. thermosphacta* produced the same unpleasant odours as spoiled products that had been contaminated naturally. This was not the case when the two products were inoculated separately. The mechanisms of this interaction had not yet been explained. The presence of *Br. thermosphacta* had already been observed in MAP fish. Furthermore, this microorganism had already been described as producing odours of butter, blue cheese or feet. *Cb. maltaromaticum* produced an unpleasant chlorine odour whereas in previous studies this bacterium was described as non-spoiling only producing a buttery odour. *Psychrobacter* sp. did not seem to play a part in the spoilage of cooked shrimps.

The potential for spoilage by *Carnobacterium* on cooked, peeled, MAP Nordic shrimps without brine varied depending on the species and the strain (Laursen et al., 2005). *Cb. divergens* and some *Cb. maltaromaticum* produced unpleasant odours whereas another group of *Cb. maltaromaticum* was not spoilt. Spoilage bacteria produced ammonia, tyramine, different alcohols, aldehydes and ketones. These authors had confirmed that *Br. thermosphacta* and *Cb. maltaromaticum* cultured together led to some bad odours (wet dog) those were not produced when these species were grown separately. However, this association did not produce new metabolites and consequently the "wet dog" odour did not come from a metabiotic phenomenon in which a bacterium produced a metabolite from a substrate formed by another bacterium. On the other hand, *Cb. maltaromaticum* decreased the formation of diacetyl by *Br. thermosphacta* and the latter reduced the activity of *Cb. maltaromaticum*. The "wet dog" smell might be due to the interaction between metabolites formed by *Cb. maltaromaticum*, *Cb. divergens*, and partly by *Cb. mobile*, and those formed by *Br. thermosphacta*. The presence of oxygen increased the potential for spoilage and the number of metabolites produced by *Br. thermosphacta*. In fact, *Br. thermosphacta* can be aerobiotic or anaerobiotic depending on the respective percentages of O_2 and CO_2. To reduce the potential for spoilage in shrimps, it is therefore recommended to use an atmosphere with less O_2 and more CO_2.

In contrast to cooked, peeled, MAP shrimps without brine, in which the specific spoilage bacteria was well defined (Mejlholm et al., 2005; Laursen et al., 2006), the characteristics of the spoilage flora of brined or brined/drained and MAP shrimps were not as clearly established. The composition of this spoilage flora depended on a large number of parameters: the initial contamination, the preservation parameters (organic acids and salt) and the storage conditions (atmosphere, temperature).

CONCLUSION

The susceptibility of fish and shrimp flesh to spoilage can be explained by physico-chemical characteristics (neutral pH, high concentration of low molecular weight nitrogenous compounds, etc.) which favour microbial growth. For most unprocessed products, the specific microorganisms responsible for sensory degradation have been identified and the influence of storage parameters such as the temperature and the storage atmosphere are well documented. Predictive models for spoilage and some other indicators are available (Seafood Spoilage and Safety Predictor: www.dfu. min.dk/micro/sssp/). The spoilage mechanisms for lightly preserved products are much more complex as they must take into account several groups of interacting bacteria. It is therefore necessary to continue research in this field to better understand these phenomena with the objective of providing relevant quality indicators, predictive models of storage and effective methods of preservation.

ABBREVIATIONS

CFU   Colony forming units
DGGE   Denaturing gradient gel electrophoresis
H_2S Hydrogen sulphide
LAB   Lactic acid bacteria
MAP   Modified atmosphere package
MRS   Mann-Rogosa-Sharp
TGGE   Temperature gradient gel electrophoresis
TMA   Trymethyl amine
TMAO   Trimethylamine oxide
TVBN   Total volatile bases nitrogen

REFERENCES

Abgrall, B. (1988). Poissons et autres produits de la mer. In: Microbiologie alimentaire. Tec et Doc, Lavoisier, Paris, France pp. 251-264.
Al Harbi, A.H. and Uddin, N. (2005). Bacterial diversity of tilapia *(Oreochromis niloticus)* cultured in brackish water in Saudi Arabia. Aquaculture 250: 566-572.
Anthoni, U., Borresen, T., Christophersen, C., Gram, L. and Nielsen, P.H. (1990). Is trimethylamine oxide a reliable indicator for the marine origin of fishes ? Comp. Biochem. Physiol. Part B: Biochem. Mol. Biol. 97B: 569-571.
Ast, J.C. and Dunlap, P.V. (2005). Phylogenetic resolution and habitat specificity of members of the *Photobacterium phosphoreum* species group. Environ Microbiol. 7 (10): 1641-1654.
Benner, R.A., Staruszkiewicz, W.F. and Otwell, W.S. (2004). Putrescine, cadaverine, and indole production by bacteria isolated from wild and aquacultured Penaeid Shrimp stored at 0, 12, 24, and 36°C. J. Food Protec. 67 (1): 124-133.
Bernadsky, G. and Rosenberg, E. (1992). Drag-reducing properties of bacteria from the skin mucus of the cornetfish (*Fistularia commersonii*). Microb. Ecol. 24: 63-74.

Brillet, A., Pilet, M.F., Prévost, H., Cardinal, M. and Leroi, F. (2005). Effect of inoculation of *Carnobacterium divergens* V41, a biopreservative strain against *Listeria monocytogenes* risk, on the microbiological, and sensory quality of cold-smoked salmon. Int. J. Food Microbiol. 104 (3): 309-324.

Burr, G.B., Gatlin, I.I.I.D. and Ricke, S. (2005). Microbial ecology of the gastrointestinal tract of fish and the potential application of prebiotics and probiotics in Finfish aquaculture. J. World Aquacult. Soc. 36 (4): 425-436.

Cahill, M.M. (1990). Bacterial flora of fishes: a review. Microb Ecol. 19 (1): 21-41.

Cambon-Bonavita, M.A., Lesongeur, F., Menoux, S., Lebourg, A. and Barbier, G. (2001). Microbial diversity in smoked salmon examined by a culture-independent molecular approach—a preliminary study. Int. J. Food Microbiol. 70: 179-187.

Cann, D.C. (1974). Bacteriological aspects of tropical shrimp. In: FAO; Technical Conference on Fishery Products. Tokyo, Japan, 4-11, December, 1973 pp. 338-344.

Cann, D.C., Houston, N.C., Taylor, L.Y., Smith, G.L., Smith, A.B. and Craig, A. (1984). Studies of salmonids packed and stored under a modified atmosphere. Report, Torry Research Station, Aberdeen, Scotland.

Cardinal, M., Cornet, J., Sérot, T. and Baron, R. (2006). Effects of the smoking process on odour characteristics of smoked herring (*Clupea harengus*) and relationships with phenolic compound content. Food Chem. 96: 137-146.

Cardinal, M., Gunnlaugsdottir, H., Bjoernevik, M., Ouisse, A., Vallet, J.L. and Leroi, F. (2004). Sensory characteristics of cold-smoked Atlantic salmon (*Salmo salar*) from European market and relationships with chemical, physical and microbiological measurements. Food Res. Int. 37: 181-193.

Chinivasagam, H.N., Bremner, H.A., Thrower, S.J and Nottingham, S.M. (1996). Spoilage pattern of five species of Australian prawns: deterioration is influenced by environment of capture and mode of storage. J. Aquat. Food Prod. Technol. 5: 25-30.

Chinivasagam, H.N., Bremner, H.A., Wood, A.F. and Nottingham, S.M. (1998). Volatile components associated with bacterial spoilage of tropical prawns. Int. J. Food Microbiol. 42: 45-35.

Civera, T., Parisi, E., Amerio, G.P. and Giaccone, V. (1995). Shelf life of vacuum-packed smoked salmon: microbiological and chemical changes during storage. Arch. Lebensmitt. 46: 1-24.

Dalgaard, P. (1995). Modelling of microbial activity and prediction of shelf life for packed fresh fish. Int. J. Food Microbiol. 26: 305-317.

Dalgaard, P. and Huss, H.H. (1997). Mathematical modelling used for evaluation and prediction of microbial fish spoilage. In: Seafood Safety, Processing and Biotechnology. F. Shahidi, Y. Jones and D.D. Kitts (eds). Technomic Publishing Co., Inc., Lancaster, UK pp. 73-89.

Dalgaard, P. and Jorgensen, L.V. (2000). Cooked and brined shrimps packed in a modified atmosphere have a shelf-life of > 7 months at 0°C, but spoil in 4-6 days at 25°C. Int. J. Food Sci. Technol. 35: 431-442.

Dalgaard, P., Gram, L. and Huss, H.H. (1993). Spoilage and shelf-life of cod fillets packed in vacuum or modified atmospheres. Int. J. Food Microbiol. 19: 283-294.

Dalgaard, P., Buch, P. and Silberg, S. (2002). Seafood spoilage predictor-development and distribution of product specific application software. Int. J. Food Microbiol. 73: 343-349.

Dalgaard, P., Vancanneyt, M., Euras Vilalta, N., Swings, J., Fruekilde, P. and Leisner, J.J. (2003). Identification of lactic acid bacteria from spoilage associations of cooked and brined shrimps stored under modified atmosphere between 0°C and 25°C. J. Appl. Microbiol. 94: 80-89.

Dalgaard, P., Madsen, H.L., Samieian, N. and Emborg, J. (2006). Biogenic amine formation and microbial spoilage in chilled garfish (*Belone belone belone*)—effect of modified atmosphere packaging and previous frozen storage. J. Appl. Microbiol. 101: 80-95.

Devaraju, A.N. and Setty, T.M.R. (1985). Comparative study of fish bacteria from tropical and cold temperature marine waters. FAO Fishery Rep. 317: 97-107.

Dodds, K.L., Brodsky, M.H. and Warburton, D.W. (1992). A retail survey of smoked ready-to-eat fish to determine their microbiological quality. J. Food Protec. 55 (3): 208-210.

Dondero, M., Cisternas, F., Carvajal, L. and Simpson, R. (2004). Changes in quality of vacuum-packed cold-smoked salmon (*Salmo salar*) as a function of storage temperature. Food Chem. 87: 543-550.

Drosinos, E.H. and Nychas, G.J.E. (1996). *Brochothrix thermosphancta*, a dominant microorganism in Mediterranean fresh fish (*Sparus aurata*) stored under modified atmosphere. Ital. J. Food Sci. 8 (4): 323-329.

Drosinos, E.H., Lambropoulo, K., Mitre, E. and Nychas, G.J.E. (1997). Attributes of fresh gilt-head seabream *(Sparus aurata)* fillets treated with potassium sorbate, sodium gluconate and stored under a modified atmosphere at $0 + /-1°C$. J. Appl. Microbiol. 83: 569-575.

Duffes, F. (1999). Improving the control of *Listeria monocytogenes* in cold smoked salmon. Trends Food Sci. Tech. 10: 1-6.

Einarsson, H. and Lauzon, L. (1995). Biopreservation of brined shrimp *(Pandalus borealis)* by bacteriocins from lactic acid bacteria. Appl. Environ. Microbiol. 61: 669-676.

Emborg, J., Laursen, B.G., Rathjen, T. and Dalgaard, P. (2002). Microbial spoilage and formation of biogenic amines in fresh and thawed modified atmosphere-packed salmon (*Salmo salar*) at 2°C. J. Appl. Microbiol. 92: 790-799.

Emborg, J., Laursen, B.G. and Dalgaard, P. (2005). Significant histamine formation in tuna (*Thunnus albacares*) at 2°C—Effect of vacuum and modified atmosphere-packaging on psychrotolerant bacteria. Int. J. Food Microbiol. 101: 263-279.

Espe, M., Kiessling, A., Lunestad, B.T., Torrissen, O.J. and Rora, A.M.B. (2004). Quality of cold smoked salmon collected in one French hypermarket during a period of one year. LWT Food Sci. Technol. 37: 627-638.

From, V. and Huss, H.H. (1990). Lightly preserved shrimps- shrimps in brine. Fiskerbladet. 36: 35-41.

Gennari, M., Tomaselli, S. and Cotrona, V. (1999). The microflora of fresh and spoiled sardines (*Sardina pilcharus*) caught in Adriatic (Mediterranean) sea and stored in ice. Food Microbiol. 16: 15-28.

Gibson, D.M. and Ogden, I.D. (1986). Estimating the shelf life of packaged fish. Proc. Int. Symp. Coordinated by the University of Alaska, Seafood Quality Determination, Alaska, USA pp. 437-450.

Gonçalves, A.C., López-Caballero, M.E. and Nunes, M.L. (2003). Quality changes of deepwater pink shrimp (*Parapenaeus longirostris*) packed in modified atmosphere. J. Food Sci. 68: 2586-2590.

Gonzales-Fandos, E., Garcia-Linares, M.C., Villarino-Rodriguez, A., Garcia-Arias, M.T. and Garcia-Fernandez, M.C. (2004). Evaluation of the microbiological safety and sensory quality of rainbow trout (*Oncorhynchus mykiss*) processed by the sous vide method. Food Microbiol. 21: 193-201.

Gopal, S., Otta, S.K. and Kumar, S. (2005). The occurrence of *Vibrio* species in tropical shrimp culture environments; implications for food safety. Int J. Food Microbiol. 102 (2): 151-159.

Gram, L. (1995). Quality and quality changes in fresh fish. FAO Fish Tech. Paper 348.

Gram, L. and Huss, H.H. (1996). Microbiological spoilage of fish and fish products. Int. J. Food Microbiol. 33: 121-137.

Gram, L. and Melchiorsen, J. (1996). Interaction between fish spoilage bacteria *Pseudomonas* sp. and *Shewanella putrefaciens* in fish extracts and on fish tissue. J. Appl. Bacteriol. 80: 589-595.

Gram, L., Trolle, G. and Huss, H.H. (1987). Detection of specific spoilage bacteria from fish stored at low (0°C) and high (20°) temperatures. Int. J. Food Microbiol. 4: 65-72.

Gram, L., Wedell-Neegaard, C. and Huss, H.H. (1990). The bacteriology of fresh and spoiling Lake Victoria Nile perch (*Lates niloticus*). Int. J. Food Microbiol. 10: 303-316.

Grimont, F. and Grimont, P.A.D. (1992). The Genus *Serratia*. In: The Prokaryotes, Volume 3, 2nd edn. A.Balows, H.G. Trüper, M.Dworkin, W. Harder and K.H. Schleifer (eds). Springer Verlag, Berlin, Germany pp. 2823-2848.

Guldager, H.S., Bøknæs, N., Østerberg, C., Nielsen, J. and Dalgaard, P. (1998). Thawed cod fillets spoil less rapidly than unfrozen fillets when stored under modified atmosphere at 2°C. J. Food Protec. 61: 1129-1136.

Hanninen, M.J., Oivanen, P. and Hirvelä-Koski, V. (1997). *Aeromonas* species in fish, fish-eggs, shrimp and freshwater. Int. J. Food Microbiol. 34: 17-26.

Hansen, G.H. and Olafsen, J.A. (1999). Bacterial interactions in early life stages of marine cold water fish. Microb. Ecol. 38: 1-26.

Herbert, R.A., Hendrie, M.S., Gibson, D.M. and Shewan, J.M. (1971). Bacteria active in the spoilage of certain sea foods. Symposium on Microbial Changes in Foods, Paper IV. J. Appl. Bacteriol. 34 (1): 41-50.

Hildebrandt, G. and Erol, I. (1988). Sensorische und mikrobiologische Untersuchung an vakuumverpackten Räucherlachs in Scheiben (Sensory and microbiological analysis of vacuum packed sliced smoked salmon). Arch. Lebensmitt. 39: 120-123.

Hobbs, G. (1983). Microbial spoilage of fish. In: Food Microbiology Advances and Prospects. T.A. Roberts and F.A. Skinner (eds). Academic Press, London, UK pp. 217-229.

Holben, W.E., Williams, P., Saarinen, M., Särkilahti, L.K. and Apajalahti, J.H.A. (2002). Phylogenetic analysis of intestinal microflora indicates a novel mycoplasma phylotype in farmed and wild salmon. Microb. Ecol. 44 (2): 175-185.

Hovda, M.B., Sivertsvik, M., Lunestad, B.T., Lorentzen, G. and Rosnes, J.T. (2007). Characterisation of the dominant bacterial population in modified atmosphere packaged farmed halibut (*Hippoglossus hippoglossus*) based on 16S rDNA-DGGE. Food Microbiol. 24: 362-371.

Hozbor, M.C., Saiz, A.I., Yeannes, M.I. and Fritz, R. (2006). Microbiological changes and its correlation with quality indices during aerobic ice storage of sea salmon (*Pseudopercis samifasciata*). LWT Food Sci Tech. 39: 99-104.

Huber, I., Spanggaard, B., Appel, K.F., Rossen, L., Nielsen, T. and Gram, L. (2004). Phylogenetic analysis and *in situ* identification of the intestinal microbial community of rainbow trout (*Oncorhynchus mykiss*, Walbaum). J. Appl. Microbiol. 96: 117-132.

Huss, H.H. (1999). La qualité et son évolution dans le poisson frais. Changements *post-mortem* dans le poisson. FAO Document Technique sur Les Pêches. 348: 1-26.

Huss, H.H., Jeppesen, V.F., Johansen, C. and Gram, L. (1995). Biopreservation of fish products—a review of recent approaches and results. J. Aquat. Food Prod. Technol. 4 (2): 5-26.

Jaffrès, E., Sohier, D., Leroi, F., Pilet, M.F., Prévost, H., Joffraud, J.J. and Dousset, X. (2009). Study of the bacterial ecosystem in tropical cooked and peeled shrimps using a polyphasic approach. Int. J. Food Microbiol. 131: 20-29.

Jaffrès, E., Prévost, H., Rossero, A., Joffraud, J.J. and Dousset, X. (2010). *Vagococcus penaei* sp. nov., isolated from spoilage microbiota of cooked shrimp (*Penaeus vannamei*). Int. J. Syst. Evol. Microbiol. 60: 2159-2164.

Jayaweera, V. and Subasinghe, S. (1988). Some chemical and microbiological changes during chilled storage of prawns (*Penaeus indicus*). FAO Fish Rep. 401: 19-22.

Jeyasekaran, G., Ganesan, P., Anandaraj, R., Jeya Shakila, R. and Sukumar, D. (2006). Quantitative and qualitative studies on the bacteriological quality of Indian white shrimp *(Penaeus indicus)* stored in dry ice. Food Microbiol. 23 (6): 526-533.

Joffraud, J.J., Cardinal, M., Cornet, J., Chasles, J.S., Léon, S., Gigout, F. and Leroi, F. (2006). Effect of bacterial interactions on the spoilage of cold-smoked salmon. Int. J. Food Microbiol. 112: 51-61.

Joffraud, J.J., Leroi, F. and Chevalier, F. (1998). Development of a sterile cold-smoked fish model. J. Appl. Microbiol. 85: 991-998.

Joffraud, J.J., Leroi, F., Roy, C. and Berdagué, J.L. (2001). Characterisation of volatile compounds produced by bacteria isolated from the spoilage flora of cold-smoked salmon. Int. J. Food Microbiol. 66: 175-184.

Jorgensen, B.R., Gibson, D.M. and Huss, H.H. (1988). Microbiological quality and shelf life prediction of chilled fish. Int. J. Food Microbiol. 6: 295-307.

Jorgensen, L.V., Dalgaard, P. and Huss, H.H. (2000a). Multiple compound quality index for cold-smoked salmon (*Salmo salar*) developed by multivariate regression of biogenic amines and pH. J. Agric. Food Chem. 48 (6): 2448-2453.

Jorgensen, L.V., Huss, H.H. and Dalgaard, P. (2000b). The effect of biogenic amine production by single bacterial cultures and metabolism on cold-smoked salmon. J. Appl. Microbiol. 89: 920-934.

Kim, D.H., Brunt, J. and Austin, B. (2007). Microbial diversity of intestinal contents and mucus in rainbow trout (*Oncorhynchus mykiss*). J. Appl. Microbiol. 102: 1654-1664.

Kim, S.H., Price, R.J., Morrissey, M.T., Field, K.G., Wei, C.I. and An, H. (2002). Histamine production by *Morganella morganii* in mackerel, albacore, mahi-mahi, and salmon at various storage temperatures. J. Food Sci. 67 (4): 1522-1528.

Kim, S.H., Wei, C., Clemens, R.A. and An, H. (2004). Histamine accumulation in seafoods and its control to prevent outbreaks of scombroid poisoning. J. Aquat. Food Prod. Technol. 13: 81-100.

Knockaert, C. (2005). Le fumage du poisson. Collection Ifremer "Valorisation des produits de la mer" 1-174.

Laursen, B.G., Bay, L., Cleenwerck, I., Vancanneyt, M., Swings, J., Dalgaard, P. and Leisner, J.J. (2005). *Carnobacterium divergens* and *Carnobacterium maltaromicum* as spoilers or protective cultures in meat and seafood: phenotypic and genotypic characterisation. Syst. Appl. Microbiol. 28: 151-164.

Laursen, B.G., Leisner, J.J. and Dalgaard, P. (2006). *Carnobacterium* species: Effect of metabolic activity and interaction with *Brochothrix thermosphancta* on sensory characteristics of modified atmosphere packed shrimp. J. Agric. Food Chem. 54: 3604-3611.

Lee, J.S. and Simard, R.E. (1984). Evaluation of methods for detecting the production of H_2S, volatile sulphides, and greening by *Lactobacilli*. J. Food Sci. 49: 981-983.

Leisner, J.J., Laursen, B.G., Prevost, H., Drider, D. and Dalgaard, P. (2007). *Carnobacterium*: positive and negative effects in the environment and in foods. FEMS Microbiol. Rev. 31 (5): 592-613.

Leroi, F. and Joffraud, J.J. (2000). Salt and smoke simultaneously affect chemical and sensory quality of cold-smoked salmon during 5°C storage predicted using factorial design. J. Food Protec. 63 (9): 1222-1227.

Leroi, F., Arbey, N., Joffraud, J.J. and Chevalier, F. (1996). Effect of inoculation with lactic acid bacteria on extending the shelf-life of vacuum-packed cold-smoked salmon. Int. J. Food Sci. Technol. 31: 497-504.

Leroi, F., Joffraud, J.J., Chevalier, F. and Cardinal, M. (1998). Study of the microbial ecology of cold smoked salmon during storage at 8°C. Int. J. Food Microbiol. 39: 111-121.

Leroi, F., Joffraud, J.J. and Chevalier, F. (2000). Effect of salt and smoke on the microbiological quality of cold smoked salmon during storage at 5°C as estimated by the factorial design method. J. Food Protec. 63 (4): 502-508.

Leroi, F., Joffraud, J.J., Chevalier, F. and Cardinal, M. (2001). Research of quality indices for cold-smoked salmon using a stepwise multiple regression of microbiological counts and physico-chemical parameters. J. Appl. Microbiol. 90: 578-587.

Liston, J. (1992). Bacterial spoilage of seafood. Quality assurance in the fish industry. Int. J. Food Sci. Technol. 41: 93-103.

López-Caballero, M., Gonçalves, A. and Nunes, M. (2002). Effect of CO_2/O_2-containing modified atmospheres on packed deepwater pink shrimp (*Parapenaeus longirostris*). Eur. Food Res. Technol. 214: 192-197.

Lyhs, U., Korkeala, H., Vandamme, P. and Björkroth, J. (2001). *Lactobacillus alimentarius*: a specific spoilage organism in marinated herring. Int. J. Food Microbiol. 64: 355-360.

Magnusson, H. and Traustadottir, K. (1982). The microbiological flora of vacuum packed smoked herring fillets. J. Food Technol. 17: 695-702.

Matches, J.R. (1982). Effects of Temperature on the Decomposition of Pacific Coast Shrimp (*Pandalus jordani*). J. Food Sci. 47 (4): 1044-1047.

Mejlholm, O., Boknaes, N. and Dalgaard, P. (2005). Shelf life and safety aspects of chilled cooked and peeled shrimps (*Pandalus borealis*) in modified atmosphere packaging. J. Appl. Microbiol. 99: 66-76.

Mejlholm, O., Kjeldgaard, J., Modberg, A., Vest, M.B., Bøknæs, N., Koort, J., Björkroth, J. and Dalgaard, P. (2008). Microbial changes and growth of *Listeria monocytogenes* during chilled storage of brined shrimp (*Pandalus borealis*). Int. J. Food Microbiol. 124 (3): 250-259.

Mendes, R., Gonçalves, A., Pestana, J. and Pestana, C. (2005). Indole production and deepwater pink shrimp *(Parapenaeus longirostris)* decomposition. Eur. Food Res. Technol. 221: 320-328.

Mudarris, M. and Austin, B. (1988). Quantitative and qualitative studies of the bacterial microflora of turbot, *Scophthalmus maximus* L., gills. J. Fish Biol. 32 (2): 223-229.

Nilsson, L., Gram, L. and Huss, H.H. (1999). Growth control of *Listeria monocytogenes* on cold-smoked salmon using a competitive lactic acid bacteria flora. J. Food Protec. 62: 336-342.

Olofsson, T.C., Ahrné, S. and Molin, G. (2007). The bacterial flora of vacuum-packed cold-smoked salmon stored at 7°C, identified by direct 16S rRNA gene analysis and pure culture technique. J. Appl. Microbiol. 103 (1): 109-119.

Paludan, Müller, C., Dalgaard, P., Huss, H.H. and Gram, L. (1998). Evaluation of the role of *Carnobacterium piscicola* in spoilage of vacuum and modified atmosphere-packed-smoked salmon stored at 5°C. Int. J. Food Microbiol. 39: 155-166.

Pond, M.P., Stone, D.M. and Alderman, D.J. (2006). Comparison of conventional and molecular techniques to investigate the intestinal microflora of rainbow trout (*Oncorhynchus mykiss*). Aquaculture 261: 194-203.

Rachman, C., Fourrier, A., Sy, A., De La Cochetiere, M.F., Prevost, H. and Dousset, X. (2004). Monitoring of bacterial evolution and molecular identification of lactic acid bacteria in smoked salmon during storage. Le Lait 84: 145-154.

Rakow, D.von (1977). Bemerkungen zum Keimgehaltvon im Handel befindlichen Räucherfischen. Arch. Lebensmittelhyg 28: 192-195.

Ratkowsky, D.A., Olley, J., Mc Meekin, T.A. and Ball, A. (1982). Relationship between temperature and growth rate of bacterial cultures. J. Bacteriol. 149 (1): 1-5.

Ringo, E. and Gatesoupe, F.J. (1998). Lactic acid bacteria in fish: a review. Aquaculture 160: 177-203.

Ringo, E. and Birkbeck, T.H. (1999). Intestinal microflora of fish larvae and fry. Aquacult. Res. 30: 73-93.

Ringo, E. and Olsen, R.E. (1999). The effect of diet on aerobic bacterial flora associated with intestine of Arctic charr (*Salvelinus alpinus L.*). J Appl Microbiol. 86: 22-28.

Ringo, E., Strom, E. and Tabachek, J.A. (1995). Intestinal microflora of salmonids: a review. Aquacult. Res. 26: 773-789.

Rudi, K., Maugesten, T., Hannevik, S.E. and Nissen, H. (2004). Explorative multivariate analyses of 16S rRNA gene data from microbial communities in modified-atmosphere-packed salmon and coalfish. Appl Environ Microbiol. 70 (8): 5010-5018.

Sakzaki, R. and Tamura, K. (1992). The Genus *Hafnia*. In: The Prokaryotes, Volume 3, 2nd edn. A.Balows, H.G. Trüper, M.Dworkin, W. Harder and K.H. Schleifer, (eds). Springer Verlag, Berlin, Germany pp.2816-2821.

Satomi, M., Vogel, B.F., Gram, L. and Venkateswaran, K. (2006). *Shewanella hafniensis* sp. nov. and *Shewanella morhuae* sp. nov., isolated from marine fish of the Baltic Sea. Int. J. Syst. Evol. Microbiol. 56: 243-249.

Satomi, M., Vogel, B.F., Venkateswaran, K. and Gram, L. (2007). Description of *Shewanella glacialipiscicola* sp. nov. and *Shewanella algidipiscicola* sp. nov., isolated from marine fish of the Danish Baltic Sea, and proposal that *Shewanella affinis* is a later heterotypic synonym of *Shewanella colwelliana*. Int. J. Syst. Evol. Microbiol. 57: 347-352.

Seibel, B.A. and Walsh, P.J. (2002). Trimethylamine oxide accumulation in marine animals: relationship to acylglycerol storage. J. Exp. Biol. 25: 297-306.

Sérot, T., Baron, R., Knockaert, C. and Vallet, J.L. (2004). Effect of smoking processed on the contents of 10 major phenolic compounds in smoked fillets of herring (*Cuplea harengus*). Food Chem. 85: 111-120.

Shamshad, S.I., Kher-un-nisa, R.M., Zuberi, R. and Qadri, R.B. (1990). Shelf life of shrimp (*Penaus merguiensis*) stored at different temperatures. J. Food Sci. 55 (5): 1201-1205.

Shewan, J.M. (1971). The microbiology of fish and fishery products-a progress report. J. Appl. Bacteriol. 34 (2): 299-315.

Shewan, J.M. (1974). The biodegradation of certain proteinageous foodstuffs at chill temperatures. In: Industrial Aspects of Biochemistry. B. Spencer (ed). North Holland Publishing Co., Amsterdam, The Netherlands pp. 475-489.

Shewan, J.M. (1977). The bacteriology of fresh fish and spoiling fish and the biochemical changes induced by bacterial action. In: Proceedings of the Conference on "Handling, Processing and Marketing of Tropical Fish", Tropical Products Institute, London, UK pp. 51-66.

Sivertsvik, M., Rosnes, J.T. and Bergslien, H. (1997). Shelf-life of whole cooked shrimp *(Pandalus borealis)* in carbon dioxide-enriched atmosphere. In: Sea Food from Producer to Consumer, Integrated Approach to Quality. J.B. Luten, T. Borrensen and J. Oehlenschlager (eds). Elsevier, Amsterdam, The Netherlands pp. 221-230.

Sivertsvik, M., Jeksrud, J. and Rosnes, J.T. (2002). A review of modified atmosphere packaging of fish and fishery products—significance of microbial growth, activities and safety. Int. J. Food Sci. Technol. 37 (2): 107-127.

Sivertsvik, M., Rosnes, J.T. and Kleiberg, G.H. (2003). Effect of modified atmosphere packaging and super-chilled storage on the microbial and sensory quality of Atlantic Salmon (*Salmo salar*) Fillets. J. Food Sci. 68 (4): 1467-1472.

Spanggaard, B., Huber, I., Nielsen, J., Nielsen, T., Appel, K.F. and Gram, L. (2000). The microflora of rainbow trout intestine: a comparison of traditional and molecular identification. Aquaculture 182: 1-15.

Stohr, V., Joffraud, J.J., Cardinal, M. and Leroi, F. (2001). Spoilage potential and sensory profile associated with bacteria isolated from cold-smoked salmon. Food Res. Int. 34: 797-806.

Tomé, E., Gibbs, P.A. and Teixeira, P.C. (2007). Could modifications of processing parameters enhance the growth and selection of lactic acid bacteria in cold-smoked salmon to improve preservation by natural means? J. Food Protec. 8: 1607-1614.

Truelstrup Hansen, L. (1995). Quality of chilled vacuum-packed cold-smoked salmon. PhD Thesis Danish Institute for Fisheries Research and The Royal Veterinary and Agricultural University of Copenhagen, Denmark.

Truelstrup Hansen, L., Gill, T. and Huss, H.H. (1995). Effects of salt and storage temperature on chemical, microbiological and sensory changes in cold-smoked salmon. Food Res. Int. 28 (2): 123-130.

Truelstrup Hansen, L., Drewes Rontved, S. and Huss, H.H. (1998). Microbiological quality and shelf life of cold-smoked salmon from three different processing plants. Food Microbiol. 15: 137-150.

Van Spreekens, K.J.A. (1974). The suitability of a modification of Long and Hammer's medium for the enumeration of more fastidious bacteria from fresh fisheries products. Arch. Lebensmittelhyg 25: 213-219.

Vanderzant, C., Cobb, B.F., Thompson, C.A. and Parker, J.C. (1973). Microbial flora, chemical characteristics and shelf life of four species of pond reared shrimp. J. Milk Food Technol. 36: 443-449.

Veciana-Nogues, M.T., Marine. Font, A. and Vidal-Carou, M.C. (1997). Biogenic amines as hygienic quality indicators of tuna. Relationships with microbial counts, ATP-related compounds, volatile amines, and organoleptic changes. J. Agric. Food Chem. 45: 2036-2041.

Vogel, B.F., Venkateswaran, K., Satomi, M. and Gram, L. (2005). Identification of *Shewanella baltica* as the most important H_2S-producing species during iced storage of Danish marine fish. Appl. Environ. Microbiol. 71 (11): 6689-6697.

Weiss, A. and Hammes, W.P. (2006). Lactic acid bacteria as protective cultures against *Listeria* spp. on cold-smoked salmon. Eur. Food Res. Technol. 222: 343-346.

Wilson, B., Danilowicz, B.S. and Meijer, W.G. (2008). The diversity of bacterial communities associated with Atlantic Cod *Gadus morhua*. Microb. Ecol. 55 (3): 425-434.

4

Microbiological Safety and Quality of Processed Shrimps and Fishes

Md. Latiful Bari,[1,]* Sabina Yeasmin,[2] Shinichi Kawamoto[3] and Kenji Isshiki[4]

INTRODUCTION

Seafood has traditionally been a part of the diet in many parts of the world and in some countries constitutes the main supply of animal protein. Today even more people are eating fish as a healthy alternative to red meat. The low fat content of many fish species (white fleshed, demersal) and the effects on coronary heart disease of the n-3 polyunsaturated fatty acids found in fatty (pelagic) fish species are extremely important aspects for health-conscious people particularly in affluent countries, where cardiovascular disease mortality is high. However, consumption of fish and shellfish may also cause diseases due to infection or intoxication. Some of the diseases have been specifically associated with consumption of seafood while others have been of a more general nature. Seafood-borne illness accounts for more than 10% of the reported food-borne diseases each year. Nevertheless, seafood-borne diseases cause a significant number of illness and deaths worldwide, and people should be concerned about them. Most health problems associated with seafood are due either to contaminants that are present in the environment where seafood (i.e., shellfish or fish) are grown or improper handling. Most seafood-borne diseases can be traced to pollution in the area where the sea creatures

[1]*Centre for Advanced Research in Science, University of Dhaka, Dhaka-1000, Bangla Desh*
[2]*Department of Genetic Engneeering and Biotechnology, Dhaka University, Dhaka-1000, Bangladesh*
[3]*Food Hygiene Laboratory, National Food Research Institute, Kannondai-2-1-12, Tsukuba 305- 8642, Japan*
[4]*Division of Marine Life Science, Research Faculty of Fisheries Science, Hokkaido University 3-1-1, Minato-cho, Hakodate, Hokkaido 041-8611, Japan*
**Corresponding author: E-mail:latiful@univdhaka.edu. Tel.: 880-2-9661920-59 Ext 4721*

are harvested. However, shrimp remains one of the most popular and highly valued seafood selections throughout the world. Current annual world production from both wild harvest and farm culture is estimated at approximately 3,000,000 metric tonnes.

The importance of aquaculture continues to expand, especially for freshwater species such as carp, and almost one third of fish used for human consumption are now produced in aquaculture (FAO, 2004). Developed countries accounted for more than 80% of total imports of fishery products in 2003 in value terms. Japan was the biggest importer of fishery products, accounting for some 26% of the global total. The European Union (EU) has increased its dependency on imports for its fish supply. The United States, besides being the world's fourth major exporting country, was the second biggest importer. Imports were growing mainly due to expanding shrimp imports. Shrimps and prawns are increasingly produced in aquaculture especially in Southeast Asia.

MICROBIAL HAZARD IN SEAFOOD

Fish bacterial flora is composed mainly of psychrotrophic bacteria. Although a wide variety of spoilage microorganisms can contaminate seafood limited analysis techniques may serve to control its microbiological quality. Mainly psychrotrophic aerobes and lactic acid and related bacteria might be responsible for seafood spoiling, coliforms and enterococci analyses might serve to verify fecal exposure. The most common food borne pathogen microorganisms are *Vibrio* spp., *Clostridium* spp. and *Listeria monocytogenes*, all other pathogenic microorganisms have their origin in human activity, *E. coli* and *Salmonella* spp. are found the most. Thus, the microbial flora of seafood directly reflects the environment from which the seafood is extracted. Microorganisms become associated with the gill, intestine, and slime of the fish. The mud attached to the bottom of the fish, crab, and shrimp is another source of microorganisms. If microbial buildup is allowed to occur in the fish hold it will further add to the microbial load of seafoods. The presence, growth, survival or death of microorganisms or destruction of toxins as influenced by processing, packaging and storage conditions will also need to be considered. The following is the list of major microorganisms threatening seafoods (Table 4.1).

Seafood Diseases and Illness

The responsibility for collection and recording of data on disease and illness when seafood is implicated on a worldwide basis is spread among a large number of agencies across many countries. In the United States, the major sources of information on seafood-borne disease and illness

Table 4.1 Classification of microorganisms affecting seafood products

Natural spoilage microorganisms	Non-indigenous spoilage microorganisms	Natural food borne associated microorganisms	Non-indigenous food-borne microorganisms
Acinetobacter spp.	Enterobacteriaceae	*Clostridium botulinum*	*Shigella* spp.
Brochothrix thermosphancta	Fecal coliforms	*Vibrio* spp.	*Staphylococcus*
Corynebacteria	*Enterococcus faecalis*	*Listeria monocytogenes*	*aureus, Escherichia*
Flavobacterium spp.			*coli, Salmonella* spp.
Lactic acid bacteria			
Micrococcus spp.			
Moraxella spp.			
Photobacterium phosphoreum			
Pseudomonas spp.			
Shewanella putrefaciens			

are the Centers for Disease Control (CDC) Food-borne Disease Outbreak Surveillance Program and a data base on shellfish-associated food-borne cases maintained by the Food and Drug Administration (FDA) Northeast Technical submitted by state health departments to the CDC. The FDA data come from books, news accounts, CDC reports, city and state health department files, Public Health Service regional files, case histories and archival reports (FAO, 1998).

The total foodborne incidences during the period 2002-2006 were recorded as 81,852. The bacteria identified was, *Salmonella* (31,635 cases), *Campylobacter* (27,253), *Shigella* (13,941), *Cryptosporidium* (3,806), STEC O157 (2,111), STEC non-O157 (355), *E.coli* O157 (486), *Yersinia* (817), *Vibrio* (610), *Listeria* (632), and *Cyclospora* (179). The incidence of *Vibrio* infections has increased to the highest level. These infections are most often associated with the consumption of raw seafood, particularly oysters.

Table 4.2 Summary of the number of total foodborne illness outbreaks and cases in USA, 2002 to 2006

Microorganisms	No of infections				
	2002	2003	2004	2005	2006
Total	16,580	15,600	15,806	16,614	17,252
Salmonella spp.	6,028	6,017	6,464	6,471	6,655
Campylobacter	5,006	5,215	5,665	5,655	5,712
Shigella	3,875	3,021	2,231	2,078	2,736
STEC O157	647	-	401	473	590
Cryptosporidium	541	480	613	1,313	859
Yersinia	166	161	173	159	158
Vibrio spp.	103	110	124	119	154
Listeria	101	138	120	135	138
HUS/*E. coli* O157	44	443	-	-	-
Cyclospora	43	15	15	65	41
STEC non-O157	-	-	-	146	209

Source: derived from (CDC, 2008).

Among 95% *Vibrio* isolates for which the species was identified, 64% were *V. parahaemolyticus*, and 12% were *V. vulnificus*. (CDC, 2008).

Seafood-borne illness data for Japan are published annually by the Ministry of Health and Welfare, as part of the annual statistics for all food-borne illnesses. The number of incidents (outbreaks), patients (cases), and deaths are recorded (MHLW, 2008).

Categories reported by the Japanese data collection system are shellfish, swellfish (puffer fish) and others, fish paste and other products of fish and shellfish. Among the reported outbreaks, 65% of the outbreaks in Japan from 2003 to 2007 were seafood borne resulting in 1,158 cases, 14,159 patients, and 11 deaths (Table 4.3). About 35% deaths occurred only with shellfish and puffer fish during the 5-yr period (Table 4.3 and 4.4). Total food-borne outbreaks from 2003 to 2008 numbered 7,576 resulting in 147,052 cases.

Table 4.3 Summary of the number of total foodborne and seafood-borne illness outbreaks and cases in Japan, 2003 to 2007

Year	Cases		Patients		Death	
	Total Food borne	Total seafood borne	Total Food borne	Total seafood borne	Total Food borne	Total seafood borne
2003	1, 585	285	29, 355	3, 889	6	3
2004	1, 666	304	28, 175	2, 894	5	2
2005	1, 545	243	17, 019	2, 926	7	2
2006	1, 491	168	39, 026	1, 708	6	1
2007	1, 289	158	33, 477	2, 742	7	3

Source: Derived from (Ministry of Health and Welfare of Japan, 2008).

When the cause of the food-borne illness is known, 15% of the outbreaks are from fish and shellfish sources resulting in 9% of the cases. Thirty five percent of food-borne illness deaths in Japan are from fish, shellfish and products of fish and shellfish (Table 4.4). For fish and shellfish, most illnesses caused by bacteria are from *Vibrio parahaemolyticus* at 75% of bacterial disease cases. *Salmonella, Staphylococcus aureus, Escherichia coli* and *Clostridium perfringens* are the other major sources of illness. For products of fish and shellfish, the same five bacteria are the leading causes of illness but the distribution of cases among the bacteria is more evenly spread. Natural poisons and chemical substances rank six and seven among both fish and shellfish-borne diseases and fish and shellfish product-borne illnesses (MHLW, 2008).

Seafood-borne disease and illness reported by each reporting area or system represent the minimum number of actual seafood-borne cases that occur. It is very likely that many seafood-borne illnesses (like all food-borne illnesses) are either not reported by the patient or recognized as a food-borne illness. However, these data collection agencies provide only reliable sets of data available (MHLW, 2008).

Table 4.4 Summary of the number of seafood-borne illness outbreaks and cases attributed to fish and shellfish in Japan, 2003 to 2007

Category	Outbreaks	Patients	Death
		Number	
Shellfish	230	2,968	0
Puffer fish (Fugu)	177	237	11
Others	141	2,413	0
Total fish and shellfish	548	5,618	11
Fish paste	1	83	0
Others	391	4, 237	0
Total fish products	392	4,320	0
Total	940	9,938	11

Source: Derived from (Ministry of Health and Welfare of Japan, 2008).

Consumer Perceptions of Seafood Safety and Quality

Safety and quality regarding seafood have very different meanings to the microbiologist and seafood technologist than they do to the consumer. Safety usually refers to the risk level associated with illness or death caused by the consumption of a seafood product that is contaminated with a microbiological or parasitic organism, a naturally occurring poison or a chemical contaminant. Quality is most often related to appearance, odor, flavor and texture. Consumers and media often mix the two concepts, when in fact low quality seafood can be quite safe to eat and seafood with low-risk safety factors, but high in quality, might not actually be considered high quality by a consumer. In addition, many sellers of seafood have a limited understanding of consumers' preferences for seafood, even if they could be recognized (FAO 1998).

Quality and safety become multi-dimensional and consumers have great difficulty in determining and observing actual seafood quality (Anderson and Anderson 1991). Some of the attributes that affect seafood quality are: nutritional value, incidence of parasites, presence of microorganisms and bacteria, shelf life, taste, level of additives, the use of certain treatments such as irradiation, the presence of pesticides or preservatives, discoloration, size, presence of bones, scars or cuts, odor and uniformity among others. A mixture of safety and quality factors can also affect the acceptance of seafood to the consumers. Therefore, in order to ensure safe seafood government intervention or regulation is required.

Microbiological Quality Assessment of Seafood

Microbiological quality assessment is an essential point of seafood processing and distribution chains. Bacterial populations found in the skin and digestive tract of seafood are considered as natural populations. This indigenous flora is composed of several microorganisms, some of them can be human pathogens, and others are spoilage microorganisms. As seafood is often processed in fish industries, contamination with non-indigenous microorganisms is a very real possibility that increases with handling and storage steps. Microbial contamination originating from human activity in specific coastal areas is also considered as non-indigenous. Seafood microbiological quality assessment depends greatly on the processing conditions within the seafood chain. Processing parameters modify the original physico-chemical properties of seafood, favoring and/or avoiding the growth of several microorganisms. Therefore, in the fish processing chain managing risks should be based on scientific knowledge of the microbiological hazards and the understanding of the primary production, processing and manufacturing technologies and handling during food preparation, storage and transport, retail and catering (Anonymous, 2007).

The primary responsibility for ensuring food safety rests with the private sector involved in the fish production and processing chain and the implementation of food safety management systems based on the principles of the Hazards Analysis and Critical Control Point (HACCP) system in the fish processing chain is one of the most effective methods of managing risks.

The implementation of microbiological risk management decisions is the responsibility of national authorities. Risk management options can take many forms. A very wide range of food safety measures may be implemented, either alone or in combination, and these include development of regulatory standards, guidelines and codes of practice. Implementation of food controls usually includes specification of the role of competent authorities in ensuring compliance with regulatory requirements, and enforcement actions that may result from noncompliance. The competent authority establishes standards e.g., performance criteria, and verifies that they are met by industry, but the industry primarily implements the measures that achieve standards.

Microbiological Quality Assessment of Bivalves

Bivalves are a special case because of their unique food intake method: they are filtering organisms that concentrate on bacteria, viruses and/or particles present in their environment. After harvesting it is a good practice to depurate bivalves in current water for several days before entering the commercial chain, depending on the quality of water they were grown on. Once again analysis of coliforms is mandatory, analysis of *Escherichia coli*, *Salmonella* spp. and *Shigella* spp. also must be done due to the risk of fecal contamination. Bivalves that live under low oxygen conditions might be analyzed for *Clostridium botulinum* because of the risk of botulinic toxin presence in the substrate. As emerging pathogen *Listeria monocytogenes* might be also analyzed. Imported warm water bivalves must be analyzed for *Vibrio* spp because of the risk of its presence (Anonymous, 2007).

Microbiological Quality Assessment of Cultured Fish

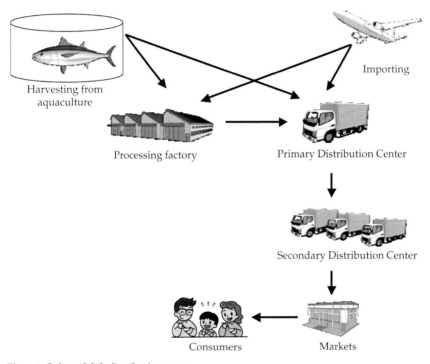

Fig. 4.1 Cultured fish distribution route.

Color image of this figure appears in the color plate section at the end of the book.

When warm water species are cultured they might be analyzed for *Vibrio* spp. It is not necessary to characterize all microorganisms; just total counts serves as an indicator of the microbiological quality of the product. If culturing waters are coliforms-free, their analysis may be overcome (if good working practices are assumed). Pathogens like *Salmonella* and *Shigella*, *Listeria monocytogenes* must be analyzed and should be absent. Microbial analyses must be done with a representative sample containing skin, gills and gut, as they are especially exposed to contamination. Specific spoilage microorganism plate count of psychrotrophs must be done as seafood is usually sold in refrigerated conditions and mesophilic aerobes are not a good indication of the real microflora present. Once cultured fish leave the production and/or processing center it goes to a Primary Distribution Center (PDC). All analyses might be repeated, but if it is assumed that they are pathogen-free, psychrotrophic aerobes (or lactic acid bacteria for modified atmosphere packaged seafood) might be analyzed in order to control the spoilage process. Seafood leaves PDC to go to Secondary Distribution Center. At least psychrotrophic aerobes (or lactic acid bacteria) might be analyzed once again. From the Secondary Distribution Center seafood is directed to retailers and then from retailers distributed to consumers. At this point, sanitary inspectors can routinely check the microbial conditions of the retail seafood products and can ensure microbiological risk to avoid food-borne pathogens in seafood products (Amárita, 2007).

Microbiological Quality Assessment of Wild Captured Fish

Once seafood arrives to the harbor microbial analyses is to be done with a representative sample containing skin, gills, and intestinal tract, as they are especially exposed to contamination, muscle also can be contaminated after fishing. Psychrotrophic aerobes must be analyzed as seafood is usually sold in a refrigerated condition and mesophilic aerobes are not a good indication of the real microflora present. Coliforms might be analyzed in order to determine fecal contamination due to unhygienic working practices and/or storage conditions; this is important in coastal fishes as they can be exposed to fecal contamination due to human proximity. Pathogens like *Salmonella* and *Shigella*, *Listeria monocytogenes* must be analyzed and should be absent. When fish is captured in warm waters the presence of *Vibrio* spp. should be verified.

When processes involve thermal treatment in vacuum or anaerobic conditions, thermophilic anaerobes must be analyzed. *Clostridium botulinum* also needs to be analyzed because of the presence of botulinic toxin. In all cases pathogens should be analyzed. In Primary and Secondary Distribution Centers, all analyses should be repeated, but if it is assumed that they are pathogen-free, psychrotrophic aerobes, lactic acid bacteria, halophilic aerobes or thermophilic anaerobes might be analyzed in order to control the spoilage process. From the Secondary Distribution Center seafood is directed to retailers and then to consumers. Sanitary authorities must ensure exhaustive controls (Amárita, 2007).

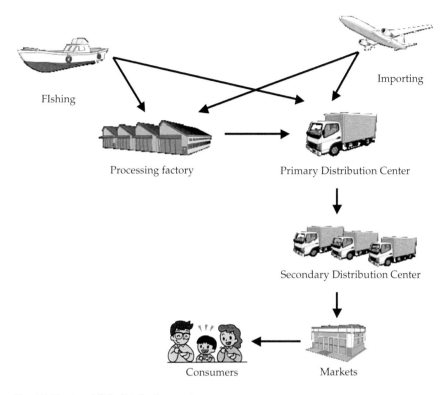

Fig. 4.2 Captured fish distribution route.

Color image of this figure appears in the color plate section at the end of the book.

Law Regulations of Quality Assessment of Seafood

New regulations clearly establish that producers are directly in charge of their own foodstuffs, in particular in UE (EC regulation 178/2002). HACCP is established, as the general proceeding, based on the Codex Alimentarius Commission principles and in accordance with correct hygienic practices, must ensure the quality of foodstuffs. However, there are no fixed microbiological criteria for seafood, except for bivalve molluscs. The European Union established regulations in the EU countries that include different aspects of the management of microbial risks (EC regulation 2073/2005) as well as strategies for control and prevention of risks for consumer protection (Reilly, 2006).

Examples of Managing Risks

Stage	How are the risks managed?	Verification that the risks are being managed
Primary production	• Classification of production areas for the harvesting of Live Bivalve Molluscs (LBM) (EC/854/2004). LBM from class B & C areas must be treated to meet the requirements of Class A areas. • Monitoring for the presence of toxin producing plankton in production and relaying waters and biotoxins in LBM (EC/854/2004) • GHP/GMP	• End product testing • Microbiological criteria*/ Biotoxin criteria**
Processing	• Implementation of a food safety management system based on the principles of HACCP (requirement of EC/852/2004 and EC/853/2004)	• Microbiological criteria (used to validate and verify HACCP based procedures—EC/2073/2005)*
Retail	• GHP (Good Hygiene Practice) and GMP(Good Manufacture Practice) (control of cold chain requirement of EC/852/2004)	• End product testing (where appropriate) • Environmental monitoring programs

*Microbiological Criteria specified in EC 2073/2005.

Type of criterion	Microorganism	Food category
Process hygiene	*Escherichia coli*	Shelled and shucked products of cooked crusta
	Coagulase positive Staphylococci	ceans and molluscan shellfish
Food safety	Salmonella *E. coli*	Cooked crustaceans and molluscan shellfish LBM and live echinoderms, tunicates and gastro pods
	Histamine	Fishery products from fish species associated with high amounts of histidine (*Scrombroid*)
		Fishery products which have undergone enzyme maturation in brine, manufactured from fish species associated with a high amount of histidine
	L. monocytogenes	RTE (Ready-to-Eat) food

** **Criteria for biotoxins specified in EC 853/2004**
Criteria established for
• PSP (Paralytic Shellfish Poisoning)
• ASP (Amnesic Shellfish Poisoning)
• Okadaic acid, dinophysistoxins and pectenotoxins
• Yessotoxins
• AZP (Azaspiracid shellfish Poisoning)

Shrimp Quality and Safety

Shrimp farmers must be aware of the current regulatory expectations in their country and in the countries where their shrimp will be sold and consumed. The regulatory authorities in most nations are assigned to protect the "safety" of their consumers. Most countries have specific regulations to assure food safety for products produced in or imported into the country. In many instances, these food safety regulations also involve or influence product quality. Regulatory expectations will be based on judgments and measures for both shrimp safety and quality (Otwell et al., 2001).

Shrimp quality and safety are closely related. A food with poor quality due to bacterial spoilage could be considered safe to eat if it is cooked to eliminate any safety concerns, but the poor quality is often considered an indirect measure for product safety. Likewise, an apparently good quality aquaculture could cause illness if it is contaminated with a potential food hazard that is not obviously based on quality judgments. Regulatory authorities should try to distinguish certain safety problems. For example, farmed shrimp could be unsafe to eat if:

1) the shrimps are contaminated with certain types or amounts of pathogenic bacteria;
2) the shrimps contain excessive amounts of food additives or improper food additives;
3) the shrimps contain pesticides, herbicides or other potential toxic chemicals introduced during pond culture; or
4) the shrimps contain improper amounts or type of therapeutic chemicals used during pond culture.

Areas of Concern for Shrimp Quality and Safety

The traditional regulatory approach has been to set various guidelines or tolerances that ensure a safe product. These standards are usually enforced by inspection of products after they are processed, combined with occasional inspections of the processing facilities to enforce good manufacture practices (GMP). The GMP's include some basic sanitation requirements that are usually designed for processing. Good aquaculture practices (GAP's) are introduced to include farming activities linked with processing.

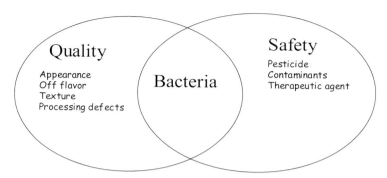

Fig. 4.3 Areas of concern for shrimp quality and safety.

As with market expectations, regulatory authorities are usually more concerned with new sources or methods of production for most foods. This concern has been most recently demonstrated by many nations adding food safety requirements based on Hazard Analysis and Critical Control Point (HACCP) programs. These requirements place more attention on "prevention" of potential food safety problems before they occur rather than the traditional approach of inspecting or trying to find problems after they have occurred. The HACCP approach does not replace the traditional regulatory approach. It is in addition to the traditional approach and depends on a solid foundation of sanitation, and the GMP's and GAP's (Otwell et al., 2001).

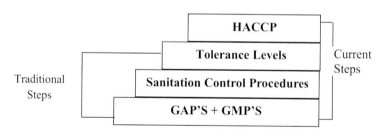

Fig. 4.4 Traditional vs. current steps to food safety.

The HACCP programs add requirements to document or record routine practices during the farm production and later processing of the shrimp. These records are evidence that proper hygiene and sanitation control procedures have been used when growing, harvesting and processing the shrimp. The key HACCP feature is monitoring of certain "critical control points" to maintain specified limits that assure the shrimps are safe to eat.

QUALITY CONCERNS OF AQUACULTURE SHRIMP

Shrimp quality is essential to maintaining product value. Poor quality does not only reduce value, but could build a poor reputation for a particular farm, processor or an entire country (EC Rapid alert system). As for product safety, certain controls must be used to maintain quality. The following list of problems and controls is based on industry experiences, buyer specifications and some related regulations for shrimps produced and sold in the world market (Table 4.5).

Table 4.5 Quality concern and preventive measures of aquaculture shrimp (Otwell et al., 2001)

Quality Concerns	Defects	Preventative Measures
Appearance	Blackspot	Proper application of sulfite or Everfresh
	Broken & damaged	Proper handling and icing
	Heat discoloration	Timely placement of product in ice
	Loose heads (whole product)	Proper handling of product in ice only
	Red heads	Stop feeding 48 h before harvest
	Soft Shell (Whole and Shell-on product	Harvest at the proper time based on periodic checks
	Yellowing	Proper use of sulfites
	Pitted or gritty shells	Proper use of sulfites
	Milky shrimp	Culling from the harvest
	Mixed Species	Separation by species at the plant
Odor/Flavor	Decomposition	Timely placement of product in ice
	Chlorine	Use proper concentration & exposure time
	Petro-chemical smell	Prevent contamination with oil, diesel, etc.
	Choclo/Earthy Smell	Sensory test before harvest
	Off-flavors in the head	Sensory test before harvest
Texture	Mushy and/or soft texture	Proper shrimp to ice ratio and timely placement of product in ice
Processing Defects	Short weight	Routine checks for proper specifications
	Off-counts	
	Uniformity	
	Dehydration	Proper glazing
	Extraneous materials	Proper culling

SAFETY CONCERNS OF AQUACULTURE SHRIMP

Shrimps remain one of the safest sources of seafood in the world. Food safety problems are rare, but certain problems can result in significant illnesses and costly damage to the industry and product reputation. The following list of potential food safety problems is based on actual market experiences and scientific evidence that indicate these problems are "reasonably likely to occur" for farmed shrimps (Table 4.6). All of these problems can be eliminated or reduced with appropriate controls (Otwell et al., 2001).

Table 4.6 Safety concern and control of aquaculture shrimp

Food Safety Concerns	Controls
Biological Pathogenic bacteria - *Salmonella* spp. - *Vibrio cholerae* - Pathogenic *Escherichia coli*	Increase culture water exchange, use of antimicrobial agents, or divert product to value added application. Determine the contamination source and apply controls.
Chemical Pesticides - Pesticides from agriculture - Insecticides, rodenticides, & other chemicals	Do not apply pesticides in the vicinity of the ponds or the feed. Be aware of application of these compounds in adjacent farms. Prevent contamination through run-offs.
Herbicides - Chlorphenoxy compounds - Triazine herbicides, & others	Do not apply herbicides in the vicinity of the ponds or the feed. Be aware of application of these compounds in adjacent farms. Prevent contamination through run-offs.
Fertilizers and Water Treatment Compounds - Ammonium compounds, Calcium phosphate, Phosphoric acid, Potassium chloride, Sodium silicate, Lime, hydrated lime, & limestone	Fertilizers and water treatment compounds are normally not considered to be a food safety problem. You should not apply fertilizers close to the harvest date.
Other contaminants - Heavy metals, i.e. methyl mercury	Not normally considered as a food safety concern in shrimp. Cadmium, lead and mercury are the most commonly found heavy metals in seafood products.
Therapeutic Agents - Oxytetracycline, Oxilinic acid, Furazolidone, Quinolona, & Terrivet	The use of these compounds is a major concern in shrimp aquaculture and controls are needed. The farmer needs to be aware of which product is approved or not for the country where the product will be shipped. Records need to be kept on the usage and recommended withdrawal times.
Food Additives - Sulfites	Sulfites are known to cause an allergic-type reaction for certain consumers and need to be controlled. If sulfites are used, product needs to be properly labeled.
Sanitizer residues	Proper labeling and proper use of cleaning and sanitizing compounds is essential to prevent any contamination on the product. Use proper concentrations and proper exposure time.
Physical - Debris - Filth	Filth and debris are normally considered to be quality defect and not a food safety issue. Both of these problems need to be minimized to avoid problems at the port of entry of the receiving country. Proper culling and sequential washes following the harvest can reduce both problems.

VIRUS

The incidence of food-borne outbreaks of viral gastroenteritis is still unknown. Viral disease transmission to humans via consumption of seafood has been known since the 1950's (Roos, 1956), and human enteric viruses appear to be the major cause of shellfish-associated disease. Presently there are more than 100 known enteric viruses, which are excreted in human feces and find their way into domestic sewage. However, only a few have shown to cause seafood-associated illness according to Kilgen and Cole (1991). These are: Hepatitis-type A (HAV), Noro (Norwalk virus, small, round structured), Snow Mountain Agent, Calicivirus, Astrovirus Non-A and Non-B.

Viruses are inert outside the living host-cell, but they survive. This means that they do not replicate in water or seafood irrespective of time, temperature or other physical conditions. Their presence on seafood is purely as a result of contamination either via infected food handlers or via polluted water. Shellfish, which are filter-feeders, tend to attract virus from the water in which they are growing. Large amounts of water pass through active shellfish (up to 1,500 L/day/oyster) according to Gerba and Goyal (1978), which means that the concentration of virus in the shellfish is much higher than in the surrounding water.

Epidemiology and Risk Assessment

The infective dose is probably much smaller than that of bacteria for causing foodborne disease (Cliver, 1988). The minimum infection dose of some enteric viruses for man is close to the minimum dose detectable in laboratory assay systems using cell cultures (Ward and Akin, 1983).

Animal/human bodies are sources of enteric viruses. The viruses are found in large quantities in the feces of infected persons a few days to several weeks after ingestion/infection depending on the virus. Direct or indirect fecal contamination is the most common source of contamination of food.

Bivalve molluscs dominate the list of food vehicles in outbreaks of viral diseases. However, another important vehicle involves ready to eat food prepared by infected food handlers. The available data show that almost any food that comes into contact with human hands and does not subsequently receive a substantial heat treatment may transmit these viruses.

With only few exceptions, all reported cases of seafood-associated viral infections have been from consumption of raw or improperly cooked molluscan shellfish (Kilgen and Cole, 1991). However, there is clear evidence that HAV has been transmitted by unsanitary practices during processing, distribution or food handling (Ahmed, 1991). These seafood-associated illnesses are very common. Each year 20,000 to 30,000 cases are reported to the Center of Disease Control (CDC) in the U.S. (Ahmed, 1991), and one of the largest outbreaks of foodborne illness ever reported, is the outbreak of hepatitis involving 290,000 cases in China in 1988. The investigation revealed that the source and mode of transmission were the consumption of contaminated and inadequately cooked shellfish (Tang et al., 1991).

The survival of viruses in the environment and in food is dependent on a number of factors such as temperature, salinity, and solar radiation, presence of organic solids as reviewed by Gerba (1988). Thus, enteric viruses are able to survive for several months in seawater at temperatures <10°C, which is much longer than, e.g., coliform bacteria (Melnick and Gerba, 1980). Thus, there is little or no correlation between the presence of virus and the usually applied indicator bacteria for fecal pollution. All enteric viruses are also resistant to acid pH, proteolytic enzymes and bile salts in the gut. Hepatitis type A virus, being one of the more heat stable viruses, has an inactivation time of 10 min at 60°C (Eyles, 1989). Thus virus is able to survive some commonly used culinary preparations (steaming, frying). Enteric viruses are also resistant to some common disinfectants (e.g., phenolics, quaternary ammonium compounds, ethanol) while the halogens (e.g., chlorine, iodine) inactivate enteric viruses in water and on clean surfaces. Ozone is highly effective in clean water (Eyles, 1989).

Contamination by food handlers can be prevented by introducing good personal hygiene and health education. Food handlers must not handle food while suffering from intestinal infections and for at least 48 h after symptoms have disappeared. In cases of doubt, disposable gloves should be worn in critical operations, as viruses are difficult to remove from hands by washing and are resistant to many skin disinfectants (Eyles, 1989).

PARASITE

The presence of parasites in fish is very common, but most of them are of little concern with regard to economics or public health. However, more than 50 species of helminth parasites from fish and shellfish are known to cause disease in humans (Healy and Juranek, 1979; Higashi, 1985; and Olson, 1987). Most are rare and involve only slight to moderate injury but some pose serious potential health risk. The most important pathogenic parasites are listed in Table 4.7. All the parasitic helminths have complicated life cycles. They do not spread directly from fish to fish but must pass through a number of intermediate hosts in their development. Very often sea-snails or crustaceans are involved as first intermediate host and marine fish as second intermediate host, while the sexually mature parasite is found in mammals as the final host. In between these hosts, one

Table 4.7 Pathogenic parasites transmitted by fish and shellfish

Parasite	Known geographical distribution	Fish and shellfish
Nematodes or round worms		
Anisakis simplex	North Atlantic others	herring sea-snails or crustaceans
Pseudoterranova dicipiens	North Atlantic and others	Cod, almost all marine fish
Gnathostoma sp.	Asia	freshwater fish, frogs
Capillaria sp.	Asia	freshwater fish
Angiostrongylus sp.	Asia, South America, Africa	freshwater prawns, snails, fish
Cestodes or tape worms		
Diphllobothrium latum	Northen hemisphere	freshwater fish
D. pacificum	Peru, Chile, Japan	seawater fish
Trematodes or flukes		
Clonorchis sp.	Asia	freshwater fish, snails
Opisthorchis sp.	Asia	freshwater fish
Metagonimus yokagawai	Far East	snails and freshwater fish
Heterophyes sp.	Middle East, Far East	snails, freshwater fish brackish water fish
Paragonimus sp.	Asia, America, Africa	snails, crustaceans, fishes
Echinostoma sp.	Asia	clams, freshwater fishes, snails

or more free-living stages may occur. Infection of humans may be part of this life cycle or it may be a sidetrack causing disruption of the life cycle.

Nematodes

Round worms or nematodes are common and found in marine fish all over the world. The anisakis nematodes *A. simplex* and *P. dicipiens* commonly known as the herring worm and the cod worm have been intensively studied (Healy and Juranek 1979, Higashi, 1985; and Olson, 1987). They are typical round worms, 1-6 cm long, and if humans ingest live worms, they may penetrate into the wall of the gastrointestinal tract and cause an acute inflammation ("herring worm disease").

A number of other nematodes are found in freshwater fish. *Gnathostoma* sp. is the most important species found in Asia. The final hosts are cats and dogs but humans may be infected. Upon ingestion the larvae migrate from the stomach to various regions, most commonly to subcutaneous sites on the thorax, arms, head and neck, where the worms induce a creeping sensation and edema (Huss, 1994).

Another nematode of public health importance is *Capillaria* sp. (e.g., *Capillaria philippinensis*). The adult worms are gut parasites in piscivorous birds and intermediate hosts are small freshwater fish. Infection in humans cause severe diarrhea and possible death attributed to fluid loss.

A well-known and common nematode in Asia is the *Angiostrongylus* sp. (e.g., *Angiostrongylus cantonensis*). The adult worm is found in the lungs of rats and the intermediate hosts are snails, freshwater prawns and land crabs. The parasite has been shown to cause meningitis in humans (Brier, 1992).

Cestodes

Only few *cestodes* or tapeworms in man are known to be transmitted by fish. However, the broad fish tapeworm *Diphyllobothrium latum* is a common human parasite reaching up to 10 m or more in length in the intestinal tract of humans. This parasite has a microcrustacean as first intermediate host and freshwater fish are required as second intermediate host. The related species (*D. pacificum*) is transmitted by marine fish and commonly occurs in the coastal regions of Peru, Chile and Japan where raw fish preparations (ceviche, sushi and others) are common.

The broad fish tapeworms, *Diphyllobothrium* sp., reach sexual maturity in the intestinal tract of mammals. Eggs may pass in the feces and develop in water into larvae that hatch and swim freely. If consumed by a copepoda or other suitable crustacean host, the larvae may then become infective for fish that consume the infected crustacean. These larvae then develop into forms that may infect other fishes, where they do not develop further, or mammals, where they may reach sexual maturity (Brier, 1992).

Trematodes

Some of the *Trematodes* or flukes are extremely common, particularly in Asia. Thus it is estimated that the *Clonorchis sinensis* (the liver fluke) is infecting more than 20 million people in Asia. In southern China human clonorchiasis rates can surpass 40% in some regions (Rim, 1982). Intermediate hosts are snails and freshwater fish, while dogs, cats, wild animals and humans are final hosts where the fluke live and develop in the bile ducts in the liver. The predominant problem in transmission is the contamination of snail-infested waters by egg-laden feces from humans (e.g., use of "night soil" as fertilizers). Two very small flukes (1-2 mm) *Metagonimus yokagawai* and *Heterophyes heterophies* differ from *Clonorchis* by living in the intestines of the final host, causing inflammation, symptoms of diarrhea and abdominal pain. Intermediate hosts are snails and freshwater fish. The adult oriental lung fluke *Paragonimus* sp. is 8-12 mm and encapsulated live in cysts in the lungs of man, cats, dogs and pigs and many wild carnivore animals. Snails and crustaceans (freshwater crab) are the intermediate hosts. All parasites of concern are transmitted to humans by eating raw or uncooked fish products.

Marinated Fish

Safe processing is primarily based on the level of NaCl in the tissue fluid. When the minimum amount of acetic acid (2.5-3.0% in the tissue fluid) is used, the following maximum, survival time of nematodes at various NaCl-levels was found: Maximum survival times of nematodes should therefore also be minimum holding time of the final product before sale.

Table 4.8 Survival of nematodes at different NaCl-levels

% NaCl in tissue fluid	Max. survival time of nematodes
4-5	6-17 wk
6-7	10-12 wk
8-9	5-6 wk

Heat-treated Fish

All nematodes are killed when heated to 55°C for 1 min. This means that hot-smoked, pasteurized, sous-vide cooked and other lightly heat-treated fish products are safe. However, some normal household cooking traditions may be on the borderline of safety.

Frozen Fish

Freezing to −20°C and maintaining this temperature for at least 24 h will kill all nematodes. The results listed above show that a number of fish products are unsafe. This applies to lightly salted fish products (< 5-6% NaCl in water phase) such as matjes-herring, gravad fish, cold smoked fish, lightly salted caviar, ceviche and several other local traditional products. A short period of freezing either of the raw material or the final product therefore must be included in the processing as a means of control of parasites.

BIOTOXINS

It has been known since ancient times that certain fish and shellfish are poisonous and can cause death when eaten. The first Chinese pharmacopoeia, dating 2800 BC, records injunctions against eating puffer fish (Kao, l966).

The chemical nature and biological basis for these food-borne intoxications have been elucidated over the last 50 yr, beginning with the pioneering work of Meyer and Sommer on the etiology of paralytic shellfish poisoning (PSP) in California (Meyer et al., l928). It is now evident that certain microscopic algae, present in phytoplankton, produce very potent toxins (phycotoxins, or algal toxins), which are chemical compounds mainly of low relative molecular mass. Concentrations of phycotoxins in the sea or in freshwater are highest during an algal bloom or red tide, a phenomenon characterized by a sudden, rapid multiplication of algal cells caused by environmental factors not yet fully understood. The phycotoxins are taken up by predators feeding on plankton, either directly as in the case of bivalve molluscs, or through several trophic levels as in fish. Man then consumes these food items. Marine biotoxins are responsible for a substantial number of seafood- borne diseases. The known toxins are shown in Table 4.9.

Table 4.9 Aquatic biotoxins

Aquatic Biotoxins		
Toxin	Where /when produced	Animal(s)/organ involved
Tetrodotoxin	in fish *ante mortem*, feed, Bacteria	pufferfish (*Tetraodontidae*) mostly ovaries, liver, intestines
Ciguatera	Marine algae	>400 tropical/subtropical fish sp.
PSP-paralytic shellfish poison	″ ″	filter feeding shellfish, mostly digestive glands and gonads
	″ ″	filter feeding shellfish
DSP-diarrhetic	″ ″	″ ″ ″
NSP-neurotoxic	″ ″	″ ″ ″ (blue mussels)
ASP-amnesic	″ ″	Shellfish and crabs edible tissue, viscera

The toxins and the diseases they can provoke have been described and reviewed by Taylor (1988), Hall (1991), WHO (1984, 1989) and Todd (1993). Some of the more important aspects are discussed below.

Tetrodotoxin

Unlike all other biotoxins algae do not produce tetrodotoxin that accumulate in live fish or shellfish. The precise mechanism in production of this very potent toxin is not clear, but quite commonly occurring symbiont bacteria are involved (Noguchi et al., 1987; Matsui et al., 1989). Tetrodotoxin is mainly found in the liver, ovaries and intestines in various species of puffer fish, the most toxic being members of the family *Tetraodontidae*, but not all species in this family contain the toxin. The muscle tissue of the toxic fish is normally free of toxin, but there are exceptions. Puffer fish poisoning causes neurological symptoms within 10-45 min after ingestion. Symptoms are tingling sensation in face and extremities, paralysis, respiratory symptoms and cardiovascular collapse. In fatal cases death takes place within 6 h.

Ciguatera

Ciguatera poisoning results from the ingestion of fish that have become toxic by feeding on toxic dinoflagellates, which are microscopic marine planktonic algae. The principal source is the benthic dinoflagellate *Gambierdiscus toxicus*, which is living around coral reefs closely attached to macroalgae. Increased production of toxic dinoflagellates are seen when reefs are disturbed (hurricanes, blasting of reefs etc.). More than 400 species of fish, all found in tropical or warm waters, have been reported to cause ciguatera (Halstead, 1978). The toxin accumulates in fish that feed on the toxic algae or larger carnivores that prey on these herbivores. Toxin can be detected in the gut, liver and muscle tissue by means of mouse-assay and chromatography. Some fish may be able to clear the toxin from their systems (Taylor, 1988).

Although the reported incidence of ciguatera poisoning is low (Taylor, 1988), it has been estimated that the worldwide incidence may be in the order of 50,000 cases yearly (Ragelis, 1984). The clinical picture varies but onset time is a few hours after ingestion of the toxin. Gastrointestinal and neurological systems are affected (vomiting, diarrhea, tingling sensation, ataxia, weakness). Duration of illness may be 2-3 d but some may also persist for weeks or even years in severe cases. Death results from circulatory collapse and 12% of fatal cases have been reported (Halstead, 1978).

Paralytical Shellfish Poisoning (PSP)

Intoxication after consumption of shellfish is a syndrome that has been known for centuries, the most common being paralytic shellfish poisoning (PSP). PSP is caused by a group of toxins (saxitoxins and derivatives) produced by dinoflagellates of the genera *Alexandrium, Gymnodinium* and *Pyrodinium*. Historically, PSP has been associated with the blooming of dinoflagellates ($>10^6$ cells/liter), which may cause reddish or a yellowish discoloration of the water. However, water discoloration may be caused by proliferation of many types of planktonic species, which are not always toxic, and not all toxic algae blooms are colored. The dinoflagellates bloom as a function of water temperature, light, salinity, presence of nutrients and other environmental conditions. However, the precise nature of factors eliciting a toxic clone is unknown. Water temperature must be >5-$8°C$ for blooms to occur. If temperatures decrease to below $4°C$, the dinoflagellates will survive as cysts buried in the upper layers of the sediments.

Mussels, clams, cockles and scallops that have fed on toxic dinoflagellates retain the toxin for varying periods of time depending on the shellfish. Some clear the toxin very quickly and are only toxic during the actual bloom, others retain the toxin for a long time, even years (Schantz, 1984). PSP is a neurological disorder, and the symptoms, include tingling, burning and numbness of lips and fingertips, ataxia, drowsiness, incoherent speech. In severe cases death occurs due to respiratory paralysis. Symptoms develop within 0.5-2 h of a meal and victims who survive more than 12 h generally recover.

Diarrhetic Shellfish Poisoning (DSP)

Thousands of cases of gastrointestinal disorders caused by diarrhetic shellfish poisoning (DSP) have been reported in Europe, Japan and Chile (WHO, 1984). The causative dinoflagellates, which produce the toxins, are within the genus *Dinophysis* and *Aurocentrum*. These dinoflagellates are widespread which means that this illness could also occur in other parts of the world. At least seven toxins have been identified, including okadaic acid. The onset of disease is within half an hour to a few hours following consumption of shellfish that have been feeding on toxic algae. Symptoms are gastrointestinal disorder (diarrhea, vomiting, abdominal pain) and victims recover within 3-4 d. No fatalities have ever been observed.

Neurotoxic Shellfish Poisoning (NSP)

Neurotoxic shellfish poisoning (NSP) has been described in people who consumed bivalves that have been exposed to "red tides" of the dinoflagellate (*Ptychodiscus breve*). The disease has been limited to the Gulf of Mexico and areas off the coast of Florida. Brevetoxins are highly lethal to fish and red tides of this dinoflagellate are also associated with massive fish kills. The symptoms of NSP resemble PSP except that paralysis does not occur. NSP is seldom fatal.

Amnesic Shellfish Poisoning (ASP)

Amnesic shellfish poisoning (ASP) has only recently been identified (Todd, 1990, Addison and Stewart, 1989). The intoxication is due to domoic acid, an amino acid produced by the diatom *Nitzschia pungens*. The first reported incidence of ASP occurred in the winter of 1987/88 in eastern Canada, where over 150 people were affected and four deaths occurred after consumption of cultured blue mussels. The symptoms of ASP vary greatly from slight nausea and vomiting to loss of equilibrium and central neural deficits including confusion and memory loss. The short-term memory loss seems to be permanent in surviving victims, thus the term amnesic shellfish poisoning.

The control of marine biotoxins is difficult and disease cannot be entirely prevented. The toxins are all of non-protein nature and extremely stable (Gill et al., 1985). Thus cooking, smoking, drying, salting does not destroy them, and from the appearance of fish or shellfish flesh it is difficult to identify whether it is toxic.

The major preventive measure is inspection and sampling from fishing areas and shellfish beds, and analysis for toxins. The mouse bioassay is often used for this purpose and confirmatory HPLC (High Performance Liquid Chromatography) is done if death occurs after 15 min. If high levels of toxin are found, commercial harvesting needs to be halted.

It seems unlikely that it will ever be possible to control phytoplankton composition in growing areas, eliminating toxigenic species, and there is no reliable way to forecast, when a particular phytoplankton will grow and thus no way to predict blooming of toxigenic species (Hall, 1991). Removal of toxin by depuration techniques may have some potential, but the process is very slow and costly. There is also a risk that a small number of individuals decline to open and pump clean water through the system and therefore retain their original level of toxicity (Hall, 1991).

BIOGENIC AMINES (HISTAMINE POISONING)

Histamine poisoning is a chemical intoxication following the ingestion of foods that contain high levels of histamine. Historically this poisoning was called scombroid fish poisoning because of the frequent association with scombroid fishes including tuna and mackerel. Histamine poisoning is a worldwide problem occurring in countries where consumers ingest fish containing high levels of histamine. It is a mild disease; incubation period is very short (few minutes to few hours) and the duration of illness is short (few hours). The most common symptoms are cutaneous such as facial flushing, urticaria, edema, but the gastrointestinal tract may also be affected (nausea, vomiting, diarrhea) as well as neurological involvement (headache, tingling, burning sensation in the mouth). Histamine is formed in the fish *post mortem* by bacterial decarboxylation of the amino acid histidine as shown in Figure 4.5.

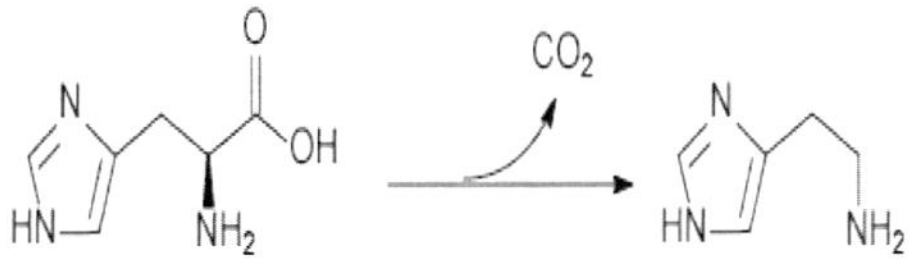

Fig. 4.5 Conversion of histidine to **histamine** by histidine decarboxylase

The fish frequently involved are those with natural high content of histidine such as those belonging to the family *Scombridae* but also non-scombroid fish such as *Clupeidae* and mahi-mahi may be involved in histamine poisoning.

The histamine-producing bacteria are certain *Enterobacteriaceae*, some *Vibrio* sp., a few *Clostridium* and *Lactobacillus* sp. The most potent histamine producers are *Morganella morganii*, *Klebsiella pneumoniae* and *Hafnia alvei* (Stratten and Taylor, 1991). These bacteria can be found on most fish, probably as a result of post-harvest contamination. They grow well at 10°C but at 5°C growth is greatly retarded and no histamine was produced by *M. morganii* when temperatures were <5°C at all times

(Klausen and Huss, 1987). However, large amounts of histamine were formed by *M.morganii* at low temperatures (0-5°C) following storage for up to 24 h at high temperatures (10-25°C) even though bacterial growth did not take place at 5°C and below. Many studies agree that histamine-producing bacteria are mesophilic. However, Ababouch et al., (1991) found considerable histamine production in sardines at temperatures < 5°C, and van Spreekens (1987) has reported on histamine production by *Photobacterium* sp., which are also able to grow at temperatures < 5°C.

The principal histamine producing bacteria *M. morganii* grow best at neutral pH, but they can grow in the pH range 4.7-8.1. The organism is not very resistant to NaCl, but optimal conditions growth can take place in up to 5% NaCl. Thus histamine production by this organism is only a problem in very lightly salted fish products. It should be emphasized that once the histamine has been produced in the fish, the risk of provoking disease is very high. Histamine is very resistant to heat, so even if the fish is cooked, canned or otherwise heat-treated before consumption, the histamine is not destroyed.

The evidence that histamine is causing disease is mostly circumstantial. High levels of histamine have consistently been found in samples implicated in outbreaks, and the symptoms noted in outbreaks are consistent with histamine as the causative agents. However, high intake of histamine does not always result in disease, even when "hazard action level" (50 mg/100 g for tunafish) is exceeded. The human body will tolerate a certain amount of histamine without any reaction. The ingested histamine will be detoxified in the intestinal tract by at least two enzymes, the diamine oxidase and histamine N-methyltransferase (Taylor, 1986). This protective mechanism can be eliminated if intake of histamine and/or other biogenic amines is very high, or if other compounds block the enzymes as shown in Figure 4.6.

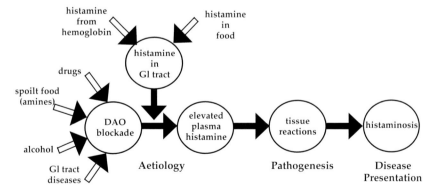

Fig. 4.6 The disease concept of food-induced histaminosis (Sattler and Lorenz, 1990).

Other biogenic amines such as cadaverine and putrescine which are known to occur in spoiled fish may therefore act as potentiators of histamine toxicity. Presumably inhibition of intestinal histamine catabolism will result in greater transport of histamine across cellular membranes and into the blood circulation.

Low temperature storage and holding of fish at all times is the most effective preventive measure. All studies seem to agree that storage at 0°C or very near to 0°C limits histamine formation in fish to negligible levels. Several countries have adapted regulations governing the maximum allowable levels of histamine in fish. Examples are shown in Table 4.10.

Table 4.10 Regulatory limits for histamine in fish

Area	Hazard action level mg/100g	Defect action level mg/100g	Maximum allowable limit g/100g
USA (FDA)	50	10-20	-
EC	-	10	20

CONCLUSIONS

Farmers, processors and buyers share responsibility for the quality and safety of farmed shrimp (Fig. 4.7). The areas of responsibility begin before harvest and continue during distribution and/or consumption. Proper controls are required during shrimp growth, pond harvest, processing, distribution and storage. Due to market and regulatory expectations, the processor usually assumes continuous responsibility from production to final sale. Farmers should work with the processor to assure proper controls are used during grow-out, pre-harvest preparations and harvest operations.

Likewise, regulatory authorities in the country where the shrimps are farmed are expected to serve as a "competent authority" or third party providing surveillance and assurances that the shrimps are produced and processed to provide safe products for domestic and foreign consumption. The authority and how it is used should support the shrimp farming industry and provide information that will attract and build buyer

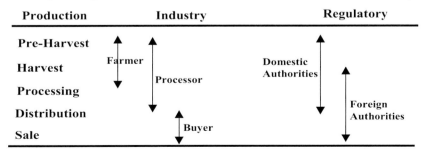

Fig. 4.7 Areas for responsible control.

confidence. Farmers and processors should work in cooperation with these authorities to ensure the market value for their shrimps.

In addition, every processor shall have to conduct, a hazard analysis to determine whether there are food safety hazards that are reasonably likely to occur for each kind of fish and fishery product processed by that processor and to identify the preventive measures that the processor can apply to control those hazards. Such food safety hazards can be introduced both within and outside the processing plant environment, including food safety hazards that can occur before, during, and after harvest. A food safety hazard that is reasonably likely to occur is one for which a prudent processor would establish controls because experience, illness data, scientific reports, or other information provide a basis to conclude that there is a reasonable possibility that it will occur in the particular type of fish or fishery product being processed in the absence of those controls. Every processor needs to implement a HACCP plan to overcome microbial safety hazards of seafood.

ABBREVIATIONS

ASP	Amnesic shellfish poisoning
AZP	Azaspiracid shellfish poisoning
CDC	Centers for disease control
DSP	Diarrhetic shellfish poisoning
FDA	Food and drug administration
GAP	Good aquaculture practices
GHP	Good hygiene practice
GMP	Good manufacturing practices
HAV	Hepatitis—type A virus
HACCP	Hazards analysis and critical control points
HPLC	High performance liquid chromatograph
LBL	Live bivalve mollusks
NSP	Neurotoxic shellfish poisoning
PDC	Primary distribution center
PSP	Paralytic shellfish poisoning
RTE	Ready-to-eat

REFERENCES

Ababouch, L., Afilal, M.E., Benabdeljelil, H. and Busta, F.F. (1991). Quantitative changes in bacteria, amino acids and biogenic amines in sardines (*Sardina pilchardus*) stored at ambient temperature (25-28°C) and in ice. Int. J. Food Sci. Technol. 26: 297-306.

Addison, R.F. and Stewart, J.E. (1989). Domoic acid and the Eastern Canadian molluscan shellfish industry. Aquaculture 77: 263-269.

Ahmed, F.E. (ed.) (1991). Seafood Safety. National Academy Press. Washington DC, USA.

Amárita, F. (2007). Methods for microbiological quality assessment of seafood. Food Research Division, A practical guide for seafood products, AZTI-Tecnalia, Spain.

Anderson, J.G. and Anderson, J.L. (1991). Seafood quality issues for consumer researchers. J. Consumer Affairs 25: 144-163.

Brier, J.W. (1992). Emerging problems in seafood—borne parasitic zoonoses. Food Control 3: 2-7.

CDC (Center for Diseases Control and Prevention)(2008). Annual Report from 2002 to 2006. URL: http://www.cdc.gov/foodnet/. Accessed on December 1, 2008.

Cliver, D.O. (1988). Virus transmission via foods. Food Technol. 42: 241-248.

Eyles, M.J. (1989). Viruses. In: Foodborne Microorganisms of Public Health ignificance, 4th edn. K.A. Buckle (ed). AIFST (NSW Branch) Food Microbiology Group. P.O. Box 277, Pymble, NSW 2073, Australia.

FAO (1998). Seafood Safety-Economics of Hazard Analysis and Critical Control Point (HACCP) Programmes. FAO fisheries technical paper 381. Rome, Italy.

FAO (2004). Assessment and management of seafood safety and quality. FAO Fisheries Technical Paper 444. Rome, Italy.

Gerba, C.P. (1988). Viral disease transmission by seafoods. Food Technol. 42: 99-102.

Gerba, C.P. and Goyal, S.M. (1978). Detection and occurence of enteric viruses in shellfish: A review. J. Food Protec. 41: 742-749.

Gill, T.A., Thompson, J.W. and Gould, S. (1985). Thermal resistance of paralytic shellfish poison in soft shell clams. J. Food Protec. 48: 659-662.

Hall, S. (1991). Natural toxins. In: Microbiology of Marine Food Products. D.R. Ward and C. Hackney (eds). Van Nostrand Reinhold, New York, USA pp. 301-330.

Halstead, B.W. (1978). Poisonous and Venomous Marine Animals of the World, Darwin Press, Princeton, USA.

Health Canada (2007). Microbiological methods used by the Health Products and Food Branch (HPFB) of Health Canada: http://www.hc-sc.gc.ca/fn-an/res-rech/analy-meth/microbio/index_e.html (English).

Healy, G.R. and Juranek, D. (1979). Parasitic infections. In: Food-Borne Infections and Toxications. H. Riemann and F.L. Bryan, (eds). Academic Press, London, UK pp. 343-385.

Higashi, G.H. (1985). Foodborne parasites transmitted to man from fish and other aquatic foods. Food Techonol. 39: 69-111.

Huss, H.H. (1994). Assurance of Seafood Quality. FAO Fishery Technical Paper No. 334. FAO, Rome, Italy.

Kao, C.Y. (1966) Tetrodotoxin, saxitoxin and their significance in the study of excitation phenomena. Pharmacol. Rev. 18: 997-1049.

Kilgen, M.B. and Cole, M.T. (1991). Viruses in seafood. In: Microbiology of Marine Food Products. D.R. Ward and C. Hackney (eds). Van Nostrand Reinhold,New York, USA pp. 197-209.

Klausen, N.K. and Huss, H.H. (1987). Growth and histamine production by *Morganella morganii* under various temperature conditions. Int. J. Food Microbiol. 5: 147-156.

Matsui, T., Taketsuyu, S., Kodama, K., Ishii, A., Yamamori, K. and Shimizu, C. (1989. Production of tetrodotoxin by the intestinal bacteria of a pufferfish *Takifugu niphobes*. Nippon Suisan Gokkaishi 55: 2199-2203.

Melnick, J.C. and Gerba, C.P. (1980). The ecology of enteroviruses in natural waters. CRC Crit. Rev. Environ. 10: 65-93.

Meyer, K.E., Sommer, K.F., and Schoenholz, P. (1928). Mussel poisoning. J. Prev. Med. **2:** 365-394.

MHLW (Ministry of Health, Labour and Welfare of Japan) (2008) URL (http://www.mhlw.go.jp/topics/syokuchu/04.html#4-2). Accessed on December 1, 2008.

Noguchi, T., Hwang, D.F. Arakawa, O., Sugita, H., Deguchi, Y., Shida, Y. and Hashimoto, K. (1987). *Vibrio alginolyticus*, a tetrodotoxin producing bacterium in the intestines of the fish *Fugu vermiculans vermi cularis*. Mar. Biol. 94: 625-630.

Olson, R.E. (1987) Marine fish parasites of public health importance. In: Seafood Quality Determination. Elsevier Science Publishers, Amsterdam, The Netherlands pp. 339-355.

Otwell, S., Garrido, L., Garrido, V. and Benner, R. (2001). Farm raised shrimp good aquaculture practices for product quality and safety. Extension Bulletin, Florida Sea Grant College Program, Gainesville, Fl., USA pp. 169-228.

Ragelis, E.P. (1984). Ciguatera seafood poisoning. Overview. In: Seafood Toxins. E.P. Ragelis (ed). ACS Symposium Series 262. Washington D.C., USA pp. 25-36.

Reilly, A. (2006). Managing microbiological risk in the fish processing chain. Presentation at the FAO-EUROFISH Workshop on Seafood Safety, 13-15, December 2006, Copenhagen, Denmark.

Rim, H.J. (1982). Clonorchiasis. In: CRC Handbook Series in Zoonosis, Section C: Parasitic Zoonoses, Volume 3 . G.V. Hillyer and C.F. Hopla, (eds). CRC Press Inc., Boca Raton, FL., USA pp. 17-32.

Roos, R. (1956). Heapatitis epidemic conveyed by oysters. Svenska Läkartidningen 53: 989-1003.

Sattler, J. and. Lorenz, W. (1990). Intestinal diamine oxidases and enteral-induced histaminosis: studies of three prognostic variables in an epidemiological model. *J. Neural Transm.* Suppl. 32: 291-314

Schantz, E.J. (1984). Historical perspective on paralytical shellfish poisoning. In: Seafood Toxins. E.P. Ragelis (ed). ACS—Symposium Series, Washington DC, USA 262: 99-111.

Stratten, J.E. and S.L. Taylor 1991. Scombroid poisoning. In: Microbiology of Marine Food Product. D.R. Ward and C.R. Hackney (eds). Van Nostrand Reinhold, New York, USA pp. 331-351.

Tang, Y.W., Wang, J.X., Xu, Z.Y., Guo, Y.F., Qian, W.H. and Xu, J.X.(1991). A serologically confirmed case-control study of a large outbreak of hepatitis A in China associated with consumption of clams. Epidemiol. Infect. 107: 651-658.

Taylor, S.L. (1986). Histamine food poisoning: Toxicology and clinical aspects. Crit. Rev. Toxicol. 17: 91-128.

Taylor, S.L. (1988). Marine toxins of microbial origin. Food Technol. 42: 94-98.

Todd, E.C.D. (1990). Amnesic Shellfish Poisoning—A new seafood toxic syndrome. In: Toxic Marine Phytoplancton. E. Graneli, B. SundstrΦ Edlar and D.M. Andersen (eds). Elsevier Science Publications, Amsterdam, The Netherlands pp. 504-508.

Todd, E. (1993). Domoic acid and amnesic shellfish poisoning—a review. J. Food Protec. 56: 69-83.

Ward, R.L. and Akin, E.W. (1983). Minimum infective dose of animal viruses. Crit. Rev. Environ. Control 14: 297-310.

WHO (1984). Aquatic (Marine and Freshwater) Biotoxins. Environ. Health Criter. 37, Geneva Switzerland.

WHO (1989). Report of WHO Consultation on Public Health Aspects of Seafood-Borne Diseases. WHO/CDS/VPH/90.86.

Van Spreekens, K.J.A. (1987). Histamine production by the psycrophilic flora. In: Seafood Quality Determinations. D. Kramer and J. Liston (eds). Elsevier Science Publications, Amsterdam, The Netherlands pp. 309-318.

5

Molecular Detection of Seafood-Borne Human Pathogenic Bacteria

Robert E. Levin

INTRODUCTION

Human bacterial infections and intoxication of seafood origin are usually derived from the consumption of raw or undercooked fish or shellfish harboring infectious organisms of mammalian intestinal origin (sewage outfalls) or from infectious organisms native to estuarine environments. Although conventional bacteriological methodology is still widely used to detect and enumerate these seafood pathogens molecular techniques involving the polymerase chain reaction (PCR) are now available allowing for rapid detection with notably high levels of sensitivity.

This chapter deals exclusively with human infectious and toxic producing bacteria that are native to marine environments and which are found to be associated with seafood and the conventional and molecular techniques available for their detection and enumeration. There are four such principle organisms that fall into this category:

- *Vibrio vulnificus,*
- *Vibrio parahaemolyticus,*
- *Vibrio cholerae*, and
- *Clostridium botulinum* type E.

Department of Food Science, Massachusetts Agricultural Experiment Station, University of Massachusetts, Amherst, MA 01003; Fax: 413-545-1262; Tel.: 413-545-0187;
E-mail: relevin@foodsci.umass.edu

VIBRIO VULNIFICUS

Overview

Vibrio vulnificus is considered the most serious and invasive of all human pathogenic vibrios in the U.S. accounting for 95% of all seafood-related deaths in this country (Oliver, 1989). Three biotypes are presently recognized and distinguished on the basis of biochemical characteristics, serology, and molecular typing. The organism requires at least 0.5% NaCl for growth and has been found to be a natural inhabitant of marine coastal waters and to be globally ubiquitous (Kaysner et al., 1987; Kelly, 1982; Oliver, 1989; O'Neill et al., 1992; Tamplin et al., 1982).

The most frequent symptoms include fever, chills, nausea, hypotension, and endotoxic shock which are usually associated with endotoxicity derived from Gram negative lipopolysaccharides (LPS) (McPherson et al., 1991). Liver damage or cirrhosis, arising from chronic alcoholism, chronic renal disease, diabetes, and immunocompromising diseases are considered major factors in susceptibility (Desenclos et al., 1991; Oliver, 1989; Johnston et al., 1986) and are thought to be responsible for the observation that outbreaks involving the consumption of oysters from a specific lot usually involve only a single susceptible individual developing symptoms (Oliver and Kaper, 1997). Secondary necrotic lesions of the extremities frequently occur (69%) often necessitating surgical debridement or limb amputation (Oliver and Kaper, 1997).

The organism produces an unusually large number of extracellular virulence factors. A number of selective agar media have been developed for isolation of the organism incorporating various levels of the antibiotics such as colistin and polymyxin-B, in addition to bile salts, and K tellurite. The heat stable hemolysin-cytolysin gene *cth* has been used most frequently for the specific PCR detection and identification of *V. vulnificus*.

Virulence Factors

Capsule Production

All virulent strains of *V. vulnificus* were found by Simpson et al. (1987) to produce opaque colonies derived from encapsulated cells whereas non-encapsulated spontaneous mutants of the same strains were found to produce translucent (non-capsular) colonies that were avirulent.

Extra-cellular Virulence Factors

V. vulnificus produces a large number of extracellular factors that are considered to contribute to its virulence including: collagenase (Poole et al., 1982; Smith and Merkel, 1982; Oliver et al., 1986), hemolysins (Moreno and Landgraf, 1998), proteases (Desmond et al., 1984; Kreger and Lockwood, 1981; Oliver et al., 1986), elastase (Kothary and Kreger, 1985; Oliver et al., 1986; Moreno and Landgraf, 1998), lipase (Desmond et al., 1984; Oliver et al., 1986; Tison et al., 1982, Moreno and Landgraf, 1998), phospholipase (Desmond et al., 1984; Testa et al., 1984; Tison et al., 1982), chondroitin sulfatase (Oliver et al., 1986), hyaluronidase (Oliver et al,. 1986), DNase (Desmond et al., 1984; Kreger and Lockwood, 1981; Oliver et al., 1986, Moreno and Landgraf, 1998), mucinase (Oliver et al., 1986; Moreno and Landgraf, 1998), fibrinolysin (Oliver et al., 1986), and an alkaline sulfatase (Kitaura et al., 1983; Oliver et al., 1986). A heat-stable hemolysin is produced that exhibits cytolytic activity against a variety of mammalian erythrocytes, cytotoxic activity to Chinese hamster ovary cells, vascular permeability to guinea pig skin, and lethality for mice (Kreger and Lockwood, 1981) and has been utilized for molecular identification of the organism. This cytolysin- hemolysin toxin is encoded by the *cth* gene.

Enrichment and Isolation Media

Alkaline Peptone Salt (APS) Broth

APS Broth is widely used for enrichment cultivation of *V. vulnificus* and is useful for MPN (Most Probable Number) enumeration followed by streaking onto one of the selective agar media described below. The medium consists of 1% peptone plus 3.0% NaCl. The pH is adjusted to 8.5 and therefore imparts some degree of selectivity for alkaline pH tolerant marine vibrios such as *V. vulnificus*.

Thiosulfate-Citrate-Bile-Salts Sucrose (TCBS) Agar

Most isolation studies from marine sources have involved the use of TCBS agar (Table 5.1) to select for *V. vulnificus* although it is widely recognized that significant numbers of non-vibrios are also capable of growth on this medium (Brayton et al., 1983; Oliver et al., 1983; West et al., 1982). Because TCBS agar does not distinguish *V. vulnificus* from other sucrose-negative vibrios, additional selective media have been developed for its isolation. The identification of lactose fermenting vibrios from seafood with TCBS is much more difficult than from clinical sources. This is due to the large number of sucrose-negative, lactose-positive vibrios in marine environments that yield colonies on selective media such as TCBS agar that are similar or identical in appearance to those produced by *V. vulnificus* (Oliver et al., 1983).

Vibrio vulnificus *Medium*

Cerdà-Cuéllar et al. (2000) proposed a more selective medium for isolation of *V. vulnificus* known as *Vibrio vulnificus* medium.This medium (Table 5.1) contains polymyxin B and colistin but in addition has a pH of 8.5.

Table 5.1 Selective agar media for isolation of *V. vulnificus*

Thiosulfate-citrate-bile-salts sucrose (TCBS) agar. (Difco)	
Yeast extract	5 g/L
Proteose peptone No. 3	10 g
Sodium citrate	10 g
Sodium thiosulphate	10 g
Oxgall	8 g
Sucrose	20 g
NACl	10 g
Ferric citrate	1 g
Brom Thymol.blue	0.04 g
Thymol blue	0.04 g
Agar	15 g

(Final pH is 8.6. Originally developed as a selective medium for *V. cholarae* El Tor which produces yellow colonies, while non sucrose fermenting *V. parahaemolyticus* and *V. vulnificus* produce blue green colonies).

Vibrio vulnificus medium (VVM) (Cerdà-Cuéllar et al., 2000)	
Cellobiose	5 g/L
NaCl	15 g
Yeast extract	4 g
$MgCl_2.6H_2O$	4 g
KCl	4 g
Cresol red	40 mg
Bromothymol blue	40 mg
Polymyxin B	10^5 U
Colistin methanesulfonate	10^5 U
Agar	15 g

All of the components are added together, heated to boiling, cooled to 50°C , and then the pH is adjusted to 8.5 with 5 M NaOH. VVM does not require boiling.

Typing of *V. Vulnificus* Isolates below the Species Level

Biogroups

The major emphasis on subspecies designations has been based on species pathogenicity (humans versus eels), and biochemical and serological observations which have led to present conclusions which contradict certain original subspecies concepts regarding the organism.

Tison et al. (1982) were the first to allocate *Vibrio* strains pathogenic for eels to the species *V. vulnificus* which at that time were not associated with pathogenicity for humans. They performed a comparative study of human clinical, environmental, and eel pathogenic isolates of *V. vulnificus* using phenotypic comparison, eel and mouse pathogenicity, and DNA-DNA hybridization studies and concluded that human clinical isolates should be designated as belonging to biogroup 1 and that eel pathogen isolates be designated as belonging to biogroup 2 of the species *V. vulnificus*. Biogroup 2 was phenotypically defined as differing biochemically from biogroup 1 in being negative for indole production, ornithine decarboxylase activity, acid production from mannitol, and sorbitol and growth at 42°C. Amaro and Biosca (1996) concluded that strains of biotype 2 (eel pathogens) are also opportunistic pathogens for humans.

During 1996-1997, 62 cases of wound infections and bacteremia due to *V. Vulnificus* were found to result from contact with purchased inland pond raised tilapia in Israel (Bisharat et al., 1999). The outbreak was due to a new marketing policy of selling live fish instead of marketing them packed in ice *post-mortem*. The isolates exhibited five atypical biochemical test results and were non-typeable by pulsed field gel electrophoresis (PFGE) and all had the same polymerase chain reaction-restriction fragment length polymorphism (PCR-RFLP) pattern derived from a 388-bp DNA fragment of the *cth* gene. Following PCR amplification of this fragment, digestion was performed with the three restriction nucleases *Dde*I, *Hha*I, and *Hpa*II. These isolates were distinguishable from biogroups 1 and 2 on the basis of negative reactions with respect to the utilization of citrate, salicin, cellobiose, and lactose. The authors designated these isolates as belonging to a newly established subspecies group biotype 3.

Molecular Methods for Detection and Typing *V. Vulnificus*

Conventional PCR

Aono et al. (1997) used the two primers VVp1 and VVp2 (Table 5.2) developed by Hill et al. (1991) for evaluating the effectiveness of the PCR in identifying isolates of *V. vulnificus* from marine environments. A total of 13,325 bacterial isolates from seawater, sediments, oysters, and goby specimens collected along the coastal regions of Tokyo Bay were metabolically screened. Among these, 713 grew at 40°C, required NaCl for growth, formed greenish colonies on TCBS agar, and were presumptively identified as *V. vulnificus*. The PCR amplified the targeted 519-bp sequence of the *cth* gene with 61 of these isolates. DNA-DNA hybridization with the type strain of *V. vulnificus* and the API 20E system confirmed the PCR results. The authors concluded that the PCR method is useful for rapid and accurate identification of *V. vulnificus* from marine sediments.

Real-time PCR (Rti-PCR)

Panicker et al. (2004) described a SYBR Green I-based real-time PCR assay for detection of *V. vulnificus* in oyster tissue homogenate. A pair of primers designated L-vvh and R-vvh (Table 5.2) were used to amplify a 205-bp sequence of the *vvh* gene. The minimum level of detection was 100 CFU (Colony Forming Units) per PCR tube. A 5 h enrichment allowed detection of 1 CFU/ml of tissue homogenate which is equivalent to 10 CFU/g of tissue. The assay required 8 h for completion.

Table 5.2 PCR primers and DNA probes for *Vibrio vulnificus*

Primer or probe	Sequence (5′ —> 3′)	Size of amplified sequence (bp's)	Gene	Reference
VVp1	CCG-GCG-GTA-CAG-GTT-GGC-C	519	*cth*	Hill et al. (1991)
VVp2	CGC-CAC-CCA-CTT-TCG-GGC-C			" " " "
RAPD–Gen 1-50-03	AGG-A(C/T)A-CGT-G	-	-	Radu et al. (1998)
RAPD–Gen 1-50-09	AGA-AGC-GAT-G	-	-	" " " "
RAPD–PSE420	TAT-CAG-GCT-GAA-AAT-CTT	-	-	Vickery et al. (1998)
M13	GAA-ACA-GCT-ATG-ACC-ATG	-	-	Arias et al. (1998)
T3	AIT-AAC-CCT-CAC-TAA-AGG	-	-	" " " "
1038 probe	UAG-CGA-AAU-UCC-UUG-UCG	-	-	" " " "
L-vvh	TTC-CAA-CTT-CAA-ACC-GAA-CTA-TGA	205	*vvh*	Panicker et al. (2004)
R-vvh	AIT-CCA-GTC-GAT-GCG-AAT-ACG-TTG			" " " "
VVAP probe	GAG-CTG-TCA-CGG-CAG-TTG-GAA-CCA	-	*vvha*	Wright et al. (1993)
RAPD	GGA-TCT-GAA-C	-	-	Warner and Oliver (1998)
D1	CCA-TCT-GTC-CCT-TTT-CCT-GC	373	*tdh*	Tada et al. (1992)
D2	CCA-AAT-ACA-TTT-TAC-TTG-G			" " " "
D5	GGT-ACT-AAA-TGG-CTG-ACA-TC	199	*tdh*	" " " "
D2	CCA-AAT-ACA-TTT-TAC-TTG-G			" " " "
D5	GGT-ACT-AAA-TGG-CTG-ACA-TC	251	*tdh*	" " " "
D3	CCA-CTA-CCA-CTC-TCA-TAT-GC			" " " "
R3	GCC-TCA-AAA-TGG-TTA-AGC-GC	210	*trh1*	" " " "
R5	TGG-CGT-TTC-ATC-CAA-ATA-CG			" " " "
R2	GGC-TCA-AAA-TGG-TTA-AGC-G	250	*trh1, trh2*	" " " "
R6	CAT-TTC-CGC-TCT-CAT-ATG-C			" " " "
P1	TGA-AAT-AAA-GCA-GTC-AGG-TG	778	*ctxB*	Koch et al. (1993)
P3	GGT-ATT-CTG-CAC-ACA-AAT-CAG			" " " "

Table 5.1 contd...

Table 5.1 contd...

Primer or probe	Sequence (5' —> 3')	Size of amplified sequence (bp's)	Gene	Reference
CTX2	CGG-GCA-GAT-TCT-AGA-CCT-CCT-G	564	*ctxA*	DePaola and Hwang (1995) from Fields et al. (1992)
CTX3 "	CGA-TGA-TCT-TGG-AGC-ATT-CCC-AC			" " " " "
STO-F STO-R	CAT-GAG-AAA-CCT-ATT-CAT-TGC TTA-ATT-TAA-ACA-TCC-AAA-GCA-AG	238	*sto*	Ripabelli et al. (1999) " " " " "
ompW-F ompW-R	CAC-CAA-GAA-GGT-GAC-TTT-ATT-GTG GAA-CTT-ATA-ACC-ACC-CGC-G	588	*ompW*	Nandi et al. 2000) " " " "
toxR-F toxR-R	ATG-TTC-GGA-TTA-GGA-CAC TAC-TCA-CAC-ACT-TTG-ATG-GC	883	*toxR*	" " " " " " " "
ctxA-F ctxA-R	CTC-AGA-CGG-GAT-TTG-TTA-GGC-ACG TCT-ATC-TCT-GTA-GCC-CCT-ATT-ACG	301	*ctxA*	" " " " " " " "
rtxC-F rtxC-r	CGA-CGA-AGA-TCA-TTG-ACG-AC CAT-CGT-CGT-TAT-GTG-GTT-GC	265	*rtx*	Fukushima et al. (2003) " " " "
CT-F (1994) CT-R	ACA-GAG-TGA-GTA-CTT-TGA-CC ATA-CCA-TCC-ATA-TAT-TTG-GGA-G	308	*ctx*	" " " from Nair et al. " " " " "
ctxA-F ctxA-R	TTT-GTT-AGG-CAC-GAT-GAT-GGA-T ACC-AGA-CAA-TAT-AGT-TTG-ACC-CAC-G	-	*ctx*	Blackstone et al. (2007) " " " "
ctxAP	FAM/TET-TGT-TTC-CAC-CTC-AAT-AGT-TTG-AGA-AGT-GCC-C-BHQ			" " " "
VHMF VHA-AS5	TGG-GAG-CAG-CGT-CCA-TTG-TG CAA-TCA-CAC-CAA-GTC-ACT-C	519 "	*lolB* "	Lalitha et al.(2008) " " " "
GF-1	AAA-AGT-CAT-ATC-TAT-GGA-TA	762	*bont/E*	Hielm et al. (1996) from Franciosa et al. (1994) " " " " "
GF-3	GTG-TTA-TAG-TAT-ACA-TTG-TAG-TAA-TCC			" "

Primer	Sequence	Product	Gene	Reference
BoTE1 BoTE2	GTG-AGT-TAT-TTT-TTG-TGG-CTT-CCG-AGA TTA-TTT-TCA-CCT-TCG-GGC-ACT-TTC-TG	307	*bont/E*	Alsallami & Kotawski (2001) " " " "
CBMLA1 CBMLA2	AGC-TAC-GGA-GGC-AGC-TAT-GTT GGT-ATT-TGG-AAA-GCT-GAA-AAG-G	782	*bont/A*	Lindström et al. (2001) " " " " " " " "
CBMLB1 CBMLB2	CAG-GAG-AAG-TGG-AGC-GAA-AA CTT-GCG-CCT-TTG-TTT-TCT-TG	205	*bont/B*	" " " " " " " "
CBMLE1 CBMLE2	CCA-AGA-TTT-TCA-TCC-GCC-TA GCT-ATT-GAT-CCA-AAA-CGG-TGA	389	*bont/E*	" " " " " " " "
CBMLF1 CBMLF2	CGG-CTT-CAT-TAG-AGA-ACG-GA TAA-CTC-CCC-TAG-CCC-CGT-AT	543	*bont/F*	" " " " " " " "
BE1430F BE1709R BE1571FP	GTG-AAT-CAG-CAC-CTG-GAC-TTT-CAG GCT-GCT-TGC-ACA-GGT-TTA-TTG-A 6-FAM-ATG-CAC-AGA-AAG-TGC-CCG- AAG-GTG-A-TAMRA	269	*bont/E*	Kimura et al. (2001) " " " " " " " "
mRNA-EF mRNA-ER	AGC-AAA-TAG-AAA-ATG-AAC GAA-TAC-TAT-TAT-TTA-GGG-TA	250	BoNT/E mRNA	Sharkey et al. (2004) " " " "
RAPD OPJ-6 RAPD OPJ-13	TCG-TTC-CGC-A CCA-CAC-TAC-C	- -		Hyytiä et al. (1999) " " " "
RAPD Primer 1 RAPD Primer 2	GGT-GCG-GGA-A CCC-GTC-AGC-A	- -		Wang et al. (2000) " " " "

Wang and Levin (2006) developed a Rti-PCR assay for quantification of *Vibrio vulnificus* seeded into clam tissue homogenates using the primers of Panicker et al. (2004). Without enrichment, the limit of detection was 1×10^2 CFU/g of tissue with a linear detection range of 1×10^2 to 1×10^8 CFU/g. With a 5 h non-selective enrichment, the limit of detection was 1 CFU/g of tissue with a linear detection range of 1 to 1×10^6 CFU/g of tissue. A 10-fold higher detection limit with seeded clam tissue homogenates occurred compared to a pure culture. After 5 h of non-selective enrichment the detection limit with a pure broth culture and seeded tissue homogenates were identical at 1 CFU/ml, however, the Ct (thermal cycles required for initial detection of amplification) value with tissue homogenates was about 3 cycles higher than with a pure culture reflecting some level of PCR inhibition from the tissue.

Lee and Levin (2008) developed a novel method for discriminating *Vibrio vulnificus* by Rti-PCR before and after γ-irradiation based on the observation that γ-irradiation results in extensive reduction in the molecular size of DNA. Irradiation of viable cells (1×10^6 CFU/ml) at 1.08 KGy [KiloGray (radiation unit measure)] resulted in 100% destruction determined by plate counts, with most of the DNA from the irradiated cells having a bp-length of less than 1000. The use of a pair of primers to amplify a 1000-bp sequence of DNA from cells exposed to 1.08 KGy failed to yield amplification. In contrast, primers designed to amplify sequences of 700, 300, and 70-bp yielded amplification with Ct values resulting in 13.4, 27.6, and 45.4% detection of genomic targets. When viable cells of *V. vulnificus* were exposed to 1.08, 3.0, and 5.0 KGy, the average molecular size of genomic DNA visualized in an agarose gel decreased with increasing dose, corresponding to an increased probability of amplification with primers targeting sequences of decreasing size.

RAPD (Random Amplified Polymorphic DNA)

Radu et al. (1998) subjected 26 biotype 1 and 10 biotype 2 isolates to RAPD analysis using two random primers designated Gen 1-50-03 and Gen 1-50-09 (Table 5.2). A total of six RAPD types were distinguished with primer Gen 1-50-03, with all six RAPD types represented by one or more strains of biotype 1. With biotype 2 strains, only three of these RAPD types were distinguished. With primer Gen 1-50-09, a total of five RAPD types were distinguished, with all five RAPD types represented by one or more strains of biotype 1. With biotype 2 strains, only four of these RAPD types were generated. Results also indicated that certain biotype 1 and biotype 2 strains yielded identical RAPD profiles with both RAPD primers, indicating a high degree of DNA sequence similarity between such strains of the two biotypes.

Vickery et al. (1998) made use of a random primer designated R-PSE420 (Table 5.2) for generating RAPD profiles of *V. vulnificus* strains. The primer yielded 15 different DNA banding profiles with 16 strains. A great deal of genomic heterogeneity was observed with strains derived from different oyster samples and even from strains derived from the same patient with wound infections.

Warner and Oliver (1999) developed an RAPD protocol for detecting *V. vulnificus* and for distinguishing this organism from other members of the genus *Vibrio*. A 10-mer primer (Table 5.2) previously described by Warner and Oliver (1998) was used. Each of 70 *V. vulnificus* strains examined produced a unique banding pattern, indicating that members of this species are highly heterogeneous. All of the clinical isolates yielded a unique band (200 to 178-bp) that was only occasionally found with environmental strains. The authors concluded that this band may be correlated with human pathogenicity. Subsequent observations by DePaola et al. (2003a) with this primer indicated that only 70% of clinical isolates possessed this amplicon and that a band of ca. 460-bp was present in 86% of these same strains.

Arias et al. (1998) determined the genetic relationships among 132 strains of *V. vulnificus* derived from human infections, diseased eels, seawater, and shellfish with the use of ribotyping and RAPD. For ribotyping, genomic DNA was digested with *Kpn*I and hybridized with the 18-mer universal digoxigenin-labeled 1038 olignucleotide probe (Table 5.2) complementary to a highly conserved sequence in the 23S rRNA gene. RAPD was performed with the universal primers M13 and T3 (Table 5.2). Both ribotyping and RAPD revealed a high level of homogeneity of diseased-eel isolates in contrast to the genetic heterogeneity of seawater-shellfish isolates of the Mediterranean. Although differentiation within diseased-eel isolates was only possible by ribotyping, the authors proposed that RAPD is a better technique than ribotyping for less laborious and rapid typing of new *V. vulnificus* isolates.

Oligonucleotide Probe

Wright et al. (1993) developed an oligonucleotide probe designated VVAP (Table 5.2) constructed from a portion of the *V. vulnificus* cytolysin gene *vvhA* and labeled with alkaline phosphatase. Naturally occurring *V. vulnificus* were detected without enrichment or selective media by plating dilutions of oyster homogenates and seawater directly onto Luria agar followed by incubation overnight at 35°C or at room temperature for 72 h. Plates with colonies were overlayed with membrane filters or filter paper discs which were microwaved and then treated with proteinase K to remove background alkaline phosphatase activity prior to hybridization. The VVAP probe was then added and hybridization was allowed to occur for

30 min at 50°C or 1 h at 56°C. After washing, alkaline phosphatase activity was then assayed with nitroblue tetrazolium plus 5-bromo-4-chloro-3-indolyl-phosphate. Color development of *V. vulnificus* colony blots was complete within 60 min. The method has the advantage of not requiring purification of colonies or metabolic characterization of isolates.

VIBRIO PARAHAEMOLYTICUS

Overview

Vibrio parahaemolyticus is a Gam-negative polarly flagellated facultatively anaerobic enteropathogenic rod indigenous to coastal marine environments and shellfish, that is capable of causing mild gastroenteritis to severe debilitating dysentery. Infections of the gasro-intestinal (GI) tract are usually due to consumption of raw shellfish. In addition, extra-intestinal infections have also been reported to be due to the organism such as eye and ear infections, and wound infections of the extremities. The first recorded outbreak of seafood infection due to *V. parahaemolyticus* occurred in Japan in 1950 (Miwatani and Takeda, 1976; Fujino, 1951). Among 272 patients with acute gastroenteritis, 20 succumbed. The incubation period with most cases was 2 to 6 h. The symptoms included acute abdominal pain, vomiting, and diarrhea which was watery and in some cases bloody diarrhea. A more severe and debilitating dysenteric form of gastrointestinal infection with bloody stools and marked leucocytosis requiring hospitalization has been observed in South East Asia, India, and Bangladesh, with several outbreaks in the U.S., due particularly to strains of the serotype O3:K6 (Bolen et al., 1974; Hughes et al., 1978; Daniels et al., 2000).

The organism is considered to be halophilic with an optimum NaCl concentration of about 3.0% (Takikawa, 1958). The characteristics of the organism were first described in detail by Fujino et al. (1953). A major distinction between *V. parahaemolyticus* and members of the genera *Aeromonas* and *Pseudomonas* is the formation of spheroplasts (Fujino et al., 1965). All strains of *V. parahaemolyticus* have been found to possess a thermolabile hemolysin encoded by the *lht* gene which is not directly related to virulence. PCR primer pairs have therefore been developed utilizing the resulting amplicons from the *lht* gene for identification of all isolates of *V. parahaemolyticus*. Virulence has been found to be associated with two principle genes that code for (1) a thermally stable direct acting hemolysin (*tdh*) and (2) a thermally stable direct acting—related hemolysin (*trh*). However, not all clinical strains have been found to possess the *trh* gene. Primer pairs targeting sequences of the *tdh* gene are therefore used to distinguish virulent from nonvirulent strains. Virulent strains are usually characterized as Kanagawa Phenomenon (KP) positive which refers to β–hemolysis on a special blood agar known as Wagatsuma blood agar (Wagatsuma, 1968) (Table 5.3). Epidemiological studies have indicated that specific clones of certain serotypes, notably 03:K6 having enhanced virulence have become endemically established in certain global locals.

Table 5.3 Composition of Wagatsuma's medium (Wagatsuma, 1968)[a]

Yeast extract	5.0g
Peptone	10.0
Mannitol	5.0
K_2HPO_4	5.0
NaCl	70.0
Agar	15.0
dH_2O	1.0L

[a]Final pH is 7.5. 5% defibrinated and washed rabbit or human red blood cells are added after sterilization of medium.

Ecology of *V. parahaemolyticus*

V. parahaemolyticus is widespread along marine coastal waters globally. The organism is considered a common inhabitant of estuaries and is infrequently found in freshwater or full-strength seawater (Joseph et al., 1982). The organism is cold sensitive and appears to be limited to inshore coastal and estuarine areas. Counts of *V. parahaemolyticus* have been found to be as high as 1,300/g in oysters (Felsenfeld and Cabirac, 1977) which is of considerable public health importance keeping in mind the widespread practice of consuming shellfish raw or undercooked.

Antigenic Properties

Three principle categories of outer antigens are produced by strains of *V. parahaemolyticus*: thermostable somatic O antigens, thermolabile K antigens, and flagellar antigens. The complete antigenic scheme for *V. parahaemolyticus* encompassing both O and K antigens is given by Twedt (1989).

Direct Acting Hemolysins

Hemolysis on Wagatsuma's blood agar (Table 5.3) is referred to as the Kanagawa phenomenon (KP) and has been found to correlate well with human pathogenicity. The KP phenomenon is characterized by the appearance on "Wagatsuma Agar" of a sizeable clear halo of hemolysis after 18 to 24 h of incubation at 37°C. The presence of 7% NaCl in Wagatsuma agar (Table 5.3) is thought to stress the cells resulting in enhanced production of hemolysin. The use of serological detection of the hemolysin has resulted in some KP- strains being designated weak KP+ (Ohashi et al., 1977).

Lu (2003) found that the incorporation of sphingomyelinase (10 units/l) to blood agar base containing 5.0% defibrinated sheep blood resulted in most shellfish isolates of *V. parahaemolyticus* exhibiting readily discernible but weak β-hemolysis. Sakazaki et al. (1968) reported that 2,655/2,720 (96.6%) of human clinical isolates were KP$^+$ and that only 7/650 (1%) of environmental isolates were KP$^+$. Thompson et al. (1976) found only 4/2,218 environmental isolates to be KP$^+$.

Chun et al. (1974) found that the addition 0.01 M CaCl$_2$ to Wagatsuma blood agar resulted in KP$^-$ strains becoming strongly hemolytic after 24 h incubation and the hemolysis of KP$^+$ strains was greatly intensified.

Nishibushi et al. (1985) developed a 406-bp gene probe specific for the Vp-TDH gene. All 66 KP$^+$ strains examined were detected with the probe, in addition to 12/14 weakly KP$^+$ strains and 10/61 KPO-strains. Among 121 other *Vibrio* isolates exclusive of *V. parahaemolyticus*, only *V. hollisae* strains reacted with the probe under stringent conditions.

Molecular epidemiological studies have revealed that not only strains carrying the *tdh* gene but strains carrying a *trh* gene or both genes are strongly associated with gastroenteritis (Okuda et al., 1997a; Shirai et al., 1990).

Shirai et al. (1990) used *tdh* and *trh* gene probes to detect the TDH and TRH producing genes in strains of *V. parahaemolyticus*. Among a total of 214 clinical strains, 112 (52.3%) had the *tdh* gene only, 52 strains (24.3%) had the *trh* gene, and 24 strains (11.2%) carried both the *tdh* and *trh* genes. Among 71 environmental strains, 5 (7.0%) exhibited weak hybridization with the *trh* gene probe and none hybridized with the *tdh* gene probe. These results suggest that the TRH as well as the TDH is an important virulence factor for *V. parahaemolyticus*.

DePaola et al. (2003b) examined oysters for the presence and density of *V. parahaemolyticus* harvested from March, 1999 through September, 2000 in Mobile Bay, Alabama, U.S.A., a coastal area known to endemically harbor the organism and to be involved in outbreaks at that time. DNA probes targeting the *tlh and tdh* genes were used for confirmation of total and pathogenic (*tdh$^+$*) strains. *V. parahaemoltyicus* was detected in all 156 samples with densities ranging from <10 to 12,000/g.

Urease (Uh) Production

The majority of clinical and environmental isolates in early studies were usually found to be urease negative (Uh$^-$). Sakazaki et al. (1963) reported only 4% of strains tested to be urease positive. Abbott et al. (1989) were the first to report the isolation of Uh$^+$ clinical strains from the West Coast of California and Mexico and indicated that by 1983, Uh$^+$ *V. parahaemolyticus* had become the predominant biotype in California outbreaks. A correlation has been found between the presence of the *trh* gene and urease production among clinical isolates, which is considered to be an unusual characteristic of *V. parahaemolyticus*. Kaufman et al. (2002) suggested that the *tdh, trh* and urease test can be used to identify and track potentially virulent strains in oysters.

Sensitivity of *V. Parahaemolyticus* to Low Temperatures

The sensitivity of *V. parahaemolyticus* to low temperatures is well documented. Temyo (1966) observed a 3 log reduction of *V. parahaemolyticus* at 4°C in peptone broth containing 3% NaCl. Similar observations have been reported when the organism was held at low temperatures in shrimp (Bradshaw et al., 1974; Vanderzant and Nickelson, 1972) oysters (Johnson and Liston, 1973; Thomson and Thacker, 1973; Goatcher et al., 1974; homogenized fish fillets (Matches et al., 1971) and crabmeat (Johnson and Liston, 1973).

Detection and Isolation of *V. Parahaemolyticus* Enrichment Cultivation

The U.S.F.D.A.(United States Food and Drug Administration) Manual (Elliot et al., 1998) recommends blending food samples in 3% NaCl or phosphate buffered saline (PBS), enriching into alkaline Peptone Water (1% NaCl) or APS (3% NaCl), and then streaking onto TCBS agar (Table 5.1). Typical colonies are bluish green. Isolates such as *V. alginolyticus* produce yellow colonies due to the fermentation of sucrose. The inability of *V. parahaemolyticus* isolates to ferment sucrose is a primary differential characteristic. Twedt (1989) listed 13 liquid media and 11 agar media for selective cultivation of *V. parahaemolyticus*.

Use of the Polymerase Chain Reaction for Detection of *V. Parahaemolyticus*

Tada et al . (1992) established PCR protocols for the specific detection of the *tdh* and *trh* genes of *V. parahaemolyticus*. The selection of primers (Table 5.2) took into consideration that the *tdh* and *trh* genes are known to have sequence divergence of up to 3.3% and 16% respectively.

Bej et al. (1999) developed a multiplex PCR assay for total and virulent strains of *V. parahaemolyticus* based on the amplification of a 450-bp sequence (Brasher et al., 1998) of the thermolabile hemolysin gene (*tlh*), a 269-bp sequence of the thermostable direct hemolysin gene (*tdh*), and a 500-bp sequence of the thermostable direct hemolysin-related (*trh*) gene (Table 5.4). All 111 *V. parahaemolyticus* isolates studied yielded the *tlh* amplicons. However, only 60 isolates yielded the *tdh* amplicon, and 43 yielded the *trh* amplicon.

The homology of the *gyrB* sequences between *V. parahaemolyticus* and *V. alginolyticus* is 86.8% (Venkateswaran et al., 1998). For this reason, Venkateswaran et al. (1998) developed a PCR procedure using primers VP-1/VP-2r (Table 5.4) targeting a 285-bp sequence of the *gyrB* gene for specific detection of *V. parahaemolyticus*.

The *toxR* gene was first discovered as the regulatory gene of the cholera toxin operon and was later found to regulate many other genes in *V. cholerae* (DiRita, 1992; Miller et al., 1987). The *toxR* gene is well conserved among species of *Vibrio*. The degree of homology of the *toxR gene* between *V. parahaemolyticus* and *V. cholerae* is 52% which is much lower than the value of 91-92% for the rRNA gene (Kita-Tsukamoto et al., 1993; Lin et al., 1993). Based on these earlier observations, Kim et al. (1999) developed a DNA colony hybridization test with the use of a 678-bp polynucleotide probe (Lin et al., 1993) for the *toxR* gene of *V. parahaemolyticus*, to confirm the identity of isolates. Kim et al. (1999) also developed a specific PCR assay for the identification of *V. parahaemolyticus* based on amplifying amplicons of the *toxR* gene. Three effective primer pairs were identified (Table 5.2). These primer sequences were selected from the regions of the *toxR* gene not conserved between *V. parahaemolyticus* and *V. cholerae* (Lin et al. 1993). A total of 373 strains of *V. parahaemolyticus* were all found to carry the *toxR* gene.

Blackstone et al. (2003) developed a Rti-PCR assay for detection of *V. parahaemolyticus* in oysters with the use of a pair of primers amplifying a 75-bp sequence of the *tdh* gene (Table 5.2) in conjunction with a dual labeled fluorogenic probe. Their procedure involved homogenizing oyster tissue at a 1:10 dilution in alkaline peptone water (pH 8.5) followed by overnight enrichment incubation at 35°C. The assay detected target DNA from 1 CFU per PCR reaction.

A PCR method for quantitative detection of *Vibrio parahaemolyticus* was developed by Wang and Levin (2004). The primers L-tdh/ R-tdh (Table 2) from Bej et al. (1999) were used. Several lysis methods were compared and a lysis solution designated TZ developed by Abolmaaty et al. (2000) proved effective.

Molecular Typing of *V. parahaemolyticus*

Wong et al. (1996) found that the restriction nuclease *Sfi*I yielded 17 clear and discernible PFGE bands with 130 clinical strains of *V. parahaemolyticus* from Thailand resulting in a total of 39 discernible PFGE patterns.

Wong et al. (1999) subjected 308 clinical isolates of *V. parahaemolyticus* derived from food outbreaks in Taiwan between 1993 and 1995 to RAPD analysis. The 10-mer primer designated 284 (Table 5.2) was used and generated 41 RAPD patterns. The patterns were grouped into 16 RAPD types, the first four of which were the major patterns and accounted for 91.25% of the domestic clinical isolates. The RAPD typing patterns were correlated with previously reported PFGE typing patterns (Wong et al., 1996) of these isolates.

The O3:K6 Pandemic Clone

Honda et al. (1987) were the first to report on the isolation of KP⁻ clinical isolates of *V. parahaemolyticus* belonging to the serovar O3:K6. The occurrence of food-borne disease outbreaks in Taiwan increased dramatically in 1996 (Chiou et al., 2000). This increase was correlated with a high rate of isolation of *V. parahaemolyticus*, which caused 69 to 71% of the total outbreaks from 1996 to 1999. Serotyping of 3,743 *V. parahaemolyticus* isolates yielded 40 serovars, the most frequent of which was O3:K6 (Chiou et al., 2000).

Okuda et al. (1997b) examined 134 strains isolated from 1994 to 1996 in Calcutta with respect to serovar, the presence of the *tdh* gene and the *tdh*-related hemolysin genes *trh1* and *trh2*. All of the serovar O3:K6 strains carried the *tdh* gene but not the *trh* genes and did not produce urease. RAPD analysis indicated that the O3:K6 serovar strains belonged to a unique clone. Clinical O3:K6 strains isolated between 1982 and 1993 from travelers arriving in Japan from Southeast Asia were found to be RAPD distinct from the Calcuttta O3:K6 clone, while strains isolated in 1995 and 1996 were indistinguishable from the Calcutta O3:K6 strains. These results suggested that the unique O3:K6 clone may have become prevalent not only in Calcutta but also in Southeast Asia. The O3:K6 serovar is considered more highly infectious than other serovars with up to 75% of exposed individuals becoming infected compared to 56% with other serovars (Daniels et al., 2000).

Meyers et al. (2003) described the development and use of a set of primers specific for a 369-bp sequence of ORF8 designated F-03MM824 and R-03MM1192 (Table 5.2). This set of primers was found to be highly effective in rapidly screening and detecting newly acquired isolates of *V. parahaemolyticus* from marine waters from the Gulf of Mexico for the O3:K6 serovar. These authors concluded that since all newly emerged O3:K6 isolates are derived from a single clone, it is likely that this strain has been transported from one geographic local to another *via* ship ballast water.

The first reported outbreak of gastroenteritis due to V. *parahaemolyticus* serovar O3:K6 in North America occurred between May 31 and July 10, 1998. The outbreak involved 416 individuals in 13 states who had eaten raw oysters harvested from Galvaston Bay, Texas and Long Island Sound (Daniels et al., 2000; Gendel et al., 2001). Matsumoto et al. (2000) showed with RAPD analysis that O3:K6 strains from six countries including the U.S. isolated from 1997 and later belong to the same clone.

Khan et al. (2002) reported that O3:K6 strains possessed a specific 850-bp sequence that was absent in other *Vibrio* species and related organisms. A set of primers VPF2/VPR2 (Table 5.4) was then developed that amplifies a 327-bp segment of this unique sequence.

VIBRIO CHOLERAE

Overview

Vibrio cholerae is a Gram-negative facultative anaerobic curved rod that is polarly flagellated, cytochrome oxidase positive and characteristically ferments glucose to acid without gas production. The generally accepted method for its isolation involves enrichment in alkaline peptone water (APW) followed by culture on selective thiosulfate-citrate-bile salts-sucrose (TCBS) agar. Sucrose positive, smooth yellow colonies on TCBS agar are submitted to the oxidase test. Positive colonies are then purified on a nonselective agar and subjected to biochemical and serological tests. Detection of cholera toxin (CTX) is usually by an ELISA (Enzyme Linked Immuno Sorbent Assay) assay.

V. cholera is considered a heterogeneous species with 206 serotypes presently recognized. However, only two serotypes are associated with epidemic infections, 01 and 0139 (Islam et al., 2004). There are strains of these two serogroups however that do not produce the cholera enterotoxin (CTX) and are not infectious. In addition, CTX negative *V. cholera* 01 strains have been implicated in occasional cases of diarrhea and extraintestinal infections. Some non-01/non-0139 strains produce a heat stabile enterotoxin designated NAG-ST (non-agglutinable *Vibrio* ST) encoded by the *sto* gene. *V. cholerae* 01 is divided into two biotypes, classical and El Tor (Keasler and Hall, 1993; Islam et al., 2004). The El Tor biotype is presently considered the most significant. In the USA crabs, shrimp, and oysters have been the most frequently implicated vehicles.

Factors Associated With the Virulence of *V. cholerae*

Cholera toxin (CTX) consists of two polypeptides; the A subunit encoded by *ctxA* is responsible for adenylate cyclase activation in enterocytes of the small intestine, inducing extensive secretion resulting in tremendous loss of water and electrolytes associated with cholera. The B subunit encoded by *ctxB* consists of five identical peptides responsible for binding to the epithelial cell surface receptor GM_1. In addition to CTX the pathogenicity of *V. cholera* O1 and O139 strains also depends on the ability to adhere and colonize the small intestine via production of pili encoded by *tcpA*.

There are a large number of genes associated with the virulence of *V. cholerae* that have been used to detect and characterize isolates. The major virulence-associated factors are present in clusters within at least three regions in the *V. cholera* genome. The first consists of the *ctxA/ctxB* genes which reside on a lysogenized phage (Waldor and Mekalanos, 1996). The second is a large pathogenicity island that encodes a toxin co-regulated pilus gene cluster involving a type IV pilus that is the primary adhesion and colonizing factor (Taylor et al., 1987) and acts as a phage receptor. The third gene cluster, the RTX gene cluster, encodes a cytotoxin active against cultured Hep-2 cells in *V. cholera* El Tor strains (Lin et al., 1999).

PCR Detection, Identification and Characterization of *V. cholerae*

In January 1991, an outbreak of cholera started in Peru and rapidly spread throughout most of Latin America. Within 15 mon over 450,000 cases occurred with about 4,000 deaths. The causative organism was toxigenic *V. cholerae* O1 of the El Tor biotype which is distinct from the U.S. Gulf Coast strains. Fields et al. (1992) reported on the use of primers CTX2/CTX3 (Table 5.2) that amplified a 564-pb sequence of the *ctxA* gene for its detection in 150 *V. cholerae* isolates derived from patients, food, and water from the 1991-1991 outbreak. One hundred forty isolates were found to be toxigenic both by PCR and immuno assay.

Koch et al. (1993) reported on the development of a PCR assay for detection of *V. cholerae* seeded onto oysters, crab meat, shrimp, and lettuce. The primers P1/P3 (Table 5.2) amplified a 778-bp sequence of the *ctxB* gene from a *V. cholerae* O1 strain. Seeded foods were homogenized or rinsed with APW followed by a 6-8 h enrichment incubation at 37°C. One ml enrichments were boiled and 2 to 5 ml added to 100 ml PCR reaction volumes. A detection limit of 1 CFU/10g of food was obtained.

DePaola and Hwang (1995) determined the optimum conditions of enrichment for detection of *V. cholerae* by the PCR. Recovery and PCR detection was significantly greater from oyster homogenates diluted 1:100 in alkaline peptone water and incubated at 42°C for 18-21 h. The primers used (Table 5.2) were from Fields at al. (1992) and amplified a 564-bp of the *ctxA* gene.

Ripabelli et al. (1999) reported on the occurrence of various pathogenic *Vibrio* species from mussels harvested from approved shellfish waters in the Adriatic Sea. *V. cholerae* O1 and O139 serotypes were not detected. However, *V. cholerae* non-01/non-O139 was found in only 1.6% of the samples compared to 32.2% for *V. alginolyticus*. PCR with primers STO-F/STO-R (Table 5.2) were used to amplify a 238-bp sequence of the *sto* gene that encodes the thermotolerant enterotoxin of *V. cholerae* and revealed the absence of the *sto* gene in all of these environmental isolates of *V. cholerae*.

Chow et al. (2001) developed a PCR assay for detection of the *rtxA*, *rtxC*, (encoding the RTX repeat in toxin) and the *ctxB* toxin genes among

166 clinical and environmental isolates of *V. cholerae*. All 166 isolates were 01 El Tor, O139 or non-O1 serotypes and all harbored the *rtxA* and *rtxC* genes which are considered specific for all *V. cholerae* isolates. Only the non-O1 serogroups failed to harbor the *ctxB* gene.

Nandi et al. (2000) assessed the distribution of genes for an outer membrane protein (OmpW) and a regulatory protein ToxR) among 254 *V. cholerae* isolates. The primers ompW-F/ompW-R (Table 5.2) amplified a 588-bp sequence of the *ompW* gene. The primers toxR-F/toxr-R (Table 5.2) amplified an 883-bp sequence of the t*oxR* gene. The primers ctxA-F/ctxA-R (Table 5.2) amplified a 301-bp sequence of the *ctxA* gene. All 254 isolates were found to harbor the *ompW* gene while 229 (98%) were found to harbor the *toxR* gene. None of the other 40 strains belonging to other *Vibrio* species produced amplicons with either *ompW*- or *toxR*-specific primers nor did 80 strains from other bacterial genera. Restriction fragment length polymorphism (RFLP) analysis and nucleotide sequence data revealed that the *ompW* gene sequence is highly conserved among *V. cholerae* strains belonging to different biotypes and/or serogroups. The authors concluded that their observations suggested that the *ompW* gene can be targeted for the species-specific identification of *V. cholera*e strains and are more suitable than the *toxR* gene. They then developed a multiplex PCR involving both the *ompW* and *ctxA* genes for screening both toxigenic and nontoxigenic strains of clinical and environmental isolates of *V. cholerae*.

V. cholerae 01 and 0139 serotypes are considered to cause noninvasive epidemic cholera in developing countries, but non-01/non 0- 0139 serotypes may be invasive and cause systemic bacteremia and septicemia. Namdari et al. (2000) reported on the consumption of raw clams by a healthy individual in Maryland, USA followed 18 h later by the development of severe profuse watery diarrhea, nausea, and vomiting with complete recovery after 72 h. Blood cultures were positive for a non-01 and non-0139 strain of *V. cholerae* that was cytotoxic to Hep-2 cell cultures. PCR confirmed that the isolate did not harbor the *ctxA* gene.

Lee et al. (2003) developed a multiplex PCR assay linked to a microwell sandwich assay for detection of *Salmonella* and three *Vibrio* species including *V. cholerae* in seeded oyster homogenates. The primers L-ctx/R-ctx (Table 5.2) amplified a 302-bp sequence of the *ctxA* gene of *V. cholerae*. Individual capture probes were then added and covalently bound to the wells. The phosphorylated capture probe PP-ctx (Table 5.2) was used for *V. cholerae*. Multiplex amplicons were denatured in the wells and incubated for hybridization to the immobilized probes. The biotinylated probe BP-ctx (Table 5.2) was then added followed by an alkaline phosphatase-avidin conjugate and then enzyme substrate added for color development. Enrichment of seeded oyster homogenates in APW allowed the detection of 10^2 CFU/g of tissue.

Fukushima et al. (2003) developed a series of duplex Rti-PCR assays for detection of 17 species of food and waterborne pathogens including *V. cholerae* in stools utilizing SYBR Green as the fluorescent reporter molecule. The primers rtxC-F/rtxC-R (Table 5.2) amplified a 265-bp sequence of the *rtx* toxin gene and were used to detect *V. cholerae* strains O1 and O139 as well as non-O1 strains, except for the classical *V. cholerae* O1 strains. The primers CT-F/CT-R (Table 5.2) amplified a 308-bp sequence of the *ctx* gene and were used to detect *V. cholerae* O139 Bengal (Nair et al., 1994). Without enrichment of seeded stool samples, the detection level with DNA purification was about 10^5 CFU/g of stool. The protocol for detection of less than 10^4 CFU/g required overnight enrichment.

Blackstone et al. (2007) developed a Rti-PCR assay for detection of *V. cholerae* harboring the *ctxA* gene. The primers ctxA-F/ctxA-R (Table 5.2) amplified a sequence of the *ctxA* gene. The probe ctxAP (Table 5.2) was labeled at the 5′-end with FAM or TET and at the 3′-end with a black hole quencher (BHQ). Shellfish tissue from Mobile Bay (3 oyster samples and 3 clam samples) were homogenized in APW and subjected to overnight enrichment incubation at 42°C. A 1-ml aliquot of enrichment was boiled for 10 min and 2 to 2.5 ml incorporated into Rti-PCR assays. All six shellfish samples were positive for *V. cholerae* and harbored the *ctxA* gene. The detection limit was 0.8 CFU per Rti-PCR reaction with clams and was less then 10 CFU per Rti-PCR with oysters.

Lalitha et al. (2008) developed a PCR assay specific for all strains of *V. cholerae* including O1, 0139, and non-O1/non-O139 serogroups and biotypes. The primers VHMF/VHA-AS5 (Table 5.2) amplified a 519-bp sequence of the *lolB* gene that encodes an outer membrane lipoprotein. All 45 *V. cholerae* strains (34 O1 El Tor, one classical, four O139, and five non-1/non-O139) were found to harbor the *lolB* gene while 40 other *Vibrio* species and 56 enteric Gram-negative reference strains of other genera did not harbor the gene. The diagnostic sensitivity and specificity with 633 clinical rectal swab samples were 98.5% and 100% respectively.

Mendes et al. (2008) developed a multiplex single-tube PCR assay for detection of the *V. cholerae* serotype. The *ctxA* gene was targeted with a pair of external primers and a pair of internally nested primers that yielded a final amplicon of 302-bp (Table 5.2). In addition, a pair of primers (Table 5.2) was added that amplified a 198-bp sequence of the *rfbN* gene that encodes the O1 serotype.

CLOSTRIDIUM BOTULINUM

Overview

C. botulinum is a Gram-positive obligately anaerobic spore-forming rod of which there are seven types A-G based on serological distinction of the respective neurotoxins produced. Human botulism is caused by types A, B, E, and rarely type F (Franciosa et al., 1994). Types C and D cause botulism in animals. Type G is not associated with neurotoxicity in humans or animals. All isolates of *C. botulinum* can be placed into one of four groups based on physiological differences: group I, all type A strains and proteolytic B and F types; group II, all type E strains and nonproteolytic B and F strains; group III, C and D type strains; and group IV, G type strains. 16s and 23S rRNA gene sequence studies (Hutson et al. 1993a, b and Hutson et al. 1994) have confirmed this grouping and have documented a high level of relatedness among strains within each group and little relatedness between members of the different groups (Hatheway, 1993).

Relationship Between Botulism and Seafood

In recent years an average of 450 botulism outbreaks have been annually reported in the international literature; 12% of the outbreaks being caused by Type E (Hatheway, 1995). Coastal marine environments usually exhibit serotype E as predominant although certain marine sediments have been found to contain predominantly serotype B. Although isolates of types B, E, and F are known to be psychrotolerant only isolates of type E are truly psychrotrophic and exhibit the ability to grow in seafood tissue under refrigerated conditions (~4°C). Type E has been the most frequent cause of botulism derived from seafood.

There is a well established history of salted fish causing type E botulism. Uneviscerated, salt-cured fish have been implicated in a number of additional botulism outbreaks (Badhey et al., 1986; Kotev et al. 1987; Telzak et al. 1990). The intestines of uneviscerated, salted fish are thought to result in a low-salt environment allowing spores of *C. botulism* to germinate, grow and produce toxin. Weber et al. (1993) reported on a massive outbreak of type E botulism associated with the consumption of traditional salted fish in Cairo, Egypt. Low levels of type E toxin are known to result in primarily gastrointestinal (GI) symptoms. Sobel et al. (2007) reported on an outbreak of clinically mild botulism type E illness among five individuals resulting in predominantly GI symptoms consisting of nausea, vomiting, abdominal pain, dry mouth, shortness of breath and in one individual diplopia (double vision). Fresh, uneviscerated whitefish with salt had been placed in a sealed ziplock bag and stored for ~1 mon at ambient temperature prior to consumption. Remnant fish tested positive for botulinum type E toxin.

Commercially produced vacuum-packaged hot-smoked fish is presently considered one of the most important botulism food vehicles. Hot-smoked Canadian whitefish was reported by Korkeala et al. (1998) to be the cause of a single family outbreak of type E botulism in 1997. The fish was smoked only 5 d before consumption indicating that toxin production had been rapid and that there had been marked temperature abuse during storage or transport of the fish. Type E toxin was confirmed by toxin neutralization and the mouse bioassay and by PCR.

PCR for Detection of *C. botulinum* Type E Strains

Hielm et al. (1996) developed protocols for combined PCR-MPN detection and enumeration of *C. botulinum* types A, B, E, and F in fish and sediment samples. Spore counts of type E in sediment samples varied from 95 to 2710 per kg of sample. Rainbow trout were seeded with spores of *C. botulinum* type E at 10^2 to 10^6 spores per kg of tissue in addition to inoculating fish intestines. Each sample was subjected to a 5-d enrichment in TPGY broth followed by transfer of 0.5 ml into 10 ml of TPGY broth with overnight incubation prior to PCR. Washed vegetative cells from such enrichment broth cultures were boiled for 10 min and 1 ml incorporated into PCR reactions. The primers GF-1/GF-3 (Table 5.2) from Franciosa et al. (1994) amplified a 760-bp sequence of the botulinum neurotoxin type E gene (*bontE*). All seeded samples were detected as positive.

Alsallami and Kotlwoski (2001) developed improved primer pairs for detection of the BoNTB and BoNTE genes. The detection limit was increased from 1 to 0.1 ng of DNA by increasing the annealing temperature from 50°C to 62°C. The primers BoTE1/BoTE2 (Table 5.2) amplified a 307-bp sequence of the *botnE* gene.

Lindström et al. (2001) developed a multiplex PCR for detection of *C. botulinum* types A, B, E, and F in food and fecal material. The primer pairs (Table 5.2) yielded amplicons of 782-, 205-, 389-, and 543-bp respectively. With a two-step enrichment the detection limit in food and fecal samples was 10 spores per 0.1g of sample material for type E.

Kimura et al. (2001) developed a Rti-PCR assay for quantifying *C. botulinum* type E in modified-atmosphere packaged fish samples (jack mackeral). The primers BE1430F/BE1709R (Table 5.2) amplified a 269-bp sequence of the *botnE* gene. The dual labeled probe BE1571FP (Table 5.2) was labeled at the 5′-end with 6-FAM and at the 3′-end with TAMRA. The quantifiable range was 10^2 to 10^8 CFU/g which allowed detection much earlier than the toxin could be detected with the mouse bioassay.

Sharkey et al. (2004) developed a competitive reverse transcription PCR assay (cRT-PCR) to quantify toxin-encoding mRNA production by a type E strain in media with either sorbic acid or sodium nitrite. The primers mRNA-EF/mRNA-ER (Table 5.2) amplified a 250-bp sequence of the BoNTE mRNA. A 10-fold reduction in toxin mRNA production and a 25-fold reduction in the proportion of mRNA to total RNA was estimated when either 1 mg/ml of sorbic acid or 100 mg/ml of sodium nitrite was added to the medium at pH 7.0.

Molecular Typing of *C. botulinum* Type E Strains

The distribution of *C. botulinum* serotypes A, B, E, and F in Finnish trout farms was assessed using PCR by Hielm et al. (1998). The PCR primers from Hielm et al. (1996) including those for *botnE* were used (Table 5.2). A total of 333 samples were tested with neurotoxin gene specific PCR assays. *C. botulinum* type E was found in 68% of farm sediment samples, in 15% of fish intestinal samples, and in 5% of the fish skin samples. No other serotypes were found. The average spore count in sediments, fish intestines, and skin were 2×10^3, 1.7×10^2, and 3×10^2 per kg respectively. PFGE with *Sma*I of 42 Finnish isolates plus 12 North American reference strains generated 28 PFGE profiles indicating extensive genetic diversity.

The genetic diversity of 92 type E. *C. botulinum* strains was assessed by Hyytiä et al. (1999). Sixty-seven were of Finnish seafood and fishery origin, 15 were from German farmed fish, and 10 from North American seafoods. PFGE performed with *Sma*I-*Xma*I resulted in 75 typeable strains which yielded 33 profiles. PFGE performed with *Xho*I allowed 91 strains to be typed yielding 51 profiles. All 92 strains were typeable with RAPD primers OPJ-6 and OPJ-13 (Table 5.2), which yielded 27 and 19 banding patterns respectively. The frequent occurrence of small fragments and faint bands made RAPD interpretation difficult. A high level of genetic diversity among the isolates was observed regardless of their source, presumably because of the absence of strong evolutionary selection factors.

Wang et al. (2000) analyzed type E botulinum toxin producing strains isolated from botulinum cases or soil specimens in Italy and China using sequencing of the *bontE* gene, RAPD (Table 5.2) PFGE, and Southern blot hybridization for the *bontE* gene. The deduced amino acid sequences of the BoNTE's of 11 *C. butyricus* isolates from China were identical and exhibited 95.0 and 96.9% identity with those of the Italian BoNTE strain of *C. butyricum* BL6340 and *C. botulinum* type E respectively. The results indicated that BoNTE-producing *C. butryicum* is clonally distributed globally.

Leclair et al. (2006) undertook a comparative typing study involving the PFGE, RAPD, and automated ribotyping of *C. botulinum* type E strains derived from clinical and food sources associated with four botulinum outbreaks that occurred in the Canadian Arctic. All type E strains previously untypeable by PFGE, even with the use of a formaldehyde fixation step, could be typed by the addition of 50 mM thiourea to the electrophoresis running buffer. Digestion with *Sma*I and *Xho*I followed by PFGE was used to link food and clinical isolates from the four different type E botulinum outbreaks and to differentiate them from among 31 recently isolated Arctic environmental group II *C. botulinum* strains. *Sma*I PFGE typing yielded 18 profiles while *Xho*I PFGE typing yielded 23 profiles. Strain differentiation was unsuccessful with the automated ribotyping system which yielded only two profiles. RAPD analysis of the group II strains was not consistently reproducible with primers OPJ-6 and OPJ-13 (Table 5.2). Primer OPJ-13 did however yield 28 profiles.

ABBREVIATIONS

APS	Alkaline Peptone Salt
APW	Alkaline Peptone Water
CFU	Colony Forming Units
CTX	Cholera toxin
ELISA	Enzyme Linked Immuno Sorbent Assay
KP	Kanagawa phenomenon
LPS	Lipopolysaccharides
PBS	Phosphate buffered saline
PCR	Polymerase Chain Reaction
PCR-RFLP	Polymerase chain reaction-restriction fragment length polymorphism
PFGE	Pulsed field gel electrophoresis
RAPD	Random Amplified Polymorphic DNA
Rti-PCR	Real-time PCR
TCBS	Thiosulfate-Citrate-Bile-Salts Sucrose
U.S.F.D.A.	United States Food and Drug Administration

REFERENCES

Abbott, S., Powers, C., Kaysner, C., Takeda, Y., Ishibashi, M., Joseph, S. and Ianda, J. (1989). Emergence of a restricted bioserovar of *Vibrio parahaemolyticus* as the predominant cause of *Vibrio*-associated gastroenteritis on the West Coast of the United States and Mexico. J. Clin. Microbiol. 27: 2891-2893.

Abolmaaty, A, Vu, C., Oliver, J. and Levin, R.E. (2000). Development of a new lysis solution for releasing genomic DNA from bacterial cells for DNA amplification by polymerase chain reaction. Microbios. 101: 181-189.

Alsallami, A. and Kotlowski, R. (2001). Selection of primers for specific detection of *Clostridium botulinum* types B and E neurotoxin genes using PCR method. Int. J. Food Microbiol. 69: 247-253.

Amaro, C. and Biosca, E. (1996). *Vibrio vulnificus* biotype 2 pathogenic for eels, is also an opportunistic pathogen for humans. Appl. Environ. Microbiol. 62: 1454-1457.

Aono, E., Sugita, H., Kawasaki, J., Sakakibara, H., Takahashi, T., Endo, K. and Deguchi, Y. (1997). Evaluation of the polymerase chain reaction method for identification of *Vibrio vulnificus* isolated from marine environments. J. Food Protec. 60: 81-83.

Arias, C., Pujalte, M., Garay, E. and Aznar, R. (1998). Genetic relatedness among environmental, clinical, and diseased-eel *Vibrio vulnificus* isolates from different geographic regions by ribotyping and randomly amplified polymorphic DNA PCR. Appl. Environ. Microbiol. 64: 3403-3410.

Badhey, H., Cleri, D., D'Amaato, R., Vernaleo, R., Veinni,V. and Hochstein, L. (1986). Two fatal cases of type E adult food-borne botulism with early symptoms and terminal neurological signs. J. Clin. Microbiol. 23: 616-618.

Bej, A., Patterson, D., Brasher, C., Vickery, M., Jones, D. and Kaysner, C. (1999). Detection of total and hemolysin-producing *Vibrio parahaemolyticus* in shellfish using multiplex PCR amplification of *tlh, tdh*, and *trh*. J. Microbiol. Meth. 36: 215-225.

Bisharat, N., Agmon, V., Finkelstein, R., Raz, R., Ben-Dror, G., Memer, L., Soboh, S., Colodner, R., Cameron, D., Wykstra, D., Swerdlow, D. and Farmer J., III. (1999). Clinical, epidemiological, and microbiological features of *Vibrio vulnificus* biogroup 3 causing outbreaks of wound infection and bacteremia in Israel. Lancet. 354: 1421-1424.

Blackstone, G., Nordstrom, J., Vickery, M., Bowen, T., Meyer, R. and DePaola, A. (2003). Detection of pathogenic *Vibrio parahaemolyticus* in oyster enrichments by real time PCR. J. Microbiol. Meth. 53: 149-155.

Blackstone, G. Nordsrom, J., Bowen, M. Myer, R., Imbro, P. and dePaola, A. (2007). Use of real time PCR assay for detection of the *ctxA* gene of *Vibrio cholerae* in an environmental survey of Mobile Bay. J. Microbiol. Meth. 68: 254-259.

Bolen, J., Zamiska, A.and Greenough, W. (1974). Clinical features in enteritis due to *Vibrio parahaemolyticus*. Amer. J. Med. 57: 638-641.

Bradshaw, J., Francis, D. and Twedt, R. (1974). Survival of *Vibrio parahaemolyticus* in cooked seafood at refrigeration temperatures. Appl. Microbiol. 27(4): 657-661.

Brasher, C., DePaola, A., Jones, D. and Bej., A. (1998). Detection of microbial pathogens in shellfish with multiplex PCR. Curr. Microbiol. 37: 101-107.

Brayton, P., West, P., Russek, E. and Colwell, R. (1983). New selective plating medium for isolation of *V. vulnificus* biogroup 1. J. Clin. Microbiol. 17: 1039-1044.

Cerdà-Cuéllar, M., Jofre, J. and Blanch, A. (2000). A selective medium and a specific probe for detection of *Vibrio vulnificus*. Appl. Environ. Microbiol. 66: 855-859.

Chiou, C., Hsu, S., Chiu, S., Wang, S. and Chao, C. (2000). *Vibrio parahaemolyticus* serovar O3:K6 as cause of unusually high incidence of food-borne disease outbreaks in Taiwan from 1996 to 1999. J. Clin. Microbiol. 38(12): 4621-4625.

Chow, K., Ng, T., Yuen, K. and Yam, W. (2001). Detection of RTX toxin gene in *Vibrio cholerae* by PCR. J. Clin. Microbiol. 39: 2594-2597.

Chun, D., Chung, J. and Tak, R. (1974). Some observations on Kanagawa type hemolysis of *Vibrio parahaemolyticus*. In: Int. Symp. Vibrio parahaemolyticus. T. Fujino, G. Sakaguchi, R. Sakazaki and Y. Takeda (eds). Saikon Publ. Co., Tokyo, Japan pp. 199-204.

Daniels, N., Ray, B., Easton, A., Maaranao, N., Kahn, E., McShan, A., Del Rosario, L., Baldwin, T., Kingsleyk, M., Puhr, N., Wells, J. and Angulo, F. (2000). Emergence of a new *Vibrio parahaemolyticus* serotype in raw oysters. J.Am. Med. Assoc. 284: 1541-1545.

DePaola, A. and Hwang, G. (1995). Effect of dilution, incubation time, and temperature of enrichment on cultural and PCR detection of *Vibrio cholerae* obtained from the oyster *Crassostrea virginica*. Molec. Cell. Probes. 9: 75-81.

DePaola, A., Nordtrom, J., Dalsgaard, A., Forslund, A., Oliver, J., Bates, T., Bourdage, K., Gulig., P. (2003a). Analysis of *Vibrio vulnificus* from market oysters and septicemia cases for virulence markers. Appl. Environ. Microbiol. 69: 4006-4011.

DePaola, A., Nordstrom, J., Bowers, J., Wells, J. and Cook. D. (2003b). Seasonal abundance of total and pathogenic *Vibrio parahaemolyticus* in Alabama oysters. Appl. Environ. Microbiol. 69: 1521-1526.

DiRita, V., (1992). Co-ordinate expression of virulence genes by ToxR in *Vibrio cholerae*. Mol. Microbiol. 6: 451-458.

Desenclos, J., Klontxz, K., Wolfe, L. and Hoecherl, S. (1991). The risk of *Vibrio* illness in the Florida raw oyster eating population, 1981-1988. Am. J. Epidemiol. 134: 290-297.

Desmond, E., Janda, J., Adams, F. and Bottone, E. (1984). Comparative studies and laboratory diagnosis of *Vibrio vulnificus* and invasive *Vibrio* sp. J. Clin. Microbiol. 19:122-125.

Elliot, E., Kaysner, C., Jackson, L. and Tamplin, M. (1998). *Vibrio cholerae, V. parahaemolyticus, V. vulnificus*, and other *Vibrio* spp. U.S.F.D.A. Bacteriological Analytical Manual., A.O.A.C. International, Gaithersburg, MD, USA 9.01: 9-27.

Felsenfeld, O. and Cabirac, H.B. (1977). A study of the ecology of *Vibrio parahaemolyticus* and *Vibrio alginolyticus* in Southeast Louisiana USA with special consideration of seafood consumption. J. Appl. Nutr. 29(1/2): 17-28.

Fields, P., Popovic, T.,Wachsmuth, K. and Olsvik, O. (1992). Use of polymerase chain reaction for detection of toxigenic *Vibrio cholerae* O1 strains from the Latin American cholera epidemic. J. Clin. Microbiol. 30: 2118-2121.

Franciosa, G., Ferreira, J. and Hatheway, C. (1994). Detection of type A, B, and E botulinum neurotoxin genes in *Clostridium botulinum* and other *Clostridium* species by PCR: Evidence of unexpressed type B toxin genes in type A toxigenic organisms. J. Clin. Microbiol. 32: 1911-1917.

Fujino, T. (1951). Bacterial Food Poisoning. Saishin Igaku. 6: 263-271. In Japanese.

Fujino, T., Okuno, Y., Nakada, D., Aoyama, A., Fukai, K., Mukai, T. and Ueho, T. (1953). On the bacteriological examination of shirasu food poisoning. Med. J. Osaka Univ. 4: 299-304.

Fujino, T., Miwatani, T., Yasuda, J., Kondo, M., Takeda, Y., Akita, Y., Kotera, K., Okada, M., Nishimune, H., Shimizu, Y., Tamura, T. and Tamura, Y. (1965). Taxonomic studies on the bacterial strains isolated from cases of "shirasu" food poisoning (*Pasteurella parahaemolytica*) and related organisms. Biken J. 8: 63-71.

Fukushima, H., Tsunomori, Y. and Seki, R. (2003). Duplex real-time SYBR green PCR assays for detection of 17 species of food- or waterborne pathogens in stools. J. Clin. Microbiol. 41: 5134-5146.

Gendel, S., Ulaszek, J., Nishibuchi, M. and DePaola, A. (2001). Automated ribotyping differentiates *Vibrio parahaemolyticus* O3 : K6 strains associated with a Texas outbreak from other clinical strains. J. Food Protec. 64(10): 1617-1620.

Goatcher L., Engler, S., Wagner, D. and Westhoff, D. (1974). Effect of storage at 5 C on survival of *Vibrio parahaemolyticus* in processed Maryland oysters (*Crassostrea virginica*). J. Milk Food Technol. 37: 74-77.

Hatheway, C. (1993). *Clostridium botulinum* and other clostridia that produce botulism neurotoxin. In: Clostridium botulinum: Ecology and Control in Foods. A. H. W. Hauschild and K.L. Dodds (eds). Marcel Dekker, Inc., New York, USA pp. 3-20.

Hatheway, C. (1995). Botulism: The present status of the disease. Curr. Top. Microbiol. 195: 55-75.

Hielm, S., Hyytiä, E., Ridell, J. and Korkala, H. (1996). Detection of *Clostridium botulinum* in fish and environmental samples using polymerase chain reaction. Int. J. Food Microbiol. 31: 357-365.

Hielm, S., Björkroth, J., Hyytiä, E. and Korkeala, H. (1998). Prevalence of *Clostridium botulinum* in Finnish trout farms: pulsed-field gel electrophoresis typing reveals extensive genetic diversity among type E isolates. Appl. Environ. Microbiol. 64: 4161-4167.

Hill, W., Keasler, S., Trucksess, M., Feng, P., Kaysner, C. and Lampel, K. (1991). Polymerase chain reaction identification of *Vibrio vulnificus* in artificially contaminated oysters. App. Environ. Microbiol. 57: 707-711.

Honda, S., Goto, I., Minematsu, I., Ikeda, N., Ishibashi, M., Kinoshita, Y., Nishibushi, M., Honda, T. and Miwatani, T. (1987). Gastroenteritis due to Kanagawa negative *Vibrio parahaemolyticus*. Lancet. i: 331-332.

Hughes, J., Boyce, J., Alem, A., Wells, A., Rhaman, A., and Curlin, G. (1978). *Vibrio parahaemolyticus* enterocolitis in Bangdladesh: Report of an outbreak. Amer. J. Trop. Med. Hyg. 27: 106-112.

Hutson R., Thompson, D. and Collins, M. (1993a). Genetic interrelationships of saccharolytic *Clostridium botulinum* types B, E and F and related clostridia as revealed by small-subunit rRNA gene sequences. FEMS Microbiol. Lett. 108: 103-110.

Hutson, R., Thompson, D., Lawson, R., Schocken-Itturino, R., Bottger, E. and Collins, M. (1993b.) Genetic interrelationships of proteolytic *Clostridium botulinum* types A, B, and F and other members of the *Clostridium botulinum* complex as revealed by small-subunit rRNA gene sequences. Anton. Leeuwen. J. Microbiol. Serol. 64: 273-283.

Hutson, R., Collins, M., East, A. and Thompson, D. (1994). Nucleotide sequence of the gene coding for non-proteolytic *Clostridium botulinum* type B neurotoxin: comparison with other clostridial neurotoxins. Curr. Microbiol. 28: 101-110.

Hyytiä, E., Hielm, S., Björkroth, J. and Korkeala, H. (1999). Biodiversity of *Clostridium botulinum* type E strains isolated from fish and fishery products. Appl. Environ. Microbiol. 65: 2057-2064.

Islam, M., Ahsan, S., Khan, S., Ahmed, Q., Rashid, M., Islam, K. and Sack, R. (2004). Virulence properties of rough and smooth strains of *Vibrio cholerae* 01. Microbiol. Immunol. 48: 229-235.

Johnson, H.C. and Liston, J. (1973). Sensitivity of *Vibrio parahaemolyticus* to cold in oysters, fish fillets, and crab meat. J. Food Sci. 38: 437-441.

Joseph, S.W., Colwell, R.R. and Kaper, J.B. (1982). *Vibrio parahaemolyticus* and related halophilic vibrios. Crit. Revs. Microbiol. 10: 77-124.

Johnston, J., Becker, S. and Mcfarland, L. (1986). Gastroenteritis in patients with stool isolations of *Vibrio vulnificus*. Am. J. Med. 80: 336-338.

Kaufman, G., Myers, M., Pass, C., Bej, A. and Kaysner, C. (2002). Molecular analysis of *Vibrio parahaemolyticus* isolated from human patients and shellfish during US Pacific north-west outbreaks. Lett. Appl. Microbiol. 34: 155-161.

Kaysner, C., Abeyta, C. Jr., Wekell, M., DePaola, A., Stott, R., Jr. and Leitch, J. (1987). Virulent strains of *Vibrio vulnificus* from estuaries of the United States West Coast. Appl. Environ. Microbiol. 53: 1349-1351.

Keasler, S. and Hall, R. (1993). Detecting and biotyping *Vibrio cholerae* 01 with multiplex polymerase chain reaction. Lancet. 341: 1661.

Kelly, M. (1982). Effect of temperature and salinity on *Vibrio (Beneckea) vulnificus* occurrence in a Gulf Coast environment. Appl. Environ. Microbol. 44: 820-824.

Khan, A., McCarthy, S., Wang, R. and Cerniglia, C. (2002). Characterization of United States outbreak isolates of *Vibrio parahaemolyticus* using enterobacterial repetitive intergenic consensus (ERIC) PCR and development of a rapid PCR method for detection of O3:K6 isolates. FEMS Microbiol. Lett. 206: 209-214.

Kim, Y., Okuda, J., Matsumoto, C., Takahashi, N., Hashimoto, S. and Nishibushi, M. (1999). Identification of *Vibrio parahaemolyticus* strains at the species level by PCR targeted to the *toxrR* gene. J. Clin. Microbiol. 37(4): 1173-1177.

Kimura, B., Kawasaki, S., Nakano, H. and Fujii, T. (2001). Rapid, quantitative PCR monitoring of growth of *Clostridium botulinum* type E in modified-atmosphere-packaged fish. Appl. Environ. Microbiol. 67: 206-216.

Kita-Tsukamoto, K., Oyaizu, H., Namba, K. and Shimidu, U. (1993). Phylogeneic relationships of marine bacteria, mainly members of the family *Vibrionaceae*, determined on the basis of 16S rRNA sequences. Int. J. Syst. Bacteriol. 43: 8-19.

Kitaura, T., Doke, S., Azuma, I., Imaida, M., Miyano, K., Harada, K. and Yabuuchi, E. (1983). Halo production by sulfatase activity of *Vibrio vulnificus* and *Vibrio cholerae* O1 on a new selective sodium dodecyl sulfate-containing agar medium: a screening marker in environmental surveillance. FEMS Microbol. Lett. 17: 205-209.

Koch, W. Payne, W., Wentz, B. and Cebula, T. (1993). Rapid polymerase chain reaction method in detection of *Vibrio cholerae* in foods. Appl. Environ. Microbiol. 59: 556-560.

Korkeala, H., Stengel, G., Hyytiä, E., Vogelsang, B., Bohl, A., Wihlman, H., Pakkala,P. and Hielm, S. (1998). Type E botulism associated with vacuum-packaged hot-smoked whitefish. Int. J. Food Microbiol. 43: 1-5.

Kotev, S., Leventhal, A., Bashary, A., Zahavi, H. and Cohen, A. (1987). International outbreak of type E botulism associated with ungutted, salted whitefish. Morbrd. Mortal. Weekly Rep. 36: 812-813.

Kothary, M. and Kreger, A. (1985). Production and partial characterization of an elastolytic protease of *Vibrio vulnificus*. Infect. Immun. 50: 534-540.

Kreger, A. and Lockwood, D. (1981). Detection of extracellular toxin(s) produced by *Vibrio vulnificus*. Infect. Immun. 33: 583-590.

Lalitha, P., Suraiya, M., Lim, K., Le, S., Halindawaty, A., Chan, Y., Ismail, A., Zainuddin, Z. and Ravichandran, M. (2008). Analysis of *lolB* gene sequence and its use in the development of a PCR assay for the detection of *Vibrio cholerae*. J. Microbiol. Meth. 75: 142-144.

Leclair, D., Pagotto, F., Farber,J., Cadieux, B. and Austin, J. (2006). Comparison of DNA fingerprinting methods for use in investigation of type E botulism outbreaks in the Canadian Arctic. J. Clin. Microbiol. 44: 1635-1644.

Lee, C., Panicker, G. and Bej, A. (2003a). Detection of pathogenic bacteria in shellfish using multiplex PCR followed by Covalink™ NH microwell plate sandwich hybridization. J. Microbiol. Meth. 53: 199-209.

Lee, C., Cheng, M., Yu, M. and Pan, M. (2003b). Isolation and characterization of a putative virulence factor, serine protease, from vibrio parahaemolyticus. FEMS Microbiol. Lett. 209: 31-37.

Lee, J. and Levin, R. (2008). New approach for discrimination of *Vibrio vulnificus* by real-time PCR before and after γ-irradiation. J. Microbiol. Meth. 73: 1-6.

Li, M., Shimada, T., Morris, Jr., J., Sulakvelidze, A. and Sozhamannan, S. (2002). Evidence for the emergence of non-O1 and non-O139 *Vibrio cholerae* strains with pathogenic potential by exchange of O-antigen biosynthesis regions. Infect. Immun. 70: 2441-2453.

Lin, W., Fullner, J., Clayton, R., Sexdton, J., Rogers, M., Calia, K., Calderwood, S., Fraser, C. and Mekalanos. (1999). Identification of a *Vibrio cholerae* RTX toxin gene cluster that is tightly linked to the cholera toxin prophage. Proc. Natl. Acad. Sci. USA 96: 1071-1076.

Lin, Z., Kumagai, K.K, Baba, K., Mekalanos, J. and Nishibushi, M. (1993). *Vibrio parahaemolyticus* has a homolog of the *Vibrio cholerae toxRS* operon that mediates environmentally induced regulation of the thermostable direct hemolysin gene. J. Bacteriol. 175: 3844-3855.

Lindström, M., Keto, R., Markkula, A., Nevas, M., Hielm, S. and Korkeala, H. (2001). Multiplex PCR assay for detection and identification of *Clostridium botulinum* types A, B, E, and F in food and fecal material. Appl. Environ. Microbiol. 67: 5694-5699.

Lozano-León, A., Tores, J., Osorio, C. and Martinex-Urtaza, J. (2003). Identification of *tdh*-positive *Vibrio parahaemolyticus* from an outbreak associated with raw oyster consumption in Spain. FEMS Microbiol. Lett. 226: 281-284.

Lu, Baofang. (2003). The isolation, enumeration, biochemical and molecular identification of *Vibrio vulnificus* and *Vibrio parahaemolyticus* from shellfish. M.S. Thesis, University of Massachusets, Amherst, MA, USA.

Matches, J., Liston, J. and Daneault, L. (1971). Survival of *Vibrio parahaemolyticus* in fish homogenate during storage at low temperature. Appl. Microbiol. 21: 951-952.

Matsumoto, C., Okuda, J., Ishibashi, M., Iwanaga, M., Garg, P., Rammamurthy, T., Wong, H., Depaola, A., Kim. Y., Albert, M. and Nishibuchi, M. (2000). Pandemic Spread of an O3:K6 Clone of *Vibrio parahaemolyticus* and Emergence of Related Strains Evidenced by Arbitrarily Primed PCR and *toxRS* Sequence Analyses. J. Clin. Microbiol. 38: 578-585.

McPherson, V.L., Watts, J.A., Simpson, L.M. and Oliver, J.D. (1991). Physiological effects of the lipopolysaccharide of *Vibrio vulnificus* on mice and rats. Microbios. 67: 141-149.

Mendes, C., Abath, F. and Leal, N. (2008). Development of a multiplex single-tube nested PCR(MSTNPCR) assay for *Vibrio cholerae* O1 detection. J. Microbiol. Meth. 72: 191-196.

Meyers, M., Panicker, G. and Bej, A. (2003). PCR detection of a newly emerged pandemic *Vibrio parahaemolyticus* O3:K6 pathogen in pure cultures and seeded waters from the gulf of Mexico. Appl. Environ. Microbiol. 69: 2194-2200.

Miller, Y., Taylor, R. and Mekalanos, J. (1987). Cholera toxin transcriptional activator ToxR is a transmembrane DNA binding protein. Cell. 48: 271-279.

Miwatani, T. and Takeda, Y. (1976). *Vibrio parahaemolyticus*—A causative bacterium of food poisoning. Saikon Publ. Co. Tokyo, Japan.

Moreno, M.L.G. and Landgraf, M. (1998). Virulence factors and pathogenicity of *Vibrio vulnificus* strains isolated from seafood. J. Appl. Microbiol. 84: 747-751.

Nair, G., Shimada, T., Kurazono, H., Okuda, J., Pal, A., Karasawa, T., Mihara, T., Uesaka, Y., Shirai, H., Garg, S., Saha, P., Mukhopadhyay, A., Ohashi, T., Tada, J., Nakayama,

T., Fukushima, S., Takeda, T. and Yoshifumi, Y. (1994). Characterization of phenotypic, serological, and toxigenic traits of *Vibrio cholerae* O139 Bengal. J. Clin. Microbiol. 32: 2775-2779.

Namdari, H., Klaips, C. and Hughes, J. (2000). A cytotoxin-producing strain of *Vibrio cholerae* non-01, non-0139 as a cause of cholera and bacteremia after consumption of raw clams. J. Clin. Microiol. 38: 3518-3519.

Nandi, B., Nandy, R., Mukhopadhyay, S., Nair, G., Shimada, T. and Ghose, A. (2000). Rapid method for species-specific identification of *Vibrio cholerae* using primers targeted to the gene of outer membrane protein OmpW. J. Clin. Microbiol. 38: 4145-4151.

Nishibushi, M., Ishibashi, M., Takeda, Y., Kaper, M. (1985). Detection of the thermostable direct hemolysin gene and related DNA sequences in *Vibrio parahaemolyticus* and other *Vibrio* species by the DNA colony hybridization test. Inf. Immun. 49: 481-486.

Nogva, H., Drømtorp, S., Nissen, H. and Rudi, K. (2003). Ethidium monoazide for DNA-based differentiation of viable and dead bacteria by 5'-nuclease PCR. BioTechniques. 34: 804-813.

Ohashi, M., Ohta, K., Tsuno, M. and Zen-Yoji, H. (1977). Development of a sensitive serological assay, reversed passive hemagglutination test, for detection of enteropathogenic toxin (Kanagawa hemolysin) of *Vibrio parahaemolyticus*, and re-evaluation of the toxin producibility of the isolates from various sources. Proc. 13th Joint Conf. Cholera, Atlanta, Georgia, USA, Dept. Environ. Health and Welfare (DEHW) Publ No. (NIH) 78-1590, pp. 403-413.

Okuda, J., Ishibashi, M., Abbott, I., Janda, J. and Nishibuchi, M. (1997a). Analysis of the thermostable direct hemolysin (*tdh*) gene and the *tdh*-related hemolysin (*trh*) genes in the urease-positive strains of *Vibrio parahaemolyticus* isolated on the West Coast of the United States. J. Clin. Microbol. 35: 1965-1971.

Okuda, J., Ishibashi, M., Hayakawa, E., Nishino, T., Takeda, Y., Mukhopadhyay, A., Garg, S., Bhattacharya, S., Nair, G. and Nishibuchi, M. (1997b). Emergence of a unique O3 : K6 clone of *Vibrio parahaemolyticus* in Calcutta, India and isolation of strains from the same clonal group from Southeast Asian travelers arriving in Japan. J. Clin, Microbiol. 35: 3150-3155.

Oliver, J.D. (1989). *Vibrio vulnificus*. In: Foodborne Bacterial Pathogens. M.P. Doyle (ed). Marcel Dekker, New York, USA pp. 569-600.

Oliver, J. and Kaper, J. (1997). *Vibrio* species. In: Food Microbiology Fundamentals and Frontiers. M.P. Doyle, L.R. Beuchat and T.J. Montville (eds). American Society of Microbiology Press, Washington, D.C., USA pp. 228-264.

Oliver, J., Warner, R. and Cleland, D. (1983). Distribution of *Vibrio vulnificus* and other lactose-fermenting vibrios in the marine environment. Appl. Environ. Microbiol. 45: 985-998.

Oliver, J., Thomas, M. and Wear, J. (1986). Production of extracellular enzymes and cytotoxicity by *Vibrio vulnificus*. Microbiol. Infect. Dis. 5: 99-111.

Omori, G., Iwao, M., Iida, S. and Kuroda, K. (1966). Studies on K antigen of *Vibrio parahaemolyticus*. I. Isolation and purification of K antigen from *Vibrio parahaemolyticus* A55 and some of its biological properties. Biken J. 9: 33-43.

O'Neill, K., Jones, S. and Grimes, D. (1992). Seasonal incidence of *Vibrio vulnificus* in the Great Bay estuary of New Hampshire and Maine. Appl. Environ. Microbiol. 58: 3257-3262.

Panicker, G., Myers, M. and Bej, A. (2004). Rapid detection of *Vibrio vulnificus* in shellfish and gulf of Mexico Water by real-time PCR. Appl. Environ. Microbiol. 70: 498-507.

Poole, M., Bowdre, J. and Klapper, D. (1982). Elastase produced by *Vibrio vulnificus*: *in vitro* and *in vivo* effects. Abstracts, Annual Meeting of the American Society for Microbiology, Washington DC, USA B155, p. 43.

Radu, S., Elhadi, N., Hassan, Z., Rusul, G., Lihan, S., Fifadara, N., Yuherman, S. and Purwati, E. (1998). Characterization of *Vibrio vulnificus* isolated from cockles (*Anadara granosa*): antimicrobial resistance, plasmid profiles and random amplification of polymorphic DNA analysis. FEMS Microbiol. Lett. 165: 139-143.

Ripabelli, G., Sammarco, M., Grasso, G., Fanelli,I., Capriola, A. and Luzzi, I. (1999). Occurrence of *Vibrio* and other pathogenic bacteria in *Mytilus galloprovinialis* (mussels) harvested from Adriatic Sea, Italy. Int. J. Food Microbiol. 49: 43-48.

Sakazaki, R., Iwanami, S. and Fukumi, H. (1963). Studies on the enteropathogenic, facultatively halophilic bacteria, *Vibrio parahaemolyticus*. I. Morphological, cultural, and biochemical properties and its taxonomic position. Jap. J. Med. Sc. Biol. 16: 161-188.

Sakazaki, R., Iwanami, S. and Fukumi, H. (1968a). Studies on the enteropathogenic, facultatively halophilic baceria, *Vibrio parahaemolyticus*. II. Serological characteristics. Jap. J. Med. Sci. Biol. 21: 313-324.

Sakazaki, R., Tamura, K., Kato, T., Obara, Y., Yamai, S. and Hobo, K. (1968b). Studies on the enteropathogenic, facultatively halophilic bacterium *Vibrio parahaemolyticus*. III. Enteropathogenicity. Jap. J. Med. Sci. Biol. 21: 325-331.

Sharkey, F., Markos, S. and Haylock, R. (2004). Quantification of toxin-encoding mRNA from *Clostridium botulinum* type E in media containing sorbic acid or sodium nitrite by competitive RT-PCR. FEMS Microbiol. Lett. 232: 139-144.

Shirai, H., Ito, H., Hirayama, T., Nakabayashi, Y., Kumagai, K., Takeda, Y. and Nishibuchi, M. (1990). Molecular epidemiological evidence for association of thermostable direct hemolysin (TDH) and TDH-related hemolysin of *Vibrio parahaemolyticus* with gastroenteritis. Infect. Immun. 58: 3568-3573.

Simpson, L., White, V., Zande, S. and Oliver, J. (1987). Correlation between virulence and colony morphology in *Vibrio vulnificus*. Infect. Immun. 55: 269-272.

Smith, G. and Merkel, J. (1982). Collagenolytic activity of *Vibrio vulnificus*: potential contribution to its invasiveness. Infect. Immun. 35(3): 1155-1156.

Sobel, J., Malavet, M. and John, S. (2007). Outbreak of clinically mild botulism type E illness from home-salted fish in patients presenting with predominantly gastrointestinal symptoms. Clin. Infect. Dis. 45: 14-16.

Tada, J., Ohashi, T., Nishimura, N., Shirasaki, Y., Ozaki, H., Fukushima, S., Takano, J., Nishibushi, M. and Takeda, Y. (1992). Detection of the thermostable direct hemolysin gene (*tdh*) and the thermostable direct hemolysin-related gene (*trh*) of *Vibrio parahaemolyticus* by polymerase chain reaction. Mol. Cell Probes. 6: 477-487.

Takikawa, I. (1958). Studies on pathogenic halophilic bacteria. Yokohama M. Bull. 2: 313-322.

Tamplin, M., Rodrick, G., Blake, N. and Cuba, T. (1982). Isolation and characterization of *Vibrio vulnificus* from two Florida estuaries. Appl. Environ. Microbiol. 44: 1466-1470.

Taylor, R., Miller, V., Furlong, D. and Mekalanos, J. (1987). Use of *phoA* gene fusions to identify a pilus colonization factor coordinately regulated with cholera toxin. Proc. Natl. Aad. Sci. USA 84: 2833-2837.

Telzak, E., Bell, E., Kautter, D., Crowell, L., Budnick, L., Morse, D. and Schulz, S. (1990). An international outbreak of type E botulism due to uneviscerated fish. J. Infect. Dis. 161: 340-342.

Temyo, R. (1966). Studies on the prevention of outbreaks of food poisoning caused by *Vibrio parahaemolyticus*. Bull. Tokyo Med. Dent. Univ. 13: 89-510.

Testa, J., Daniel, L. and Kreger, A. (1984). Extracellular phospholipase A_2 and lysophospolipase produced by *Vibrio vulnificus*. Infect. Immun. 45: 458-463.

Thompson, W. and Thacker, C. (1973). Effect of temperature on *Vibrio parahaemolyticus* in oysters at refrigerator and deep freeze temperatures. Can. Inst. Food Sci. Technol. J. 6: 156-158.

Thompson, C., Jr., Vanderzant, C. and Ray, S. (1976). Serological and hemolytic characteristics of *Vibrio parahaemolyticus* from marine sources. J. Food Sci. 41: 204-205.

Tison, D., Nishibuchi, J. Greenwood, J. and Seidler, R. (1982). *Vibrio vulnificus* biogroup 2: new biogroup pathogenic for eels. Appl. Environ. Microbiol. 44: 640-646.

Torii, M. (1974). Extraction and antigenic specificity of o-antigens of *Vibrio parahaemolyticus*. In: International Symposium on *Vibrio parahaemolyticus*. T, Fugino, G. Sakaguchi, R. Sakazaki, and Y. Takeda (eds). Saikon Publ. Co., Tokyo, Japan pp. 187-192.

Torii, M., An, T., Igarashi, K., Sakai, K. and Kuroda, K. (1969). Immunochemical studies on O-antigens of *Vibrio parahaemolyticus*. 1. Preparation, specificity and chemical nature of the antigens. Biken J. 12: 77-84.

Twedt, R.M. (1989). *Vibrio parahaemolyticus*. In: Foodborne Bacterial Pathogens. M.P. Doyle (ed). Marcel Dekker, New York, USA pp. 543-600.

Vanderzant, C. and Nickelson, R. (1972). Survival of *Vibrio parahaemoyticus* in shrimp tissue under varius environmental conditions. Appl. Microbiol. 23: 34-37.

Venkateswaran, K., Dohmoto, N. and Harayama, S. (1998). Cloning and nucleotide sequence of the *gyrB* gene of *Vibrio parahaemolyticus* and its application in detection of this pathogen in shrimp. Appl. Environ. Microbiol. 64: 681-687.

Vickery, M., Smith, A., DePaola, A., Jones, D., Steffan, R. and Bej, A. (1998). Optimization of the arbitrary-primed polymerase chain reaction (AP-PCR) for intra-species differentiation of *Vibrio vulnificus*. J. Microbiol. Meth. 33: 181-189.

Wagatsuma, S. (1968). On a medium for hemolytic reaction. Media Circle. 13: 159-162 (in Japanese).

Waldor, M. and Mekalanos, J. (1996) Lysogenic conversion by a filamentous phage encoding cholera toxin. Science 272: 1910-1914.

Wang, S. and Levin, R.E. (2004.) Quantitative determination of *Vibrio parahaemolyticus* by polymerase chain reaction. Food Biotechnol. 18: 279-28.

Wang, S. and Levin, R. (2006). Rapid quantification of *Vibrio vulnificus* in clams (*protochaca staminea*) using real-time PCR. Food Microbiol. 23: 757-761.

Wang, X., Maegawa, T., Karasawa, T., Kozaki, S., Tsukamoto, K., Gyobu, Y., Yamakawa, K., Oguma, K., Sakaguchi, Y. and Nakamura, S. (2000). Genetic analysis of type E botulinum toxin-producing *Clostridium butyricum* strains. Appl. Environ. Microbiol. 66: 4992-4997.

Warner J. and Oliver J. (1998). Randomly amplified polymorphic DNA analysis of starved and viable but nonculturable *Vibrio vulnificus* cells. Appl. Environ. Microbiol. 64: 3025-3028.

Warner, J. and Oliver, J. (1999). Randomly amplified polymorphic DNA analysis of clinical and environmental isolates of *Vibrio vulnificus* and other *Vibrio* species. Appl. Environ. Microbiol. 65: 1141-1144.

Weber, J., Hibbs, R. and Sarswish, A. (1993). A massive outbreak of type E botulism associated with traditional salted fish in Cairo. J. Infect. Dis. 167: 451-454.

West, P., Russek, E., Brayton, P. and Colwell, R. (1982). Statistical evaluation of a quality control method for isolation of pathogenic vibrio species on selected thiosulfate-citrate-bile salts-sucrose agars. J. Clin. Microbiol. 16: 1110-1116.

Wong, H., Liu, K., Pan, T., Lee, D. and Shih. Y. (1996). Subspecies typing of *Vibrio parahaemolyticus* by pulsed-field gel electrophoresis. J. Clin. Microbiol. 34: 1536-1539.

Wong, H., Liu, C., Pan, T., Wang T. , Lee, D. and Shih. Y. (1999). Molecular typing of *Vibrio parahaemolyticus* isolates obtained form patients involved in food poisoning outbreaks in Taiwan by random amplified polymorphic DNA analysis. J. Clin. Microbiol. 37(6): 1809-1812.

Wright, A., Micelli, G., Landry, W., Christy, J., Watkins, W. and Morris, J. Jr. (1993). Rapid identification of *Vibrio vulnificus* on nonselective media with an alkaline phosphatase-labeled oligonucleotide probe. Appl. Environ. Microbiol. 59: 541-546.

6

Fermented Fish and Fish Products: An Overview

Smita H. Panda,[1,] Ramesh C. Ray,[2] Aly F. El Sheikha,[3] Didier Montet [4]* and *Wanchai Worawattanamateekul*[5]

INTRODUCTION

Fish is an important source of protein in the human diet. However, fish also has the disadvantage that if it is not salted, dried, smoked or preserved in some way or another, it will quickly spoil. In warm regions such as tropical countries, access to fresh fish can be a problem mainly in rural areas owing to the shortage of ice and lack of refrigeration (Anihouvi et al., 2007). Fermentation is the most important way of preserving fish. Fermented fish is used both as flavouring and as a source of protein. Fish fermentation in the Southeast Asian region normally lasts for 3-9 mon and the fish flesh may be liquefied or turned into a sauce or paste. Some of the products include *Nuoc-mam* of Vietnam and Cambodia, *Nam-pla* of Thailand, *Sushi* of Japan and *Patis* of the Philippines. No African fishery products are mentioned in the FAO Fisheries Report No. 100 (FAO, 1971) on fermented fish; however *fessiekh* from Egypt and Sudan is mentioned as a Mediterranean product.

[1]*Agri-Bioresource Research Foundation, 81, District Centre, Chandrasekharpur, Bhubaneswar 751 010, Orissa, India; Tel/Fax: 91-674-2351046; E-mail: agribiores6@rediffmail.com*
[2]*Regional Centre, Central Tuber Crops Research Institute, Bhubaneswar 751 019, Orissa, India; Tel/Fax: 91-674- 2470528; E-mail: rc_rayctcri@rediffmail.com*
[3]*Minufiya University, Faculty of Agriculture, Department of Food Science and Technology, 32511 Shibin El Kom, Minufiya Government, Egypt; Tel.: 33 4 67 61 57 28; Fax: 33 4 67 61 44 44; E-mail: elsheikha_aly@yahoo.com*
[4]*Centre de Coopération Internationale en Recherche Agronomique pour le Développement, CIRAD, UMR Qualisud, TA 95B/16, 34398 Montpellier Cedex 5, France; Tel.: 33 4 67 61 57 28; Fax: 33 4 67 61 44 44; E-mail: didier.montet@cirad.fr*
[5]*Department of Fishery Products, Faculty of Fisheries, Kasetsart University, Bangkok 10900, Thailand; Tel.: 66-2-5790113 press 4088; Fax: 66-2- 9428644 press 12; E-mail ss: ffiswcw@ku.ac.th*
*Corresponding author

In Africa salting and drying of fish for preservation is accompanied by fermentation, but the period is short and the product is not transformed into a paste or sauce. Although fermented fish products are a good source of protein, they can be consumed only in limited quantities because of the high salt content of these products. Fermentation of fish is especially used in situations where drying of fish is not possible because the climate is too wet and where cooling and sterilization of the product is too expensive. In this chapter, three important traditionally fermented fish and fish products for human consumption, i.e., fish sauce, fish paste and dried/salted fish have been discussed with special emphasis on the microbiology, biochemical compositions and food safety.

FERMENTATION

During fermentation of fish, protein is broken down to peptide and amino acids in the presence of a high salt concentration. The fish protein is mainly broken down by enzymes which come from the fish itself. These enzymes are mainly present in the gut (Paludan-Muller et al., 2002). In traditional fermentation methods where the intestines are removed from the fish, fermentation will often be slower as there are fewer enzymes present in the flesh.

Role of Microorganisms

Microorganisms probably play no role in the breaking down of protein during fermentation (Tanasupawat et al., 2006, 2007). However, micro-organisms which can tolerate salt (because of the high concentrations of salt which are used during fermentation of fish) do seem to contribute to the specific taste and smell (described further on this chapter) of the fermented product (Lopetcharat and Park, 2002; Sanni et al., 2002). In some traditional fermentation techniques, such as in the production of *sushi*, a fermentable source of carbohydrates such as boiled rice is added to the fermented fish product. This combination stimulates the growth of lactic acid bacteria (LAB). The rice is a source of sugars for the LAB. Due to the formation of lactic acid, which is desirable in these products, the pH of the fish mixture is lowered making the product safer and easier to keep (Gelman et al., 2001). In addition, spices such as garlic, pepper or ginger are added to the safety of the products. Also, in some products garlic may serve as a carbohydrate source for lactic acid fermentation (Paludan-Muller et al., 1999).

Salt

Salt is used to draw liquid out of the fish and to control the fermentation. Thus, the high salt content (20-30%) ensures that spoilage due to bacteria is prevented and that the number of bacteria present drops as quickly as possible during fermentation. From a nutritional point of view, however, it would be best to use as little salt as possible. The high salt concentration also slows down the fermentation speed. (Berkel et al., 2004; Rao and Stevens, 2006).

Traditional (Natural) Fermentation

The fermentation methods described in this chapter are mostly traditional methods. This means that the fermentation is allowed to take place naturally (spontaneous fermentation) and is guided by experience. No control is exerted over the fermentation. If enough salt is added, about 30% of weight of the fish, and there is no influx of air during the fermentation process (anaerobic environment), the fermentation will proceed by itself. The fermentation methods are more or less standard for a given region. Local adaptations or changes in the procedure can, of course, be found. Experience will help determine whether or not the fermentation has gone well. If the product is different than is normal, for example if it has a different colour or smell, the product should not be eaten. Traditional fermented fish products are divided into two groups:

- Products which, in the presence of salt, are fermented by the enzymes present in the fish flesh and intestines;
- Products which are fermented in the presence of boiled or roasted rice. Usually in South-East Asia boiled rice is added to the fish-salt mixture.

Fermentation Using Starter Culture

The use of starter culture in food fermentation has become a means to increase processing rates and product consistency. Starters are used to improve the sensory characteristics and microbiological quality and to shorten the fermentation time of fermented foods (Visessanguan et al., 2006). Of late, the LAB starter such as *Lactobacillus plantarum*, *L. brevis*, *L. fermentum*, *Pediococcus acidilactici*, *P. pentosaceus*, etc are being used commercially for the production of traditional fish products like *som fug* (Riebroy et al., 2008) and other fish-based food products (Gelman et al., 2001).

Types of Fermented Products

There are three kinds of fermented fish products:

- **Fish sauce**. The fish flesh is converted into a liquid fish sauce;
- **Fish paste**. The fish is converted into a paste;
- **Dry/salted fish**. The fish, whole or in pieces, retains as much as possible of its own structure.

Fermented fish products such as fish sauce and fish paste are eaten mainly in South-East Asia, whereas dried fish are consumed in many parts of Asia and Africa. Fermented fish products are an important protein supplement. They contain a number of essential amino acids which can form an important addition to the daily diet. For example, fish sauce contains a lot of the amino acid lysine (Khem, 2009; Dincer et al., 2010). This amino acid is found only in small quantities in rice. The quality of the resulting product depends on the fat content of the fish, the enzyme activity in the fish flesh, contaminations in the salt used and the temperature.

Fish Used

Often the surplus or the side catch of the main catch are fermented. These fish would otherwise be lost to spoilage. Mostly small kinds of fish are used. Table 6.1 lists the different kinds of fish used in South-East Asia for fermentation.

Table 6.1 Saltwater and freshwater fish and crustaceans which are mainly used in the fermentation methods of South-East Asia

Product group	Species
Saltwater fish	African moonfish, anchovies, croaker, herring, deep-bodied herring, fimbriated herring, mackerel, round scad, sea bream, sole, tuna
Freshwater fish	Carp, catfish, climbing perch, gourami, mudfish
Shellfish and crustaceans	Shrimp, mussels, oysters, octopus

FISH SAUCE

Fermented fish sauces have been consumed since ancient times and the earliest reported is *garum,* which was highly prized in the Roman era (Badham, 1854). In Asia and in particular those areas with an extensive coastal line and high ambient temperature and relative humidity as found in South-East Asian countries such as Thailand, Malaysia, the Philippines and Indonesia, the use of fermentation method for fish and shrimp has been of great value from ancient times (Lopetcharat and Park, 2002). These methods have been refined over the centuries. Some fish sauces are made from raw fish, others from dried fish, some from only a single species, others from whatever is dredged up in the net, including some shell fish;

some from whole fish, others from only the blood or viscera. Some fish sauces contain only fish and salt, and others a variety of herbs and spices. A number of fish sauces that are manufactured by traditional methods are listed in Table 6.2.

Table 6.2 Fish Sauces

Country	Name	Fish species	Method (fish: salt) and time of fermentation
Japan	*Shottsuru* *Uwo-shoytu* *Ikashoyu*	*Astroscopus japanicus* (sandish) *Clupea pichardus* (sardine) *Omnastrephis sloani* (squid) *Omnastrephis pacificus* (squid)	5:1 salt + malted rice and *koji* (3:1) added; 6 months
Korea	-	-(shrimp)	Salt 4:1; 6 months
Khmer Republic	*Nuoc-mam*	*Stolephorus* spp. *Ristrelliger* spp. *Engraulis* spp. *Decapterus* spp. *Darasoma* spp. *Clupea* spp.	3:1-3:2 salt; 3-12 months
	Nuoc-mam-gau-ca	*Clarius* spp. *Ophicephalus* spp.	Livers only 10:1 salt; for 6 days then boiled and filtered
Cambodia	*Nuoc-mam*	*Stolephorus* spp. *Ristrelliger* spp. *Engraulis* spp. *Decapterus* spp. *Darasoma* spp. *Clupea* spp.	3:1-3:2 salt; 3-12 months
Thailand	*Nam-pla*	*Stolephorus* spp. *Ristrelliger* spp. *Cirrhinus* spp.	5:1-1:1 salt; 5-12 months
Malaysia	*Budu*	*Stolephorus* spp.	5:1-3:1 salt + palm sugar and tamarind; 3-12 months
Myanmar	*Ngapi*	-	5:1 salt 3-12 months
Philippines	*Patis*	*Stolephorus* spp. *Clupea* spp. *Decapterus* spp. *Leionathus* spp.	3:1-4:1salt, 3-12 months
Indonesia	*Ketjap-ikan*	*Stolephorus* spp. *Clupea* spp. *Leiagnathus* spp. *Osteochilus* spp. *Puntius* spp. *Ctenaps* spp.	6:1 salt; 6 months
	Bakasang	*Stolephorus* spp.	5; 1.5-3.5 fish; salt ratio; 3-6 weeks

Table 6.2 contd...

Table 6.2 contd...

Country	Name	Fish species	Method (fish: salt) and time of fermentation
India and Pakistan	*Colombo cure*	*Ristrelliger* spp. *Cybium* spp. *Clupea* spp.	Gutted fish with gills removed and tamarind added 6:1 salt; up to 12 months
Hong Kong	*Yeesui*	*Sardinella* spp. *Jelio* spp. *Carangidae* spp. *Engraulis pupapa* *Teuthis* spp.	4:1 salts; 3-12 months
Ghana	*Momoni*	*Caranx hippos*	10:3 fish: salt ratio; 1-5 days
Greece	*Gaross*	*Scomber colias*	Liver only 9:1 salt; 8 days
France	*Pissala*	*Aphya pellucida*	4:1 salt; 2-8 weeks (depending on size)
		Gobius spp.	
		Engraulis spp.	
		Atherina spp.	
		Meletta spp.	
	Anchovy	*Engraulis encrasicholus*	Beheaded and gutted fish 2:1 salt; 6-7 months

Source: Beddows (1985); Ijong and Ohta (1995); Sanni et al. (2002).

Garum

Garum, also called *liquamen*, is a type of fish sauce condiment that was popular in ancient Roman society. While it was very popular among the Romans, it originally came from the Greeks, getting its name from the Greek words *garos* or γάρον *gáron*, which named the fish whose intestines were originally used in the condiment's production. (Badham, 1854) *Garum* appears in most of the recipes featured in *Apicius*, a Roman cookbook. The sauce was usually made by crushing and fermentation in brine of the innards of various fish such as tuna, eel, and others.

Nuoc-mam

This is the most common fish sauce produced in the Indo-China Peninsula (Vietnam, Laos, and Cambodia) and is mostly consumed entirely within the country of manufacture. The basic principle of *nuoc-mam* preparation is the breaking down of the fish protein by enzymes in the presence of large amounts of salts. The sauce is a clear brown liquid with a distinctive mealy/sharp aroma. The sauce is prepared as follows: The fish, usually anchovies or mackerel, which are not usually cleaned, are kneaded and pressed by hand. They are then placed in layers with salt in an appropriate ration of 3:1 fish to salt (w/w) in earthenware jars that are almost buried in the ground. After filling, the jars are tightly sealed and left for several (16-18) mon. After fermentation the pots are carefully removed and within a few days of setting the supernatant liquor are decanted off carefully (Beddows, 1985). The liquid is called *nuoc-mam*.

Often caramel, roasted rice or molasses are added to fish to get a dark colour and a certain taste. This improves the keeping quality of the qualitatively inferior *nuoc-mam*. At a fermentation temperature higher than 45°C, the *nuoc-mam* loses its characteristic taste. It is therefore best to keep the vats somewhere cool.

Nam-pla

This is the equivalent of *nuoc-mam* and is reported to be very similar (Beddows, 1985). The major species of fish used are *Stolephorus* and *Sardinella* but smaller *Scomber* and *Ristrelliger* may be used as well as certain *Clupeids*. The process is more commercialized than that of *nuoc-mam*. The fermentation takes 6-12 mon. The sauce is 'run off' and exposed to the sun for 1-3 mon (ripening) (Saisithi, 1967). The liquid is blended with *meiki*, a concentrated by-product obtained from the bacterial production of monosodium glutamate. Either dark or light *meiki* may be used and both significantly affect both flavour and aroma of the sauce finally offered for sale.

In general, less salt is used in the production of *nam-pla* than with nuoc-mam (4:1 ratio fish to salt). *Padack*, a more aromatic version of *nam-pla*, is also produced.

Ngan byar yay

Similar to *nam-pla* and is indigenous to Myanmar.

Patis

A sauce called *patis* is produced in the Philippines which is comparable to *nuoc-mam*. The procedure for making *patis* is more or less the same as that for *nuoc-mam*. After the first *patis* yield, which has a characteristic taste, a saturated brine solution is used to obtain the second yield of *patis* of an inferior quality. *Patis* is usually made of small fish (*Stolephorus*, *Sardinella*, *Leignathus* or *Decapterus macrosoma*). Small shrimp or alamang, goby fry, herring fry and anchovies give the best results. Enough salt must be added to saturate the moisture which oozes from the fish. One kg of salt to 3.5-4 kg of fish gives a final product with 20-25% salt content.

Patis is also a by-product of the preparation of fish paste *bagoong* (described further on this chapter).

Budu

This is a fish sauce similar to *nam-pla* and *nuoc mam* and is produced in the North Eastern states of Malaysia (Beddows, 1985). The species of fish used is mostly *Stolephorus* (*Ikanbilis*). *Budu* is prepared by fermentation of fish with the addition of salt (3:2 fish: salt ratio; w/w) in circular earthen pots covered with a plastic sheet. Weights are placed on the top which help the fish mass to become immersed in the pickle produced by osmotic dehydration. The pickle is picked up at irregular intervals. Usually in *budu* manufacture, tamarind and caramelized palm sugar are added to the pickle and this sweetens the product and gives it a darker appearance.

Bakasang

Bakasang is a traditional fish sauce produced in Indonesia by fermenting small whole sardines (*Sardinella* spp. or *Stelophorus* spp.). It is prepared by mixing fish and salt (5:1.5-3.5 fish: salt ratio; w/w), packed in small bottles and placed in the kitchen near the fire place. The temperature ranges from 30-60°C and fermentation is allowed for about 3-6 wk.

The product has become closely integrated with the eating habits of the eastern Indonesians, especially the Manadonese people (in the north Colebes Island). It is usually used as a flavouring in many dishes or mixed with red chillies, tomato, red onion and garlic and then sautéd with coconut oil. The sautéd sauce is eaten with hot porridge mixtures of rice and vegetables called *tinutuan* (Ijong and Ohta, 1995).

Momoni

In Ghana one type of fermented fish product, *momoni* is popularly used as a condiment for preparing sauces for the consumption of yam, cocoyam and *apetum* (boiled unripe plantain). *Momoni* is similar to *fessiekh* of Egypt which is prepared from *bouri* (*Mugil cephalus*). For the preparation of *momoni*, different types of freshwater fish can be used; usually African jack mackerel (*Caranx hippos*) is used. They can be scaled and gutted followed by washing in tap water and salting (294-310 g/kg) with the gill and gut regions being heavily salted. The fish are arranged in baskets covered with aluminum trays or jute bags and fermentation is allowed for 1-5 d. Before retailing, the fermented fish are washed in brine water, rubbed with salt and cut into small pieces. The cut pieces are sun-dried on a wooden tray in the open air for a few hours. *Momoni* is a solid product that is added to boiling stew consisting of ground red pepper, tomato, onion and a small amount of palm oil. The finished product is usually of low quality with a high salt concentration and deteriorates rapidly during retailing and storage (Sanni et al., 2002).

Fessiekh

In Egypt, *Fessiekh* is a traditional fish dish consisting of fermented salted and dried grey mullet, of the Mugil family, a saltwater fish that lives in both the Mediterranean and the Red Seas.

The traditional process of preparing it is to dry the fish in the sun before being preserved in salt. It has a distinctive smell that only its true lovers would appreciate. The process of preparing *fessiekh* is quite elaborate; the information is passed from father to son in certain families. The occupation has a special name in Egypt, *fasakhani. Fessiekh* is traditionally eaten during Sham El Nese ("Smelling the Breeze"), which is a spring celebration in Egypt from ancient times. Some consider *fessiekh* as a part of the good things of Egypt (El-Tahan et al., 1998).

Other Traditional Sauces

Southeast Asians use fish sauce as a cooking sauce. However, there is a sweet and sour version of this sauce, which is used more commonly as a dipping sauce. In Thailand, fish sauce is used in cooking and is also kept in a jar at the table for use as a condiment. This jar often contains a mixture of fish sauce and chopped hot chillies, called *nam-pla prik*. In Korea, it is called *ack jeot*, and is used as a crucial ingredient in *Kimchi* (usually from *myul chi* or *kanari*, meaning anchovies), both for taste and fermentation. *Sae woo jeot* (shrimp) is also popular as a side sauce.

Shottsuru is made in Japan from sandfish. Sardines, anchovies and mollusks can also be used as starting material. The fluid is filtered and boiled and can be kept for years. Soy bean sediments or *koji*, which is fermented with wheat, can be added to *shottsuru.*

Novel Fish Sauces

A new method for producing fermented fish sauce was developed to improve the taste of the fish sauce. The method involved the production of a fermented sauce, seasoning liquid from salmon that was enzymatically hydrolyzed with the wheat gluten *koji*, LAB and yeast (*Saccharomyces cerevisiae*). During fermentation, the total nitrogen, lactic acid, and ethanol contents in the fish sauce increased continuously and then plateaued 3 mon after fermentation had begun. The chemical composition and the sensory properties of the fish sauce prepared by this 3-mon fermentation were compared with those of the fish sauce from salmon similarly hydrolyzed with soy sauce *koji* prepared from soybean and wheat. The fish sauce produced using wheat gluten was very light-coloured and had a higher content of free amino acids, especially glutamic acid. The peptides consisted mainly of *Glx, Asx* and *Pro*, compared with *Asx, Glx,* and *Gly* in the liquid from soy sauce *koji.* Sensory evaluation revealed that the fish sauce derived from the wheat gluten *koji* had an intense umami taste and a fine flavour better than that of soy sauce *koji* (Indoh et al., 2006).

FISH PASTE

Fish pastes are far more widely produced and eaten than fish sauces. They are mostly consumed in relatively small quantities with rice dishes, and often contribute quite significantly to the nutrition of poor households. Shrimp paste or shrimp sauce, is a common ingredient used in Southeast Asian and southern Chinese cuisine. It is known as *terasi* (also spelled *trassi*, *terasie*) in Indonesian, *ngapi* in Burmese, *kapi* (¡ Đ) in Thai, Khmer and Lao language, *belachan* (also spelled *belachan*, *blachang*) in Malay, *Mắm Tôm* in Vietnamese, *bagoong alamang* (also known as *bagoong aramang*) in Filipino and *Hom Ha/Hae Ko* (POJ: *hê-ko*) in Min Nan Chinese.

There are two kinds of fish and shrimp pastes in South- East Asia:

1) Fish or shrimp-salt mixtures
2) Produces, which are fermented in the presence of cooked or roasted rice on which yeasts and moulds are present.

The general method of preparation of fish and shrimp pastes is the same as that described for fish sauces. Only the fermentation time is short, as not all of the fish flesh needs to be broken down. Fish paste must be mixed regularly to help the salt distribute evenly.

Bagoong

Bagoong, a fish paste from the Philippines, is made by fermenting well-cleaned whole or minced fish, shrimp, fish or shrimp eggs in the presence of salt (3:1 fish:salt ratio; w/w). The fish used for *bagoong* include anchovies, sardines, herring, silverside, shrimp, oysters, clams, and other shellfish. The fish-salt mixture is put into earthenware pots and covered with cheesecloth for 5 d. The covered pots are then put in the sun for 7 d. After that, the product is fermented for a further 3 to 12 mon.

As a by-product, the fish sauce *patis* can be harvested by separating the liquid above from the paste. The paste is sometimes coloured by adding '*angkak*'—a coloured rice which has been treated with the red yeast-like organism *Monascus purpureus*. *Bagoong* can be stored for several years (Heen and Kreuzer 1962; http://en.wikipedia.org/).

Pra-hok

In Kampuchea, *pra-hoc* is prepared as follows: after the fish (cyprinids) are beheaded they are kneaded by hand so that the scales and intestines come loose. The fish are placed in a basket and covered with banana leaves and stones for 24 h in order to drain. The fish are salted and, after leaving them for half an hour, they are dried on mats for 1 d in the sun. The fish are then pounded into a paste. The paste is put into open jars and placed in the sun. The liquid which appears on top is removed. The paste can be eaten when no more liquid comes out (Heen and Kreuzer, 1962; http://en.wikipedia.org/wiki/Fish_sauce).

Balao Balao

Balao balao, which has its origin in the Philippines, is a fermented rice-shrimp (*Penaeus indicus* or *Macrobrachium* spp.) product. It is prepared by mixing boiled rice, whole raw shrimps and salt (5:1; shrimp:salt ratio). The product is stored in jars and is fermented for 7-10 d. The mixture becomes less sour the longer the fermentation takes place. The shells of the shrimp become red and soft, and the mixture including the rice, becomes liquid. In the general preparation it is fried with garlic and onion after fermentation. It is eaten as a sauce or as a complete meal in itself (Steinkraus, 1992, 1997).

Belachan

Belachan is a Malay-Indonesian shrimp paste made of small shrimps to which a relatively small amount of salt (20:1 shrimp:salt ratio) and chillie pepper, *belasan* is added. The mixture is dried on mats on the ground and kept in the sun. After 4-8 h of drying, during which 50% of the moisture is lost, any contaminants in the shrimp are removed. The shrimp are then chopped up and squeezed into wooden vats so that no more air is present. The paste which results is fermented for 7 d. After 7 d the substance is taken out of the barrel and is dried for 3-5 h in the sun. The paste is again ground after which it is put back in the wooden vats. The paste is fermented for a further 1 mon (www.amazines.com/Belachan_related.html).

Ngapi Yay

A watery dip or condiment that is very popular in Myanmar, especially with the Burmese and Karen ethnic groups. The *ngapi* (either fish or shrimp, but mostly whole fish *ngapi* is used) is boiled with onions, tomato, garlic, pepper and other spices. The result is a greenish-grey broth-like sauce, which makes its way to every Burmese dining table. Fresh, raw or blanched vegetables and fruits (such as mint, cabbage, tomatoes, green mangoes, green apples, olives, chillies, onions and garlic) are dipped into the *ngapi yay* and eaten. Sometimes, in less affluent families, *ngapi yay* is the main dish, and also the main source of protein (Heen and Kreuzer, 1962; en.wikipedia.org/wiki/Shrimp_paste).

Trassi(Terasi)

Trassi, an Indonesian variant of dried shrimp paste, is usually purchased in dark blocks, but is sometimes also ground. The colour and aroma of *terasi* varies depending on which village produced it. The colour ranges from a soft purple-reddish hue to darkish brown. In Cirebon, a coastal city in West Java famous for producing fine quality *terasi*, it is made from tiny shrimp called *"rebon"*, the origin of the city's name. In Sidoarjo, East Java, *terasi* is made from the mixture of ingredients such as fish, small shrimp, and vegetables. *Terasi* is an important ingredient in *sambal terasi*, also many other Indonesian dishes, such as *sayur asam* (fresh sour vegetable soup), *lotek* (also called *gado-gado*, Indonesian style salad in peanut sauce), *karedok* (similar to *lotek,* but the vegetables are served raw), and *rujak* (Indonesian style hot and spicy fruit salad) (Heen and Kreuzer, 1962; en.wikipedia. org/wiki/Shrimp_paste).

Bagoong Alamang

Bagoong alamang is a Filipino shrimp paste, made from minute shrimp or krill (*Alamang*) and is commonly eaten as a topping on green mangoes or used as a major cooking ingredient. *Bagoong* paste varies in appearance, flavour, and spiciness depending on the type. Pink and salty *bagoong alamang* is marketed as "fresh", and is essentially the shrimp-salt mixture left to marinate for a few days. The paste can be sautéed with various condiments, and its flavour can range from salty to spicy-sweet.

Cincalok is the Malaysian version of 'fresh' *bagoong alamang*. (Heen and Kreuzer, 1962; www.encyclopedia.com/doc/1O39-bagoong.html)

Hom Ha

This Chinese shrimp paste is popular in southeastern China. This shrimp paste is lighter in colour than many Southeast Asian varieties and is often used to cook pork. The shrimp paste industry has historically been important in the Hong Kong region. (www.clovegarden.com/ingred/ seaprod.html)

Hae Ko

Hae Ko means prawn paste in the Hokkien dialect. It is also called *petis udang* in Malay. This version of shrimp/prawn paste is used in Malaysia, Singapore and Indonesia. This thick black paste has a molasses like consistency instead of the hard brick like appearance of *Belachan*. It also tastes sweeter because of the added sugar. It is used to flavour common local street foods like *popiah* spring rolls, *laksa* curry, *chee cheong fan* rice rolls and *rojak* salad. (Heen and Kreuzer, 1962).

The list of fish and shrimp pastes from countries in Asia and Africa is given in Table 6.3.

Table 6.3 Some fish pastes of Asian and African continents

Country	Name	Ingredients
Cambodia	*Pra-hoc*	Fish (cyprinid, ophiocephalid)/salt
	Phaak, Paak or mam-chao	Fish/salt + glutinous rice
	Mam-ca-sat	Fish/salt + roasted rice
	Mam-ca-sak	Fish/salt roasted rice + pineapple or papaya
	Mam-ca-lok	Fish/salt +roasted rice + sugar + ginger + pineapple + colour
	Mam-ruot	Fish entrails/salt
	Mam-seing	Fish eggs+ salt + roasted rice
	Mam-ruoc	Freshwater shrimps/salt
Thailand	*Kapi*	Marine shrimps/salt
	Pla-mam	Freshwater fish/salt + roasted rice + pine apple
	Pla-chao	Freshwater fish/salt + glutinous rice
	Kung-chao	Marine or freshwater shrimp/salt + colour (and sometimes roasted rice or sesame)
Malaysia	*Blachan*	Shrimp/salt
Philippines	*Bagoong*	Fish or shrimp/salt (some times + colour)
Indonesia	*Trassi*	Fish/salt + sun drying, e.g., *Trassi-udung* from shrimp. *Trassi-ikan* from small fish
Myanmar	*Nga-Ngapi*	Fish/salt
Japan	*Shiokara tyupe*	Squid or skipjack/salt + malted rice, e.g., *Unishiokara*-ovary of sea urchin + salt, kakishiokara-oyster + salt
Pakistan and North Eastern India	*Sidal*	Small fish (*Bambus* spp.)—salting and drying, then crude fish oil is added.

Source: Beddows, 1985, modified.

FERMENTED DRIED/ SALTED FISH

Apart from fish sauces and pastes, different processing techniques are employed in fish fermentation. This is greatly influenced by factors such as availability of salt and the food habits of the local people. Three main techniques have emerged as methods commonly practised by people in Asian and African countries for fish fermentation. These are:

- Fermentation with salting and drying;
- Fermentation and drying without salting; and
- Fermentation with salting but without drying.

Asian Fish Products

1.Thailand

Plaa som

It is a low salt product, typically composed of freshwater fish, salt, boiled rice and garlic (Adams, 1986) and is mainly produced in the central and north-eastern part of Thailand. However, in the Songkhla Province in southern Thailand, a local variety of *plaa som* is produced, in which garlic and boiled rice are replaced by palm syrup and from time to time by roasted rice, thus resembling *plaa uan*, another type of Thai fermented fish (Phithakpol et al., 1995).

For *plaa so*m, the whole fish is mixed with salt (8:1; fish: salt ratio, w/w) and left overnight. Cooked rice and minced garlic are added (ratio 20fish/salt: 4 rice: 1 garlic; w/w), then the mixture is packed in jars and fermented at ambient temperature for 5-7 d. The shelf-life is reportedly 3 wk (Phithakpol et al., 1995).

Som-fug

Som-fug is Thai traditional fermented minced fish, which is composed of fish mince salt (2.5%), ground steamed rice (2-12%) and minced garlic (4%). The mixture is tightly packed in banana leaves or plastic bags and left to ferment for 2-5 d at 30°C. *Som-fug* can be served either as a main dish or as a snack with vegetables. *Som-fug* is highly nutritious and is an excellent source of protein. The fish used are mainly freshwater rather than marine fish. The fish species include giant snake-head fish (*Ophicephalus micropeltes*), Rohu (*Labeo rohita*), spotted feather back (*Notopterus chitala*) and grey featherback (*Notopterus notopterus*) (Lopetcharat et al., 2001; Riebroy et al., 2008).

Som fak

For *som fak* (similar to *som fug*), fish fillets are minced. Cooked rice, minced garlic and salt (ratio 120 fish mince; 20 rice: 7 garlic; 7 salt, w/w) are added to the minced fish and the mixture is divided into small portions and packed in banana leaves or plastic sheets. The product ferments for 3-5 d at ambient temperature. The shelf-life is reported to be 2 wk; however, the product is best consumed within a few days of fermentation (Phithakpol et al., 1995).

Plara (Plaa-raa, Plaa-ra)

Plara is popular among the Thai people especially in the north and northeast (Phithakpol et al., 1995). There are several kinds of freshwater fish normally used: *Channa striatus* or *Ophicepharus striatus* (striped snake-head fish; pla chon); *Puntius gonionotus* (silver barb, pla ta-pian); *Trichogaster trichopterus* (gouramy, *pla kra-dee*); *Cirrhina jullieni* (jullien's mud carp, *pla soi*); *Cyclocheilichthys* sp. (soldier river barb, *pla ta-kok*) and *Oreochromis niloticus* (tilapia, pla nin). Recently marine fish (*Rastrelliger neglectus* and *Rachycentron canadus*)—based *plaara* was also prepared (Sangjindavong et al., 2008).

Generally, freshwater fish are descaled, deheaded, eviscerated and washed with tap water and then rapidly dried. The prepared fish is mixed with salt in a fish to salt ratio of 3-5 : 1 by weight and then left at ambient temperature for 12-24 h before packing in an earthenware jar and left to ferment for 1 mon. Then, salted fish is added with roasted rice or rice bran in a ratio of 4-5 : 1 by weight of salted fish to roasted rice or rice bran. It is put in earthenware jars and left for at least 6 mon. The shelf life of *plara* is approximately 6 mon to 3 yr, depending on the fermentation period and good handling, such as tight packaging and occasionally mixing to exchange the upper and lower portions in the jar.

Plara can be consumed as it is by chopping into pieces and adding some vegetables; chillies, *Citrus hystrix* leaves, lemon grass or consumed after cooking by wrapping in banana leave, and roasting is also used for curries.

Hoi dorang

It is a high salt product in Thailand produced from sea mussel meat washed in brine (10% salt) and water. After drainage, sea salt is added (7 : 1; meat:salt ratio, w/w) and mixed well. The product matures in 4-5 d and is packed in sealed glass jars. It has a shelf-life of 3-6 mon (Phithakpol et al., 1995).

Pla-paeng-daeng (Red fermented fish)

It is prepared by fermenting fish and *ang-kak* rice (red rice) with the mould *Monacus purpureus* which gives it a red colour and specific flavour (Phithakpol et al., 1995).

Some of the fermented fish products are given in Figure 6.1.

2. Korea

Jeotkals

Jeotkals are the traditional Korean salted and fermented fish products and are popular as side dishes, and also as ingredients in preparing *kimchi* . To prepare *jeotkals*, salt should be added at the level of 5-20% to raw fish and then allowed to ferment for a long period of time to develop taste. The fermentation period varies depending on the salt concentration and fermentation temperature; 2 mon for most *jeotkals* with low salt level (6-18%). The favourable taste and flavour may develop gradually during fermentation by several enzymatic reactions and microbial degradation processes. Thus, fermented fish products contain relatively high amounts of amino acids, the degradation products of fish proteins. (Mah et al., 2002; Cheorun et al., 2004).

A. Plara

B. Som-fug (Plaa som)

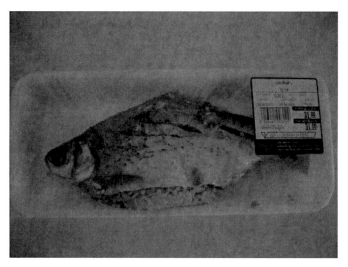

C. Som-fug (Plaa som)

D. Hoi dorang (Mussel)

E. Tilapia (Som-fug)

Fig. 6.1 Traditional fermented fish products of Thailand.

Color image of this figure appears in the color plate section at the end of the book.

3. The Philippines

Burong isda

This product prepared by fermenting cooked rice and freshwater fish (about 20%, w/v) is the typical food of central Luzon in the Philippines. Previously consumed as a condiment, it is now a main dish because of economic conditions. The fish is scaled, eviscerated, and filleted before mixing with cold cooked rice. Fermentation is carried out for 7-10 d at room temperature (Olympia et al., 1995).

Burong isda is available in two forms depending on the consumers' preferences in a particular area. One is called white *burong isda*, which has the natural product colour and the other is red *burong isda*, which is coloured by the addition of *angkak*, culture of *Monascus purpureus* grown on rice. Also, there are several varieties of *burong isda* based on the type of fish used in preparation. One example is *burong dalang*, a fermented rice-fish mixture using mud fish, *Ophicephalus striatu* (Olympia et al., 1995).

4. India, Bangladesh and Sri Lanka

Colombo cure

It is a fermented fish product of India and Sri Lanka. The intestines and gills are removed from mackerel or non-fatty sardines after which the fish are washed in drinking water. The fish are mixed with salt (3 : 1; fish: salt ratio, w/w) and put into jars. Dried fruit pulp or tamarind (a tropical fruit) is added to the salt and fish to lower the pH (8 kg of tamarind to 100 kg of fish). The fish are kept covered in brine with the help of weighted mats and are fermented for 2 to 4 mon. They are transferred to wooden barrels and care is taken to keep them covered with brine. The fermented fish can be kept for 1 yr (Prajapati and Nair, 2003; Lopetcharat et al., 2001).

Lona Ilisha

Lona ilish is a traditional salt fermented fish product widely consumed in Bangladesh and eastern and North-Eastern parts of India, prepared exclusively from *Hilsha* fish, *Tenualosa ilisha,* a high fat fish. It is prepared by dry salting the diagonally cut *hilsha* chunks followed by fermentation in saturated brine (previously brined and cooled) till desirable flavour and taste develop. The fermentation period is usually for 4-6 mon. The fish is kept in the fermentation medium (saturated brine) till consumption (Majumdar and Basu, 2010).

Very little is known on the biochemical composition and microbiology of *Lona ilish*. The moisture, salt and pH of the products available in the market averaged 54%, 15% and 5.7%, respectively. The bacterial flora comprised of *Micrococccus* and *Bacillus* species (Majumdar and Basu, 2010).

5. Myanmar

Pedah-siam

This Myanmar product is made of salted mackerel. During the preparation, the intestines are removed through the mouth. The fish are then salted, 3 kg of fish to 1 kg of salt, and stored for 24 h. Ripening takes place under anaerobic conditions. The brine formed is removed regularly. A red colour appears after ripening. (Berkel et al., 2004).

6. Japan

Sushi

Sushi is a group of preserved fish products in Japan which are formed through the addition of boiled rice to fermented fish and salt. The low pH which results from the growth of lactic acid bacteria contributes to the preserving effect. The general preparation is as follows. The intestines of the fish are removed and the fish is mixed with 20 to 30% salt. After being stored for 1 to 2 mon the fish are desalted and the liquid is removed. Boiled rice and *koji* (fermented wheat) are placed at the bottom of a basket and the desalted fish are alternated in layers with boiled rice or *koji*. The amount of boiled rice added is equal to 40 or 50% of the weight of the fish, the amount of *koji* is half the amount of boiled rice (rice: fish: *koj*i = 2 : 4 : 1). The fermentation continues for another 10 d. (Berkel et al., 2004).

Latin American Fish Products

Anchoa

*Ancho*a is a product found in a few South American countries, including Peru, Chile and Argentina. Whole anchovies are mixed with 35% fermenting fish salt and placed in barrels. The fermentation, a result of enzyme activity, takes place for a period of 3 to 4 mon (Berkel et al., 2004; Nilsson and Gram, 2002).

African Fish Products

In Africa, there are many dried and salted fish products (Table 6.4); nevertheless the two popular fish products are *momone* and *lanhouin*.

Momone

Momone is product from Ghana. In its general preparation, the intestines and gills of the fish are removed and the fish are washed in water. They are then rubbed with salt and packed in layers in barrels, alternating with layers of salt. The salt: fish ratio is 1:9. Fermentation takes place for 7 d. After that the fish are dried for 1 to 3 d on mats in the sun (Kerr et al., 2002; Berkel et al., 2004).

Lanhouin

Lanhouin is one such fermented fish used for flavour and is mostly produced in the coastal regions of the Gulf of Benin. *Lanhouin* is widely used as a condiment in urban and rural areas in southern Benin and in the neighbouring countries of Togo and Ghana. *Lanhouin* is produced by spontaneous fermentation of fish in salt. The fermentation of fish is brought about by autolytic enzymes from fish and microorganisms in the presence of salt (FAO, 2000; Anihouvi et al., 2007).

Table 6.4 Dried and salted fish of Africa

Region:	East Africa
Country:	Burundi
Product:	*Ndagala*
Raw Material:	*Limnothrissa miodon, Stolothrissa tanganicae*
Process:	The fish is sun-dried soon after harvest or purchase without any pre processing preparations. Slight fermentation takes place during sun-drying but this is pronounced when drying is delayed due to low ambient temperatures or during the rainy season.
Characteristics:	The product is hard, dry and brittle with a silvery colour. Fermentation is not the objective but it occurs during drying. The product can keep for about three months. It is susceptible to insect infestation and mould growth and therefore requires periodic re-drying to maintain a good quality.

Region:	Central Africa
Country:	Chad
Product:	Fermented dried fish (*salanga*)
Raw Material:	Alestes Nile perch, parch
Process:	Fish for fermentation may sometimes be of poor quality and unsuitable for smoking. The fish is scaled, gutted and washed. Larger species of fish are split dorsally and opened up. In one process, the dressed fish is dried immediately after washing and fermentation takes place during drying. In another, the fish is left to ferment for 12-24 hours before drying. The fish is dried on mats or grass spread on the ground. Fish may also be dried by hanging vertically or horizontally on drying lines, either by passing a stick through the head or by tying the fish with a thread along the stick. Drying takes about three to six days depending on the weather. Previous experimental trials to introduce salt curing were not successful due to the scarcity and high cost of salt.
Characteristics:	Fermented and sun-dried fish are light brown in colour with a dry, firm texture. Products made from partially spoiled fish have a characteristic strong smell. A well-dried product can be stored for at least three months. The product is susceptible to insect infestation and for this reason, some processors indiscriminately apply insecticides such as Lindane, Gammalin20, Shelltox, DDT and Gardona to the product. This practice is a health hazard to consumers.
Region:	West Africa
Country:	Côte d'Ivoire
Product:	*Adjonfa or gyagawere*
Raw Material:	Catfish, croaker, meagre, shark, mullet, skate, rays, triggerfish, horse mackerel, octopus, tuna, sole, Spanish mackerel, sea bream, herring, African moonfish
Process:	(a) **Semi-dried**: The fish is scaled, gutted and cut into pieces, when necessary, or split dorsally. It is then washed and dry-salted. The exudate from the fish is retained in the vats and the fish is immersed in this for fermentation from 12 hours to two days depending on the ambient temperatures. The fish may also be left to stand for about 12-24 houra before salting. At the end of the salting process the fish is removed and dried for two to five days to achieve a final product with a moisture content ranging from 45 to 55 percent.
	(b) **Hard-dried**: The raw fish is scaled and cut into pieces or split dorsally. It is washed and dry-salted for 12-24 hours. The fish is then removed and is sun-dried for two to seven days or until a fairly dry product with a moisture content of about 35-40 percent and salt level of 7-19 percent is obtained.
Characteristics:	(a) The semi-dry products usually have a very strong pungent smell and soft texture. They are susceptible to insect infestation and mould growth, with a shelf-life of up to two months.
	(b) Dry fermented fishery products have a mild smell and are very susceptible to fragmentation. Under humid storage conditions, mould growth occurs and so requires periodic redrying to maintain quality. These products can be stored for at least six months if properly dried.

Table 6.4 contd...

Region:	West Africa
Country:	Gambia and Senegal
Product:	*Guedj, tambadiang, yeet*
Raw Material:	(a) *Guedj*—Mackerel, sea bream, threadfin, croaker, mullet, catfish, meagre, herrings, skate, rays, shark
	(b) *Tambadiang*—Bonito
	(c) *Yeet*—Sea snails (gastropods)
Process:	(a) *Guedj*—The raw fish is often scaled, gutted and may or may not be split. It is then washed, salted and allowed to ferment for about two to three days. The salted fermented fish is dried on raised platforms for about three to five days. In another process the raw fish is dressed, washed and left overnight to ferment before salting for 12-24 hours and drying.
	(b) The fish is washed and then placed in concrete vats with alternate layers of salt. It is left to ferment for one to three days before being des-caled, washed and dried for three to five days on raised platforms.
	(c) The flesh is removed from the shell, separated from the guts and split into two or four parts. This is placed in fermentation tanks, jute bags or polysacks and allowed to ferment for two to four days before being washed and dried on raised platforms for two to four days.
Characteristics:	(a) *Guedj* is a semi-dry product with a strong pungent smell and a light brown colour. This product is susceptible to insect infestation, mould growth and fragmentation when poorly handled. It requires redrying to avoid spoilage and has a shelf-life of about three months if well dried.
	(b) *Tambadiang* is a product with a greyish appearance and mild smell. It requires occasional redrying to ensure a longer shelf-life.
	(c) *Yeet* is a semi-dry light brown product with a strong smell. It can be stored for about two months and is susceptible to a lot of mould growth. Redrying results in a hard product which is dark brown in colour.
Region:	West Africa
Country:	Ghana
Product:	*Momone, kako, koobi, ewurefua* (triggerfish)
Raw Material:	(a) *Momone*—Various species of fish such as catfish, barraccuda, sea bream, threadfin, croaker, grouper, bonito, mackerel, herrings, squid, octopus, bumper, snapper, ribbon fish
	(b) *Kako* and *Koobi*—Shark, skates, ray (for kato) and tilapia (for koobi)
	(c) *Ewurefua*—Triggerfish
Process:	The fish may or may not be scaled, gutted and have the gills removed.
	(a) *Momone*—Larger species of fish (e.g., shark) may be cut into smaller pieces or split dorsally. The dressed fish is thoroughly washed with sea or fresh water. The raw fish is either left overnight before salting or dry salted immediately after washing. Salting and fermentation lasts for one to six days after which the fish is dried on the ground, grass, nets, stones or raised platforms for one to three days.
	(b) *Koobi*—The raw fish is dressed and thoroughly washed. Dry salt is then rubbed into the gills, the belly cavity and on the surface. It is then arranged in alternate layers with salt and is allowed to ferment for two to three days before being dried for two to four days. More salt may be sprinkled on the fish during drying.

	(c) *Ewurefua*—The fish is gutted by making a slit from the belly to the lower jaw. It is then washed and transferred into brining tanks containing a small quantity of saturated brine from previous salting or from sea water. Dry salt is poured over the fish and stirred into the brine to ensure it comes into contact with the fish. The tank is then covered with jute sacks or polythene sheets and weighted down with stones or a wooden board and allowed to ferment for 12-24 hours. After fermentation, the fish is removed and is sundried for two to three days, either on the ground, grass, nets or raised racks.
Characteristics:	(a) *Momone*—This is a soft product with a very strong pungent and sometimes offensive smell. It is susceptible to larvae infestation (maggots), mould growth and bacterial spoilage especially if the salt level is low. It can be stored for at least two to three months.
	(b) & (c) *Koobi/Kaka/Ewule*—These are dried products with a mild to strong odour. They are susceptible to fragmentation, insect infestation and mould growth. For long-term storage (four to six months) they have to be redried periodically in order to maintain quality.
Region:	**West Africa**
Country:	Mali
Product:	Fermented dried fish, *djege, djadan*
Raw Material:	Tilapia, Clarias spp., Alestes spp., Schilbe spp. and Hydrocynus spp.
Process:	The fish is de-scaled and, with larger species, split open dorsally and headed. For smaller species (clupeids), the fish is dried immediately after washing and fermentation occurs during drying. Medium and larger species are put into water in an earthenware pot or oil drum and allowed to ferment for 12 hours. The fish is then removed and immersed in a solution of Gardona or K-Othrine for a few minutes to prevent it from being attacked by blowflies during drying. The fish is then dried on grass or mats for three to seven days depending on weather conditions.
Characteristics:	Fermented and sun-dried fish are light brown in colour with a mild fermented odour. During storage, they are arranged on a platform and covered with a mat. The outer surface of the mat is first treated with the insecticide malathion to prevent insect attack. A well-dried product can be stored for more than six months.
Region:	**Central Africa**
Country:	The Sudan
Product:	*Fessiekh, terkeen/mindeshi, kejeick*
Raw Material:	(a) *Fessiekh*—*Alestes* spp. and *Hydrocynus* spp.
	(b) *Terkeen/mindeshi*—Tilapia, Nile perch, *Alestes* spp.
	(c) *Kejeick*—Mixed fish, usually small species of *Alestes* spp. and *Hydrocynus* spp.

Table 6.4 contd...

Table 6.4 contd...

Process:	(a) *Fessiekh*—Processing of fessiekh is typically carried out in temporary sheds (*batarikh*) to provide shade or a cool environment. *Alestes* spp. is the fish species commonly used to produce fessiekh. Whole, fresh fish is washed, covered with salt and arranged in alternate layers with salt either on a mat, in a basket or in perforated drums to ferment for three to seven days depending on ambient temperatures. Liquid exudate from the fish is allowed to drain off. After this period the fish is transferred into larger fermentation tanks where more salt and new batches of fish are added. The fermentation tanks are covered with jute sacks or polythene sheets and weights are placed on top to press the fish. Fermentation is carried out in sheds at temperatures ranging from 18°C to 20°C. The fish is allowed to ferment for a further 10-15 days after which they are transferred into vegetable oil cans or kerosene cans (1015 kg) and sealed for export. The cans are covered with polythene sheets before closing. Fessiekh is also packed in polythene bags (2-5 kg) for sale in local markets and shops.
	(b) *Terkeen/Mindeshi*—The fish is washed and salted by arranging it in alternate layers with salt in an earthenware pot or barrel. Pepper or other spices may be added. The salt to fish ratio is 1:5. Pickling and fermentation takes place for 9-15 days after which the fermentation vat or pot is placed in the sun or near a fireplace to speed up fermentation. This process continues for about four days and the fish is stirred daily in order to break up the fish into a pasty mixture of muscle and bones which has a dark colour and strong odour.
	(c) *Kejeick*—Large fish species are gutted, split dorsally, scaled or headed and washed. Smaller species of fish are dried whole or split at the belly to remove the guts. In northern Sudan, the fish is dried immediately after it has been washed by hanging it on sticks or laying it on stones, grass or mats. In the south, the fish may be sprinkled with salt or dipped into a strong salt solution and fermented for two to three days before drying for up to one week.
Characteristics:	(a) *Fessiekh* is a wet salted product, soft in texture with a strong pungent smell and a shiny silvery appearance. It can be stored for more than three months. Spoilage is often due to poor quality raw fish especially if the fish is damaged through poor handling. A processing temperature above 25°C also results in a poor product. Fessiekh is highly susceptible to microbial attack and maggot infestation.
	(b) *Terkeen/Mindeshi*—This is a wet pasty mixture of fish muscle and bones which loses moisture during storage and becomes more viscous and dark.
	(c) *Kejeick* is a hard dried product and the colour depends on the species of fish used. It can be stored for about three months or more. The product is susceptible to insect infestation especially if it absorbs moisture. If fatty fish is used, there is rancidity due to lipid oxidation which adversely affects the flavour of the product.

Region:	East Africa
Country:	Uganda
Product:	(a) Salted dried tilapia, *Alestes* spp., Nile perch, *Hydrocynus* spp.
	(b) Sun-dried *Haplochromis* spp., sun-dried *Rastrineobola* spp. (dagaa)
Process:	(a) The fish may be gutted, split dorsally or headed and washed. Smaller species may be processed whole or split at the belly to remove the guts. The fish may be dried without salting, but when salt is used only small quantities are sprinkled on the surface. It is left to stand for three to six hours and then dried for three to seven days. Fermentation occurs during salting and drying. The fish may be dried on rocks, mats, the ground or, occasionally raised platforms.
	(b) *Haplochromis* spp. are usually dried by passing a stick through the eyes of 10-12 individual fish. Ten such sticks of fish are joined to form a mat which is then hung in the open air for the fish to dry. Rastineobola spp. are spread on the ground or a mat to dry. The drying process takes three to five days. Fermentation occurs during drying.
Characteristics:	(a) The products may be very dry or semi-dry with a mild to moderately strong smell. They are susceptible to insect infestation and mould growth and require periodic drying to maintain quality. They can be stored from one to three months.
	(b) The final products are greyish in colour, with a very hard and dry texture. They have a very mild smell which becomes pronounced if drying is inadequate. They are brittle and susceptible to insect infestation with a high incidence of fragmentation. Contamination with sand is a major problem when the fish is dried on the ground.

Source: Essuman, 1992.

MICROBIOLOGY AND BIOCHEMISTRY OF FERMENTED FISH PRODUCTS

Microbiology

General Studies

Fish in its natural environment has its own microflora in the slime on its body, in its gut and in its gills. These microorganisms (Table 6.5), as well as the enzymes in the tissues of the fish, bring about putrefactive changes in fish when it dies. Furthermore, the microorganisms generally present in the salt used for salting also contribute to the degradative changes in the fish. Microorganisms require water in an available form for growth and metabolism. Halophiles grow optimally at high salt concentrations but are unable to grow in salt-free media. Halotolerant organisms grow best without significant amounts of salt but can also grow in concentrations higher than that of seawater. Xerophiles are those organisms which grow rapidly under relatively dry conditions or below A_w of 0.85, while osmophiles can grow under high osmotic pressure. Most food-borne bacterial pathogens are not able to grow in an A_w range of 0.98-0.93 (Table 6.6).

Table 6.5 Genera of bacteria most frequently associated in fish and seafood (Lopetcharat et al., 2001)

Genus	Gram Reaction	Frequency
Acinetobacter	−	x[b]
Aeromonas	−	x
Alcaligenes	−	x
Bacillus	+	x
Corynebacterium	+	x
Enterobacter	−	x
Enterococcus	+	x
Escherichia	−	x
Flavobacterium	−	x
Lactobacillus	+	x
Listeria	+	x
Microbacterium	+	x
Moraxella	+	x
Psychrobacter	−	x
Shewanella	−	xx[c]
Vibrio	−	x
Pseudomonas	−	xx

[b]x indicated known to occur
[c]x indicates most frequently reported
Source: Lopetcharat et al. (2001).

Table 6.6 Growth of microorganisms in salted fish (Gram and Huss, 2000)

Water Activity (A_w)	Sodium Chloride Concentrate (%)	Microorganisms	
		Pathogens	Spoilage Organisms
0.98	< 3.5	All known food-borne	Most micro-organisms pathogens of concern in foods particularly the gram negative rods
0.98-0.93	3.5-10	*Bacillus cereus, Clostridium botulinum, Salmonella* spp., *Clostridium perfringens, Vibrio parahaemolyticus*	*Lactobacilliaceae, Enterobacteriaceae, Bacilliaceae, Micrococcaceae*, moulds
0.93-0.85	10-17	*Staphylococcus*	Cocci, yeasts, moulds
0.85-0.60	> 17	Mycotoxic, xerophilic moulds (no mycotoxin is produced at A_w less than 0.80)	Halophilic bacteria, yeasts, moulds (dun = *Wallemia sebi*)

Source: ICMSF, 1980a.

Various types of salts are used for salting and fermentation of fish. They include solar salt, rock salt and vacuum salt and have their own microflora. Solar salt, which is the most widely used in fish curing, has been found to contain the largest amount of microorganisms. The general bacterial flora of solar salt mostly comprises *Bacillus* types (75%) with the remainder being *Micrococcus* and *Sarcina* types. The most important spoilage organisms always present in solar salt are the red halophilic bacteria. Krieg and Holt (1984) noted that the red halophiles belong to two genera of bacteria, namely *Halobacterium* and *Holococcus*. *Halobacterium* consists of rod-shaped bacteria and requires at least 10-15% salt concentration for growth while *Halococcus* can thrive at 5-10% salt concentration.

Both genera are strictly aerobic and grow optimally at about 37°C and also produce red carotenoid pigments. In situations where brine is reused a number of times, the chemical composition of the salt solution is altered. Significant amounts of organic material are introduced and the bacterial load of the brine becomes extremely high, especially the red halophiles and the osmophilic moulds. Two common defects of salted fermented fishery products called pink and dun are the result of spoilage by red halophilic bacteria and a highly osmophilic fungus respectively. The red halophilic bacteria grows in brine solutions at temperatures ranging from 15°C to 55°C. Table 6.5 shows that only a few pathogenic organisms can proliferate at salt concentrations higher than 10%. However, it is known that many of these organisms survive in saturated salt solutions. For instance, *Typhus* bacteria can survive in saturated salt solutions for 3 to 6 mon, salmonella in 10% salt solutions for 1 to 3 mon. *Escherichia coli* and *Staphylococcus aureus* can survive for many weeks in salted fish (ICMSF, 1980b).

In general, the types of microorganisms identified in fermented fish products are:

1) Gram-negative aerobic halophilic rods (70%)
2) Gram-positive aerobic halotolerant cocci (20%)
3) Yeasts (3%).

Microorganisms in Fish Sauce and Paste

Fish sauces contain high concentrations of salt (25-30% NaCl, w/v), thus microorganisms found during sauce production are generally classified as halotolerant or halophilic bacteria (Tanasupawat et al., 2009). Earlier studies have indentified the following microorganisms from samples of *nam-pla*; mostly *Bacillus* spp.; *Bacillus cereus*, *B. circulans*, *B. licheniformis*, *B.megaterium*, *B. pumilus* and *B. subtitils* (Crisan and Sands, 1975). The important roles of bacteria in fish sauce/paste are protein degradation and flavour-aroma development. Bacteria involved in fish sauce and paste can be classified into two major groups (Lopetcharat et al., 2001).

- Bacteria that produce proteolytic enzymes. These include *Bacillus, Pseudomonas, Micrococcus, Staphylococcus, Halococcus, Halobacterium.* Highly concentrated NaCl (25%) does not have any effect on the proteolytic activity of enzymes from *Halobacter.*
- Bacteria that relate to flavour and aroma development. Ten out of 17 *Bacillus* type isolates produced a measurable amount of volatile acids in *nam-pla. Staphylococcus* strain 109 also produced a significant amount of volatile acid in *nam-pla.*

The *patis* liquid contained strains *Bacillus pumilus, Micrococcus colpogenes, M. varians* and *Candida clausenii* (Crisan and Sands, 1975). The *patis* residue contained single strains of *Bacillus coagulans, B. licheniformis* and *Achromobacter thalassius.* Fungi and obligatory anaerobic bacteria were absent (Crisan and Sands, 1975).These organisms are halo-tolerant rather than halophiles, as they could grow on 10% NaCl, not beyond that level (Crisan and Sands, 1975). When microbiological changes during *bakasang* processing were monitored, a variety of bacteria grew during the first 10 d of fermentation; however after 20 d, *Streptococcus, Pediococcus, Micrococcus* were dominant (Ijong and Ohta, 1996). The microflora found from Korean anchovy sauce in the final stage of fermentation included *Bacillus cereus, B. megaterium, B. pumilus, Pseudomonas halophilus* and *Serratia marcesens* (Sands et al., 1974).

In some recent studies, strains of aerobic, spore forming, gram positive, moderately halophilic bacteria were isolated from *nam-pla* and *budu.* They grow optimally in the presence of 10% NaCl at 37°C and pH 7.0. Three strains were identified as *Lentibacillus salicampi* and the remaining strains were proposed as *Lentibacillus juripiscarius* based on 16s rRNA-based phytogenetic analysis (Namwong et al., 2005). Some other strains isolated from fish sauce in Thailand were: *Lantibacillus halophilus, Halococcus thailandensis. Filobacillus* sp. RP 2-5, *Piscibacillus salipiscarius, Tetragenococcus halophilus, T. muriaticus, Halobacterium salinarum* (Namwong et al., 2007; Tanasupawat et al., 2006, 2007, 2009).The bacteria were identified based on phenotypic and chemotaxonomic characteristics including DNA-DNA relatedness and phylogenetic properties. Likewise, *Lentibacillus kapialis* sp. Nov. and *Oceanobacillus kapialis* were isolated from fermented shrimp paste in Thailand (Pakdeeto et al., 2007; Namwong et al., 2009).

A total of 67 microbial strains were isolated from *momoni* (Ghanian fermented fish sauce) obtained from retail outlets. The strains belonged to nine genera of microorganisms, i.e., *Bacillus, Lactobacillus, Pseudomonas, Pediococcus, Staphylococcus, Klebsiella, Debaryomyces, Hansenula* and *Aspergillus,* with *Bacillus* having a predominant occurrence of 37.7% (Sanni et al., 2002).

Microorganisms in Fermented/Salted Fish

Most fermented fishery products are made from fatty fish. Lean fish has sometimes been noted to give a less acceptable texture and flavour. The role of fats in the fermentation process has not, however, been studied in any detail. Fish oils are highly unsaturated and hence very prone to oxidation. Certain pro-oxidants, such as *haem*, in the proteins catalyze the oxidation reaction. Similarly, iron impurities in the crude solar salt used for curing also accelerate auto-oxidation (Saisithi, 1967). Oxidized fish oils have a characteristic taste and paint-like smell, but the acceptability of products having the typical taste and flavour of oxidized fats depends very much on local preferences. The products of fat oxidation take part in further reactions especially with amines (Saisithi, 1967) and with other decomposition products of proteins (Bal and Dominova, 1967) to produce coloured compounds as well as substances with odour (Jones, 1966). Lipases present in the fish flesh also hydrolyze the lipids (Lovern, 1962), but the extent is dependent on the level of salting and fermentation (Amano, 1962; Lee et al., 2008).

Pathogens rarely multiply at high salt concentrations; however, Karnop (1988) demonstrated that *Pediococcus halophilus* is able to produce histamine during long storage at ambient temperatures of 20 to 25°C. Toxins produced by *Clostridium botulinum* in poor quality fish before salting may be stable in the salted product (Huss and Rye-Petterson, 1980).

A study by Knochel and Huss (1984) on the microbiology of barrel salted herrings revealed that both aerobic and anaerobic viable counts (in media containing 15% NaCl) were low, i.e. not more than 3 x 105/g of fish. Villar et al. (1985) also showed that *Pediococcus halophilus* is the dominating organism in salted anchovies. Other halophilic bacteria isolated from fermented fish products were: *Lentibacillus jeotgali* (Korean fermented seafood) (Jung et al., 2010), *Gracilibacillus thailandensis* and *Salinivibrio siamensis* (from *Plara*) (Chamroensaksri et al., 2009, 2010), *Piscibacillus salipiscarius* (from *Plara*) (Tanasupawat et al., 2007). Anihouvi et al. (2007) studied the microbiological changes in naturally fermented cassava fish (*Pseudotolithus* sp.) for *lanhouin* fermentation. A total of 224 isolates belonging to the genera *Bacillus*, *Staphylococcus*, *Micrococcus*, *Streptococcus*, *Corynebacterium*, *Pseudomonas*, *Achromobacter* and *Alcaligenes* were isolated from the fermenting fish samples.

Lactic acid bacteria (LAB) are also found as the dominant microorganisms in many fermented fish products (Paludan-Muller et al., 2002). The primary role of LAB is to ferment the available carbohydrates and thereby cause a decrease in pH. The combination of low pH and organic acids (mainly lactic acid) is the main preservation factor in fermented fish products. In *plaa-som*, major LAB isolated were *Pediococcus pentosaceus*, *Lactobacillus alimentarius*, *L. planatrum*, *L. garviae* and *Weisella confusa* (Paludan-Muller et al., 2002), 42% isolated strains were *P. pentosaceus*. In *som-fak* (a Thai low-salt fermented fish product), a succession of aciduric, homo-fermentative lactobacillus species, dominated by *Lactobacillus plantarum, L. pentosus, Lactococcus lactis, Leuconostoc citreum* was found during fermentation (Paludan-Muller et al., 1999). Further, at the start of fermentation, *Leuconostoc* spp., *Leuconostoc lactis* subsp. *lactis* and *Lactobacillus brevis* were dominant, followed by more acid tolerant species of *L. planatrum*, the latter dominating the LAB microflora towards the end of fermentation.

Biochemistry

General Studies

Several studies have been carried out to study the biochemical pathways followed during the degradation process of fish fermentation. Pearson (1970) identified the following five chemical changes in deteriorating fish:

- Enzymic degradation of nucleotides and nucleosides in the flesh leading to the formation of inosine, hypoxanthine, ribose, etc.
- Bacterial reduction of trimethylamine oxide (TMAO), a non-volatile and non-odouriferous compound, to volatile trimethylamine (TMA) which has an amoniacal smell
- Formation of dimethylamine (DMA)
- Breakdown of protein with subsequent formation of ammonia (NH_3) indole, hydrogen sulphide, etc.
- Oxidative rancidity of the fat.

In the degradative changes occurring during fermentation no significant changes were observed in the amino acids particularly the essential ones. The degradation process, however, brings out certain characteristic flavours which are essential for the quality of the final product.

Hiltz et al. (1976) reported that the volatile bases particularly TMA, DMA and NH_3, are associated with changes in the organoleptic and textural quality of fish. The development of a specific aroma in fermented fish sauces and pastes may not be due to the action of microorganisms. Adams (1986) advocated that microorganisms play little or no part in aroma production. Beddows (1985) isolated halotolerant organisms, *Bacillus* spp. and used them in pure culture but none of them produced the typical fish sauce aroma.

Biochemistry of Fish Sauce and Paste

Amino acids are considered the major contributors to the taste of fish sauce (Lopetcharat et al., 2001). The flavour and aroma of fish sauce is thought to arise in part from glutamic acid, histidine and proline (Saisithi et al., 1996; Raksakulthai and Haard, 1992). For example, glutamic acid and lysine were found as predominant amino acids in *bakasang*, a fish sauce from Indonesia (Ijong and Ohta, 1995). Similarly, glutamic acid was the predominant amino acid in *momoni* (Sanni et al., 2002). Anti-oxidant activity has also been found in a number of fermented fishery products such as fermented blue mussels (Jung et al., 2005), fish sauces (Harada et al., 2003; Michihata, 2003) and fermented shrimp paste (Peralta et al., 2005). In a recent study, the antioxidant activity and nutritional components of the Philippine salt-fermented shrimp paste were improved through prolonged fermentation (90, 180, and 360 d). The antioxidant ability against 1, 1-diphenyl-2-picryhydrazyl (DPPH) radical, hydrogen peroxide, and lipid peroxidation increased significantly with prolonged fermentation. Polyunsaturated fatty acids like eicosapentaenoic acid (EPA) and docosahexaenoic acid (DHA) in the shrimp paste were not substantially damaged for 360 d, while free amino acid content dramatically increased at 90 d. These results suggest that properly prolonged fermentation would improve antioxidant ability and some nutritional value in the salt-fermented shrimp paste (Peralta et al., 2008).

NUTRITIONAL AND FUNCTIONAL EVALUATION OF FERMENTED FISHERY PRODUCTS

The primary objectives of fermenting fish are to preserve the fish and develop a desirable flavour. However, processing sometimes tends to affect the nutritional value of food products. Table 6.7 shows the proximate analysis of different types of fresh and cured fishery products. Considering fish as a major source of protein, a general observation of Table 6.6 shows that curing (smoking, fermentation and salting) does not adversely affect the crude protein content of fishery products. The moisture content of fermented fish varies from about 12% in smoked dried anchovies to 65% in wet salted cured fish (*momone*). e salt level is low. The protein content of fermented fish ranges from about 20% to nearly 50% depending on the water content. This makes the product a good source of animal protein.

Table 6.7 Proximate compositions of raw and cured fish

Type of Fish	Energy in Calories	Moisture %	Protein %	Fat %	Ash %
Anchovy					
Fresh	92	73.8	18.4	1.5	7.3
Dry smoked	380	11.9	68.6	3.5	16.0
Sun-dried	308	16.2	58.4	4.0	18.5
Bigeye Snapper (*Priacanthus tayenum*)					
Fermented (*somfug*)		70-89	11.4-16.0	1.2-1.6	2.5-3.6

Table 6.7 contd...

Table 6.7 contd...

Type of Fish	Energy in Calories	Moisture %	Protein %	Fat %	Ash %
Cassava Croaker (*Pseudotolithus senegalensis*)					
Fresh	104	78.6	18.4	2.8	1.0
Dried, salted (*kako*)	161	43.9	34.9	1.3	20.8
Fermented (*momone*)	134	57.6	26.2	2.4	14.6
Grey Snapper (*Lutjanus agennes*)					
Fresh	101	76.4	14.4	1.5	1.1
Fermented (*momone*)	-	64.9	18.2	1.4	-
Scad (*Decapterus rhonchus*)					
Fresh	125	72.5	22.6	2.8	1.3
Hot smoked	212	56.9	33.9	7.4	2.4
Fermented	-	54.5	18.9	-	1.3
Seabream (*Pagrus* spp., *Dentex* spp.)					
Fresh	104	77.3	18.3	2.7	1.2
Smoked	179	61.7	29.1	4.7	1.3
Salted, fermented, dried	168	62.5	27.6	4.1	2.5
Threadfin (*Galeoides decadactylus*)					
Fresh	113	73.9	18.9	2.0	1.5
Fermented	149	55.9	28.7	2.9	16.8
Triggerfish (*Balistes* spp.)					
Salted, fermented, dried	196	40.1	40.7	2.5	22.5
Raw	-	79.1	19.4	0.3	1.4
Squid (*Sepia* spp.)					
Salted, dried	182	42.1	38.1	2.1	16.5
Tiger Fish (*Hydrocynus* spp.)					
Smoked	369	15.4	73.3	6.2	5.1
Salted, fermented, sun-dried	157	43.1	32.6	2.0	22.3
Tilapia					
Fresh	123	73.4	16.6	5.8	6.8
Salted, fermented, sun-dried (koobi)	189	39.9	38.1	0.6	16.1
Smoked	368	20.9	67.5	8.8	2.7
Unsalted, fermented, sun-dried	357	12.6	71.9	3.1	6.8

Source: Riebroy et al., 2007.

HEALTH HAZARDS AND HYGIENIC ASPECTS OF FERMENTED FISH PROCESSING

Processing practices for fermented fish products constitute health hazards to consumers, processors and the environment. These practices relate to dressing, washing, salting/fermentation, drying and waste disposal of the fish, as well as the use of poor quality salt and curing containers.

Dressing

It was observed in African countries such as Ghana and Senegal that fish may be held under the foot on the ground during dressing. This practice can cause microbial contamination of the fish and injury to the processor either from the spines of the fish or by the knife (Essuman, 1992).

Washing

This is a very essential stage in fish processing. In most of the processing sites potable water is often not available. Therefore, water from lagoons, rivers, lakes or seas is used to wash the fish. These water bodies are often polluted by domestic waste, making them a possible source of chemical and microbial poisoning (Essuman, 1992).

Salting/Fermentation

The non-use of salt results in uncontrolled fermentation. Under such conditions, the fish muscle becomes ideal for the growth of pathogenic organisms and the product may decay within a short period. Such products could pose a health hazard to consumers. The reuse of salt may also lead to contamination of fresh batches of fish with microorganisms (Essuman, 1992).

Drying

It is generally observed that fish is often dried on the ground except in commercial practices where raised drying racks are used. Drying fish on the ground is a source of contamination with sand and microorganisms (UNIFEM, 1988).

Waste Disposal

At most of the processing sites in the countries of the survey, and indeed in many other African countries, there are no efficient ways of disposing the offal and other waste matter. These items are indiscriminately thrown away, thus polluting the environment and the available water body as well as serving as suitable breeding grounds for flies and mites which eventually infest the cured fishery product (Essuman, 1992).

Salt Quality

Solar salt which is predominantly used for curing is often dirty and stored in heaps or in sacks on the bare ground without being covered. Solar salt is a major source of halophilic bacteria which causes poor quality fermented fishery products notably "pink" (Brennan, 2006).

QUALITY EVALUATION OF PRODUCTS

The quality of fermented fish is assessed subjectively by visual and/or organoleptic inspection. The main quality parameters are texture, colour, odour and fragility.

Texture

Two main types of textural characteristics are usually identified in fermented/salted fish products: (1) Fermented fish intended to be used as food fish is hard dried or semi-dried but firm. This is the main type of fermented product found in many African countries (e.g., *guedj, koobi, ewule*). Due to the firm texture it remains whole in the sauce after cooking. (2) In Côte d'Ivoire, Gambia, Ghana and also in the Sudan, fermented fishery products with a soft texture are produced. These products are commonly added to soups and sauces in small quantities as a condiment. *Fessiekh* is used both as a food fish and condiment. It is either minced, mashed before adding to the vegetable sauce or broken up completely during cooking (Abdalla, 1989).

Colour

The colour of the product depends on the species of fish used as well as the processing method. For whole products such as *fessiekh*, a silvery appearance close to the fresh product is considered high quality (Essuman, 1992). Poorly fermented products tend to be greyish or dark. Long storage and further drying darkens the product. *Yeet*, for instance, becomes dark brown after weeks of exposure to the sun (Essuman, 1992).

Odour

The odour of fermented fishery products varies from mild to very pungent. Soft, semi-dry products usually have a strong smell but very dry fermented fishery products have a mild odour (Steinkraus, 1996).

Fragility

When fermented fish is dried to very low moisture content, it tends to break up during packaging and storage. Soft or semi-dry products become flaky especially if the raw fish is putrid or fermentation is unduly prolonged. *Momon*e, for example, breaks up if poorly cured (Fellows and Hampton, 1992).

Shelf-life

The shelf-life of fermented fish is an important quality characteristic. At high moisture content or low salt levels, insects tend to lay eggs on the product which eventually develop into maggots and destroy the fish. Dry fish can be stored for nearly 6 mon, but the soft or semi-dry ones have a shelf life of up to 3 mon (Ebeling, 2002).

FOOD SAFETY

Hygiene of Processing

Boats, premises, fishermen, processors, equipment, water, fish and ingredients invariably have low hygienic standards at the artisanal level of fish handling and processing. This may be due to lack of economic support from the quality products' market.

Biogenic Amine

Biogenic amine including tyramine, histamine, cadaverine, spermine and spermidine are produced during fermentation of fish and in fermented products (Mah et al., 2002). These biogemine amines were detected in salted and fermented fish products, *jeotkal*s in the range of 0-75 mg/ kg (Mah et al., 2002) and in Thai fermented product, *som fug* (Riebroy et al., 2004).

Histamine poisoning has historically been referred to as scombroid poisoning because of the frequent association of the illness with the consumption of spoiled scombroid fish such as mackerels and tuna (Tsai et al., 2004, 2005).

Histamine production in foods is by the decarboxylation of histidine through a reaction catalyzed by the enzyme histidine decarboxylase (Ababouch et al., 1991a) produced by microorganisms such as *Clostridium, Morganella morganii, Klebsiella pneumoniae, Proteus vulgaris, Hafnia alvei* and *Lactobacillus* (Tsai et al., 2006). Also, various species of fish are known to have large amounts of free histidine in their muscle tissues as substrate for histidine decarboxylase (Ababouch et al., 1991b).

Clostridium Poisoning

Since the fermentation process involves bacterial activity, it is likely that if conditions are not properly set to control pathogenic bacteria they may remain in the fermented products. This is because fermentation takes place at low temperatures. The two main factors which control the growth of pathogens such as *Clostridium botulinum* are high salt concentration and pH. All types of *C. botulinum* are inhibited by 10-12% salt and a pH below 4.5. *C. botulinum* types E, F and non-proteolytic type B are able to grow between 8°C and 10°C but are inhibited below 4°C. From observations of the production methods of fermented fishery products, the low level of incidence of *C. botulinum* poisoning may be mainly attributed to the high level of salt usage (Wiriyacharee, 1992).

Salmonella Poisoning

There is very little information on *Salmonella* food poisoning arising from the consumption of fermented fish despite the unsanitary fish processing practices observed in many countries. In a study conducted by Nerquaye-Tetteh et al. (1978) to isolate various microorganisms, no *Salmonella* spp. were isolated from samples of fermented fishery products obtained from the open markets in Ghana. The absence of *Salmonella* from fermented fishery products could be attributed to the high salt level and low water activity of the products.

Mould Infestation

Moulds are able to grow under dry conditions better than bacteria. For this reason, moulds are often associated with dried fermented fishery products. The moulds commonly associated with dried cured fish in storage are *Aspergillus halophillus, A. restrictus, A. glaucus, A. candidus, A. ochraceus, A. flavus* and *Penicillum* spp. (www.fao.org/docrep/t0685e/T0685E04.htm).

Insect and Mite Infestation

In the early stages of curing, raw fish is particularly susceptible to infestation from blowfly maggots (larvae). Fully or partially cured fishery products are also attacked by a wide range of insect pests which include species of flies, mites and both adults and larvae of certain beetles (Essuman, 1992).

CONCLUSIONS AND FUTURE PERSPECTIVES

Fermented fish, fish products and manufacturing methods vary from country to country due to culture, social and geographical position and weather/temperature. In fermented fish sauce and paste, volatile compounds contribute a unique aroma and flavour and are developed during fermentation. Amino acids (glutamic acid and aspartic acid), peptides, nucleotides and organic acids (succinic acids) also contribute to the taste of fish sauce and paste. In general, fish type, salt and fish: salt ratio, oxygen level and minor ingredients such as garlic and chilli influence the taste and quality. In comparison to the number of studies made on fish sauce and paste, less research has been conducted on fried/salted fish. Unlike Southeast Asian products, fermented fishery products in Africa usually remain whole and firm after processing. The major issues of safety of fish as well as fermented fish products include histamine poisoning, Salmonella and Clostridium poisoning, aflatoxin contamination, etc., which need serious attention.

ABBREVIATIONS

DHA   Docosahexaenoic acid
DMA   Dimethylamine
DPPH   1, 1-diphenyl-2-picryhydrazyl
EPA   Eicosapentaenoic acid
LAB   Lactic acid bacteria
TMA   Trimethylamine
TMAO   Trimethylamine oxide

REFERENCE

Ababouch, L., Afilal, M.E., Rhafiri, S. and Busta, F.F. (1991a) Identification of histamine-producing bacteria isolated from sardine (*Sardina pilchardus*) stored in ice and at ambient temperature (25°C). Food Microbiol. 8: 127-136.

Ababouch, L., Afilal, M.E., Rhafiri, S. and Busta, F.F. (1991b). Quantitative changes in bacteria, amino acids and biogemine amines in sardine (*Sardina pilchardus*) stored at ambient temperature (25-28°C) and in ice. Int. J. Food Sci. Technol. 26: 297-306.

Abdalla, M.T. (1989). Microbiology and biochemistry of *fessiekh* fermentation. M.Sc. Thesis in Food Microbiology. Faculty of Agriculture, University of Khartoum, Sudan.

Adams, M.R. (1986). Fermented flesh foods [sausages, fish products]. Progr. Indust. Microbiol. 23: 159-98

Amano, K. (1962). The influence of fermentation on the nutritive value of fish with special reference to fermented fish products of Southeast Asia. In: Fish in Nutrition. E. Heen and R. Kreuzer (eds). Fishing News (Books) Ltd., London, UK pp. 180-200.

Anihouvi, V.B., Sakyi-Dawson, E., Ayernor, G.S., and Hounhouigan, J.D. (2007). Microbiological changes in naturally fermented cassava fish (*Pseudotolithus* sp.) for *lanhouin* production. Int. J. Food Microbiol. 116 (2) 10: 287-291.

Badham, C.D. (1854). Prose Halieutics or Ancient and Modern Fish Tattle, Parker and Sons, London, UK.

Bal, V.V. and Dominova, S.R. (1967). Changes of fish fat during curing. II. Interaction of fish lipids and protein decomposition products. Isv. Vyssh Uched.Zaved Pishch. Tekhnol. (2): 36.

Beddows, C.G. (1985). Fermented fish and fish products. In: Microbiology of Fermented Foods, Volume 2. B.J.B. Wood (ed). Elsevier Applied Science, London, UK pp. 1-39.

Berkel, B.M.V., Boogaard, B.V.D. and Heijnen, C. (2004). Salting fish and meat. In: Preservation of Fish and Meat, 3rd edn. Marja de Goffau-Markusse (ed). Agromisa Foundation, Wageningen, The Netherlands pp. 25-36.

Brennan, J.G. (2006). Food Processing Handbook, Wiley-VCH, Verlag GmbH & Co., KGaA, Weinheim, Germany pp. 4-5.

Chamroensaksri, N., Tanasupawat, S., Akaracharanya, A., Visessanguan, W., Kudo, T. and Itoh, T. (2009). *Salinivibrio siamensis* sp. nov. from fermented fish (*Plara*) in Thailand. Int. J. Syst. Evol. Microbiol. 59: 880-885.

Chamroensaksri, N., Ttanasupawat, S., Akaracharanya, A., Visessanguan, W., Kudo, T. and Itoh, T. (2010). *Gracilibacillus thailandensis* sp. Nov., from fermented fish (*Plara*). Int. J. Syst. Evol. Microbiol. 60: 944- 948.

Cheorun, Jo., Dong, H.K., Hee, Y.K., Won, D.L., Hyo, K.L. and Myung, W.B. (2004). Studies on the development of low-salted, fermented, and seasoned *Changran Jeotkal* using the intestines of *Therage chalcogramma*. Rad. Phys. Chem. 71(1-2): 123-126.

Crisan, E.V. and Sands, A. (1975). Microflora of four fermented fish sauce. Appl. Microbiol. 29: 206.

Dincer, T., Cakli, S., Kilinc, B. and Tolasa, S. (2010). Amino acids and fatty acid composition content of fish sauce. J. Am. Vet. Adv. 9(2): 311-315.

Ebeling, W. (2002). Pests of stored food products. In: Urban Entomology, University of California, Division of Agricultural Sciences. Entomology UC Riverside, USA pp. 275-309. [Online: http://www.entomology.ucr.edu/ebeling]

El-Tahan, M.H., Hassan, S.A., El-Awamry, Z.K., and Hamza, A.S. (1998). Studies on microorganisms contaminated salted fish in Egypt. Central Laboratory for Food & Feed (CLFF), Agricultural Research Central, Cairo, Egypt. J. Union Arab Biol. 6 (B): 339-352.

Essuman, K.M. (1992). Fermented fish in Africa: A study on processing, marketing and consumption. FAO Fisheries Technical Paper. No. 329, FAO, Rome, Italy.

FAO (1971). Fermented fish products. FAO Fishery Report. I.M. Mackie, R. Hardy and G. Hobbs (eds). FAO, Rome, Italy.

FAO. (2000). United Nations Food and Agriculture Organization. FAO Year Book. Fishery Statistics Capture Production 86(1): 99-100.

Fellows, P. and Hampton, A. (1992). Small- scale food processing: A guide to appropriate equipment. Practical Action Publishing—formerly the Intermediate Technology Development Group,London, UK.

Gelman, A., Drabkin, V. and Glatman, L. (2001). Evaluation of lactic acid bacteria, isolated from lightly preserved fish products, as starter cultures for new fish-based food products. Innov. Food Sci. Emerg. Technol. 1 (3): 219-226.

Gram, L. and Huss, H.H. (2000). Fresh and processed fish and fish products. In: The Microbiological Safety and Quality of Food. B.M. Lund, T.C. Balrd-Parker and G.W. Gould (eds). Aspen Publishers, New York, USA pp. 472-506.

Harada, K., Okano, C., Kadoguchi, H., Okubo, Y., Ando, M., Kitao, S., et al. (2003). Peroxyl radical scavenging capability of fish sauces measured by the chemiluminescence method. Int. J. Mol. Med. 12: 621-625.

Heen, E. and Kreuzer, R. (1962). Fish in Nutrition. Fishing News Books Limited, Technology Branch, Fisheries Division, Food and Agriculture Organization of the United Nations, Roiqe, Italy. [Online: http://www.archive.org/details/fishinnutrition034834mbp]

Hiltz, D.F., Lall, B.S., Lemon, D.W. and Dyer, W.J. (1976). Deteriorative changes during frozen storage in fillets and minced flesh of silver hake (*Merluccius bilinearis*) processed from round fish held in ice and refrigerated seawater. J. Fish. Res. Board Can. 33: 256.

Huss, H. H. and Rye-Petersen, E. (1980). The stability of *Clostridium botulinum* type E toxin in salty and/or acid environment. J. Food Technol. 15 (6): 619-627.

ICMSF (1980a). Microbial Ecology of Foods, Volume 1: Factors Affecting the Life and Death of Microorganisms, Academic Press, London, UK pp. 180-184.

ICMSF (1980b). Microbial Ecology of Foods, Volume 2: Food Commodities, Academic Press, New York, USA pp. 731-751.

Ijong, F.G. and Ohta, Y. (1995). Microflora and chemical assessment of an Indonesian traditional fermented fish sauce "*Bakasang*". J. Fac. Appl. Biol. Sci. 34: 95-100.

Ijong, F.G. and Ohta, Y. (1996). Physico-chemical and microbiological changes associated with *Bakasang* processing—a traditional Indonesian fermented fish sauce. J. Sci. Food Agric. 71: 59-74.

Indoh, K., Nagata, S., Kanzaki, K., Shiiba, K. and Nishimura, T. (2006). Comparison of characteristics of fermented Salmon fish sauce using wheat gluten *Koji* with those using Soy Sauce *Koji*. Food Sci. Technol. Res. 12: 206-212.

Jones, N.R. (1966). Fish flavours. In: Chemistry and Physiology of Flavours. H.W. Schultz, E.A. Day and L.M. Libbey (eds). Avi Publishing Co., Connecticut, USA.

Jung, M.J., Roh, S.W., Kim, M.S. and Bae, J.W. (2010). *Lentibacillus jeotgali*, a halophilic bacterium isolated from traditional Korean fermented seafood. Int. J. Syst. Evol. Microbiol. 60: 1017-1022.

Jung, W.K., Rajapakse, N. and Kim, S.K. (2005). Antioxidative activity of a low molecular weight peptide derived from the sauce of fermented blue mussel, *Mytilus edulis*. Food Res. Technol. 220: 535-539.

Karnop, G. (1988). Verderb von salzsardellen durch histaminbildende Pediokokken. (Spoilage of salted sardine products by histamine forming pediococci.). Inf. fur die Fischwirtschaft 1: 28-31.

Kerr, M., Lawicki, P., Aguirre, S. and Rayner, C. (2002). Effect of storage conditions on histamine formation in fresh and canned Tuna. State Chemistry Laboratory—Food Safety Unit, Department of Human Service, Werribee, 2002: 5-20.

Khem, S. (2009). Development of model fermented fish sausage from New Zealand marine species. M.Sc. Thesis, Auckland University of Technology, New Zealand.

Knochel, S. and Huss, H.H. (1984). Ripening and spoilage of sugar-salted herring with and without nitrate. I. Microbiological and related chemical changes. J. Food Technol. 19 (2): 203-213.

Krieg and Holt (1984). Bergey's Manual of Systematic Bacteriology, Volume1. Williams and Wilkins, Baltimore, London, UK.

Lee, S.Y., Choi, W.Y., Oh, T.K. and Yoon, J.H. (2008). *Lentibacillus salinarum* sp. Nov., isolated from a marine solar salt in Korea. Int. J. Syst. Evol. Microbiol. 58: 45-49.

Lopetcharat, K. and Park, J.W. (2002). Characteristics of fish sauce made from Pacific whiting and surumi by-products during fermentation stage. J. Food Sci. 7 (2): 511-516.

Lopetcharat, K., Choi, Y.J., Park, J.W. and Daeschel, M.A. (2001). Fish sauce products and manufacturing: a review. Food Rev. Int. 17: 65-88.

Lovern, J.L. (1962). The lipids of fish and changes occurring in them during processing and storage. In: Fish in Nutrition. E. Heen and R. Kreuzer (eds). Fishing News (Books) Ltd., London, UK pp. 86-111.

Mah, J-H., Han, H-K., Oh, Y-J., Kim, M-G., and Hwang, H-J. (2002). Biogenic amines in *Jeotkal*s, Korean salted and fermented fish products. Food Chem. 79 (2): 239-243.

Majumdar, R.K. and Basu, S. (2010). Characterization of traditional fermented fish products of North-east India, *Lona ilish*. Indian J. Tradit. Knowl. 9: 453-458.

Michihata, T. (2003). Components of fish sauce *Ishiru* in Noto peninsula and its possibilities as a functional food. Foods Ingred. J. Jpn. 208: 683-692.

Namwong, S., Tanasupawat, S., Smitinont, T., Visessanguan, W., Kudo, T. and Itoh, T. (2005). Isolation of *Lentibacillus salicampi* strains and *Lentibacillus juripiscarius* sp. Nov. from fish sauce in Thailand. Int. J Syst. Evol. Microbiol. 55: 315-320.

Namwong, S., Tanasupawat, S., Visessanguan, W., Kudo, T. and Itoh, T. (2007). *Halococcus thailandensi*s sp. Nov, from fish sauce in Thailand. Int. J. Syst. Evol. Microbial. 57: 2199-2203.

Namwong, S., Tanasupawat, S., Lee, K.C. and Lee, J.S. (2009). *Oceanobacillus kapialis* sp. Nov., from fermented shrimp paste in Thailand. Int. J. Syst. Evol. Microbiol. 59: 2254-2259.

Nerquaye-Tetteh, G., Eyeson, K.K. and Tette-marmon, K. (1978). Studies on *"Bomone"*—a Ghanian fermented fish product. Food Research Institure, Accra, Ghana.

Nilsson, L. and Gram, L. (2002). Improving the control of pathogens in fish products. Lactic acid bacteria in fish preservation. In: Safety and Quality Issues in Fish Processing. H.A. Bremner (ed). Woodhead Publishing Ltd., CRC Press, New York, USA pp. 54-84.

Olympia, M., Fukuda, H., Ono, H., Kaneko and Takano, M. (1995). Characterization of starch-hydrolyzing lactic acid bacteria isolated from a fermented fish and rice food, *"burong isda"*, and its amylolytic enzyme. J. Ferment. Bioeng. 80(2): 124-130.

Paludan-Müller, C., Huss, H.H. and Gram, L. (1999). Characterization of lactic acid bacteria isolated from a Thai low-salt fermented fish product and the role of garlic as substrate for fermentation. Int. J Food Microbiol. 46 (3): 219-229.

Paludan-Müller, C., Madsen, M., Sophanodora, P., Gram, L. and Møller, P.L. (2002). Genotypic and phenotypic characterization of garlic-fermenting lactic acid bacteria isolated from *som-fak*, a Thai low-salt fermented fish product. Int. J Food Microbiol. 73 (1): 61-70.

Pakdeeto, A., Tanasupawat, S., Thawai, C., Moonmangmee, S., Kudo, T. and Itoh, T. (2007). *Lentibacillus kapialis* sp. Nov., from fermented shrimp paste in Thailand. Int. J. Syst. Evol. Microbial. 57: 364-369.

Pearson, D. (1970). The Chemical Analysis of Food, 6th edn. Churchill Publisher, London, UK.

Peralta, E., Hatate, H., Watanabe, D., Kawabe, D., Murata, H., Hama, Y. (2005). Antioxidative activity of Philippine salt-fermented shrimp paste and variation of its contents during fermentation. J. Oleo Sci. 54: 553-558.

Peralta, E.M., Hatate, H., Kawabe, D., Kuwahara, R. and Wakamatsu, S., Murata, H. (2008). Tamami Yuki improving antioxidant activity and nutritional components of Philippine salt-fermented shrimp paste through prolonged fermentation. Food Chem. 111 (1): 72-77.

Phithakpol, B., Varanyanond, W., Reungmaneepaitoon, S. and Wood, H. (1995). The Traditional Fermented Foods of Thailand. ASEAN Food Handling Bureau Level, Kuala Lumpur, Malaysia.

Prajapati, J.B. and Nair, B.M. (2003). The history of fermented foods. In: Handbook of Fermented Functional Foods. E.R. Farnworth (ed). CRC Press, Florida, USA pp. 1-26.

Raksakulthai, N. and Haard, N.F. (1992). Correlation between the concentration of peptides and amino acids and the flavour of fish sauce. ASEAN Food J. 7: 86-90.

Rao, M.S. and Stevens, W.F. (2006). Fermentation of shrimp biowaste under different salt concentrations with amylolytic and non-amylolytic *Lactobacillus* strains for chitin production. Food Technol. Biotechnol. 44 (1): 83-87.

Riebroy, S., Benjakul S., Visessanguan W., Kijrongrojana, K. and Tanaka, M. (2004). Some characteristics of commercial *som fug* produced in Thailand. Food Chem. 88: 527-535.

Riebroy, S., Benjakul S., Visessanguan W., Munehiko, B. and Tanaka, M. (2007). Effect of iced storage of bigeye snapper (*Priacanthus tayenus*) on the chemical composition, properties and acceptability of *Som-fug*, a fermented Thai fish mince. Food Chem. 102: 270-280.

Riebroy, S., Benjakul S. and Visessanguan W. (2008). Properties and acceptability of *Som-fug*, a Thai fermented fish mince, inoculated with lactic acid bacteria starters. LWT—Food Sci. Technol. 41 (4): 569-580.

Saisithi, P. (1967). Studies on the origin and development of the typical flavour and aroma of Thai fish sauce. Ph.D. Thesis. Department of Food Technology, University of Washington, USA.

Saisithi, P., Kasemsarn, B., Liston, J. and Dollar, A.M. (1996). Microbiology and chemistry of fermented fish. J. Food Sci. 31: 105-110.

Sands, A., Grism, E. and and Crisan, V. (1974). Microflora of fermented Korean seafoods. J. Food Sci. 39: 1002-1005.

Sangjindavong, M., Chuapoehuk, P., Runglerdkriangkrai, J., Klaypradit, W. and Vareevanich, D. (2008). Fermented fish Product (Pla-ra) from marine fish and preservation. Kasetsart J. (Nat. Sci.) 42: 129-136.

Sanni, I., Asiedu, M. and Ayernor, G.S. (2002). Microflora and chemical composition of *Momoni*, a Ghanaian fermented fish condiment. J. Food Composit. Analysis. 15 (5): 577-583.

Steinkraus, K.H. (1992). Lactic acid fermentations. In: Application of Biotechnology to Traditional Fermented Foods. Ad Hoc Panel (eds). National Academy Press, Washington, D.C., USA pp. 43-51.

Steinkraus, K.H. (1996). Introduction to indigenous fermented foods. In: Handbook of Indigenous Fermented Foods. K.H. Steinkraus (ed). Marcel Dekker, New York, USA pp. 1-5.

Steinkraus, K.H. (1997). Classification of fermented foods: worldwide review of household fermentation techniques. Food Control. 8 (5-6): 311-317.

Tanasupawat, S., Pakdeeto, A., Namwong, S., Thawai, C., Kudo, T. and Itoh, T. (2006). *Lentibacillus halophilus* sp. nov., from fish sauce in Thailand Int. J. Syst. Evol. Microbiol. 56: 1859-1863.

Tanasupawat, S., Namwong, S., Kudo, T. and Itoh, T. (2007). *Piscibacillus salipiscarius* gen. nov., sp. nov., a moderately halophilic bacterium from fermented fish (*pla-ra*) in Thailand. Int. J. Syst. Evol. Microbiol. 57: 1413-1417.

Tanasupawat, S., Namwong, S., Kudo, T. and Itoh, T. (2009). Identification of halophilic bacteria from fish sauce (*Nam-pla*) in Thailand. J. Cult. Collect 6 (1): 69-75.

Tsai, Y.H., Kung, H.F., Lee, T.M., Lin, G.T., and Hwang, D.F. (2004). Histamine related hygienic qualities and bacteria found in popular commercial scombroid fish fillets in Taiwan. J. Food Protec. 67: 407-412.

Tsai, Y.H., Lin, C.Y., Chang, S.C., Chen, H.C., Kung, H.F., Wei, C.I., et al. (2005). Occurrence of histamine and histamine-forming bacteria in salted mackerel in Taiwan. Food Microbiol. 22: 461-467.

Tsai, Y.H., Lin, C.Y., Chien, L.T., Lee, T.M., Wei, C.I., and Hwang, D.F. (2006). Histamine contents of fermented fish products in Taiwan and isolation of histamine-forming bacteria. Food Chem. 98: 64-70.

UNIFEM (1988). Fish Processing. The United Nations Development Fund for Women (UNIFEM) for Food Cycle Technology Source Book. Project GLO/85/W02. 9/pp.

Villar, M., De Ruiz Holgado, A. P., Sanchez, J. J., Trucco, R. E., and Oliver, G. (1985). Isolation and characterization of *Pediococcus halophila* from salted anchovies (*Engraulis anchoita*). Appl Environ Microbiol. 49 (3): 664-666.

Visessanguan, W., Benjakul, S., Smitinont, T., Kittikun, C., Thepkasikul, P. and Panya A. (2006). Changes in microbiological, biochemical and physico-chemical properties of *Nham* inoculated with different inoculum levels of *Lactobacillus curvatus*. LWT Food Sci. Technol. 3: 814-826.

Wiriyacharee, P. (1992). Using mixed starter culture for Thai *Nham*. In: Applications of Biotechnology in Traditional Fermented Foods. National Academy Press, Washington D. C., USA pp. 119-150.

7

Bioprospecting of Marine Microalgae, Corals and Microorganisms

V. Venugopal

INTRODUCTION

The ocean is the richest reservoir of living and non-living resources. The food resources of the oceans are potentially greater than those of the land because of larger area and diverse animals that are able to survive in the oceanic environment. Apart from serving as food, many of the marine organisms synthesize a number of secondary metabolite compounds, which are required for their growth and survival in the adverse environment of the ocean. Many of these compounds have unique therapeutic functions and therefore can be of use for treatment of a number of ailments.

The Marine Ecosystem

The marine ecosystem has been located at varying depths of the ocean from 200 m to 10,000 m supporting life at diverse levels. Many organisms that form part of the ecosystem are faced with extreme environments such as high pressure, low temperature, and low nutrient levels. Pressure increases by about one atmosphere for every 10 m depths and therefore organisms growing at 5000 m must be able to withstand pressures as high as 500 MPa. Below depths of about 100 m, ocean water stays at a constant temperature of 2-3°C. On the other hand, temperatures exceeding 100°C are found in the hydrothermal vents in the ocean bottoms. Salinities as high as 6N have been found in salt marshes and mines. Availability of nutrients for microbial growth decreases with depth and therefore microbial levels

Former Scientific Officer, Food Technology Division Bhabha Atomic Research Centre, Mumbai 400 076, India; E-mail: venugopalmenon@hotmail.com reduce with increasing depth. Thus, compared with about 3×10^5 microbial cells/ml of surface waters, at 2000 m deep, the cell counts may be as low as 3000 microbial cells/ml. These diverse environmental conditions have a profound influence on biodiversity of organisms in the ocean, each organism having a metabolism adapted to such conditions, leading to typical size and proximate compositions, particularly lipid (Haard, 1998). Many marine organisms form symbiotic associations among themselves with a view to survive the harsh conditions of the ocean. Most marine organisms also produce a number of bioactive secondary metabolites, which, although, are not directly involved in central physiological functions, yet contribute to their fitness and survival under the extreme habitats. These secondary metabolites (also called natural products) are organic compounds that are not directly involved in the normal growth, development or reproduction of organisms. There is a vast diversity in such secondary metabolites, which consist of alkaloids, terpenoids, sulfated polysaccharides, peptides, nitrogen heterocyclics, nitrogen-sulfur heterocyclics, sterols, and other novel compounds. Absence of these secondary metabolites, although may not result in immediate death of these organisms, but causes long term impairment of their survivability and fecundity. The functions of these compounds are usually of an ecological nature as they are used as defenses against predators, parasites and diseases for inter-species competition and to facilitate reproductive processes (attracting mates by color, smell etc.). These compounds are produced by marine species such as algae, corals, and microorganisms. Extracts of these organisms, which contain these compounds have been found to have excellent therapeutic effects (Bhakuni and Rawat, 2005; Kim et al., 2008). Table 7.1 gives the biological activities of diverse marine organisms, which can be

Table 7.1 Different bioactive compounds from marine sources as cure for various diseases

Analgesic agents
Anti-Alzheimer agents
Anti-asthma agents
Anti-bacterial agents
Anti-cancer agents
Anti-fungal agents
Anti-inflammatory compounds
Anti-malarials
Anti-tumor agents
Anti-viral agents
Apoptosis induction compounds
Immuno-stimulatory agents
Immuno-suppressive agents
Osteoarthritis suppressors

cures for various diseases. This chapter will discuss in brief, potentials of bioprospecting of microalgae and other marine organisms as sources of valuable nutraceuticals and other compounds.

MICROALGAE

Microalgae are single celled microscopic plants, which are responsible for the primary productivity in the open oceans. The sea, composing about three-fourth of the surface of the earth, absorbs significant amounts of solar radiation, which is utilized by autotrophic microalgae for photosynthesis, providing food for diverse oceanic creatures including sponges, crustaceans and other animals (Matsunaga et al., 2005; Irigoien et al., 2004; Falkowski, 2002). It has been estimated that the world's oceans absorb nearly half of the carbon dioxide that is released into the atmosphere from burning fossil fuels. When these organisms die, they sink down to the seabed, and take the carbon with them. Concentrations of these algae usually range between 10^5 and 10^6 cells/ml of open ocean waters. Microalgal cells are more in inshore ocean areas, which are nutritionally richer than open waters. Marine bays and inlets receiving sewage or industrial waste can have very high phytoplankton and bacterial population, which, in turn, supports higher densities of chemotrophic bacteria and aquatic animals such as fish and shellfish (Madigan and Martinko, 2005; Prabhudas et al., 2009).

Major Species

There are more than 40,000 different species of microalgae, which are divided into several phylae Some of the major phylae include Dinoflagellates, Cryptophytes, Euglenophyta, Haptophyta, Heterokontophyta, among others (Nybakken, 1997; Irigoien, et al., 2004). Dinoflagellates (Dinophyceae) form a significant part of primary planktonic production in both oceans and lakes. These microalgae could be autotrophs, mixotrophs, osmotrophs, phagotrophs, or parasites and are usually unicellular, flagellated, often photosynthetic protests. Approximately 130 genera of dinoflagellates consisting of about 2000 living and 2000 fossil species have been described, most of them belonging to marine habitat. Extreme abundance of dinoflagellates (2 to 8×10^6/l water) results in "red tide" leading to significant toxin formation with its devastating effect on the mortality of fish and invertebrates. Cryptophytes, a major species of phytoplankton, are unicellular flagellates with 12 to 23 genera comprised of 200 species. They are distributed both in freshwater and marine environments. A few species are colorless heterotrophs, but most possess various colored plastids with chlorohylls, carotenoids and phycobiliprotein and alloxanthin, a xanthophyll that is unique to these algae. The phylum

Heterokontophyta is the most diverse algal group with huge commercial and biotechnological potentials. They range in size from microscopic single cells to giant kelp averaging several meters. They form one of the most diverse algal groups with huge commercial and biotechnological potentials. Euglenophyta encompasses unicellular flagellate organisms, and comprise of 40 genera and 900 species. The chlorophast originating from the green algae contains chlorophylls-a and -b and carotenoids such as neoxanthin, diadinoxanthin and β-carotene. The phylum Haptophyta is a group of unicellular flagellates, having a brownish or yellowish-green color due to chlorophylls-a and -c_1/c_2 and carotenoids such as β-carotene, fucoxanthin, and others. The cells are commonly covered with scales made mainly of carbohydrates or calcium bicarbonate, and hence many species produce calcified scales. About 70 genera and 300 species have been isolated to date, most being tropical marine species, forming food for aquatic communities (Nybakken, 1997).

Blue Green Algae or Cyanobacteria

Blue green algae also referred as cyanobacteria, are oxygenic phytosynthetic prokaryotes that show large diversity in their morphology, physiology, ecology, biochemistry and other characteristics. Currently, more than 2000 species of cyanobacteria are recognized. These organisms are distributed widely not only in salt water but also in freshwater, brackish water, polar areas, and hot springs. In tropical and subtropical oceans, the marine cyanobacterium, *Trichodesmium*, form tufts of filaments that constitute a significant fraction of the biomass suspended in the waters. The cyanobacteria of the genus *Prochlorococcus* are the smallest (0.6 μm dia) and most numerous of the photosynthetic marine organisms. It has been estimated that a drop of seawater contains up to 20,000 cells of the organisms belonging to *Prochlorococcus*. These and another marine pelagic *Synechococcus* contribute largely to global oxygen production. Cyanobacteria are generally considered to be associated with marine plants and animals. Some of these organisms also exist in symbiotic association with sponges, ascidians, echiuroid worms, planktonic diatoms and dinoflagellates in order to survive in highly stressful marine environments giving rise to several secondary metabolites, having potentials as drugs and other bioactive compounds (Matsunaga et al., 2005; Falkowski, 2002).

Marine Microalgae as Food

Marine microalgae are hailed as the new "super food", because they are significantly rich in nutrients. Ingestion of relatively small quantities of microalgae can satisfy the requirements for some vitamins in animal and human nutrition, while supplementing others (Fabregas and Herrero, 1990). These phytoplankton, therefore, constitute the initial component of the food web, providing food for various oceanic creatures, which include shellfish species, consisting of crayfish, shellfish, crab, shrimp and lobster, mollusks including bivalves such as mussel, oyster, and scallop, univalve creatures such as abalone, snail and conch, and the cephalopods, squid, cuttlefish and octopus. The high protein contents of many microalgae, which range from 39-54% of dry matter, make them useful as single cell protein. Besides, they also contain up to 29% carbohydrates and 18% lipids and are also rich in many minerals including calcium, sodium, magnesium, zinc, iron, manganese, copper and nickel. They also contain saturated, mono- and poly-unsaturated fatty acids (Madigan and Martinko, 2005). Some of the important microalgae, which have been well studied in this respect, include species of *Chlorella*, *Spirulina* and *Dunaliella* (Prabhudas et al., 2009; Adams, 1997). *Chlorella* is a genus of single-celled green algae, belonging to the phylum, having a spherical shape of about 2 to 10 μm dia. The alga contains the green photosynthetic pigments chlorophyll-a and chlorophyll-b in its chloroplasts. It multiplies rapidly through photosynthesis, requiring only CO_2, water, sunlight and minerals for growth. Chlorella could serve as a potential source of food because of its high contents of proteins and other essential nutrients. *Spirulina* are unicellular, photosynthetic blue green algae (cyanobacteria). They have no cellulose in its cell walls, which are composed of soft muco-polysaccharides. The algae are microscopic in nature and occur naturally in warm, alkaline, salty, brackish lakes. Its color is derived from the green pigment of chlorophyll, and from the blue color of a protein called phycocyanin. The common *Spirulina*, which are used as human and animal food supplements are the species, *S. maxima* and *S. platensis*. Regular consumption of *Spirulina* may lower cholesterol, serum lipids, and low-density lipoprotein. *Spirulina* reduce the severity of strokes and improves recovery of movement after a stroke; reverses age-related and learning; and prevents and treats hay fever (Prabhudas et al., 2009; Sen and Sarkar, 2004). Regular intake of *Spirulina* was also shown to potentially limit brain damage from strokes and other neurological disorders (Wang et al., 2005a). The genus *Dunaliella* includes halo-tolerant, unicellular, motile green algae with exceptional morphological and physiological properties belonging to the family Chlorophyceae. It grows in high (1.5 M) salt concentration (Adams, 1997).

Drugs, Carotenoids and Other Compounds from Marine Microalgae

The unique metabolic capabilities of microalgae offer the potential for novel and promising natural products, which can fight various diseases providing support to human life. These compounds include antioxidants, antimicrobials, antitumor drugs, bio-adhesives, antifouling compounds, and immuno-active agents, enzymes, polysaccharides, enzyme inhibitors, herbicides, antimalarials, antimyotics, multi-drug resistance reverses etc. (Kurano and Miyachi, 2004; Plaza et al., 2009). *Dunaliella* spp., for example, contain compounds that exhibit various biological activities such as antihypertensive, bronchodilator, analgesic and muscle relaxant activities (Dufosse et al., 2005; FAO, 2004). The potential of cyanobacteria as untapped sources of nutraceuticals have only recently been recognized. Possibilities of their genetic modification and mass cultivation under controlled conditions offer scope to enhance yield of their bioactive compounds in substantial amounts. Some of the novel compounds isolated from cyanobacteria include plant growth regulators such as gibberellic acid that promote re-differentiation, germination and platelet formation, tyrosinase inhibitors, UV-absorbing compounds, sulfated polysaccharides showing anti-HIV activity and novel antibiotics with light-regulated activity (Otero and Vincenzini, 2003; Herrero et al., 2006). *Chlorella* (*C. vulgaris*) and *Spirulina* (*S. platensis*) contain peptides that inhibit angiotensin I-converting enzyme, and thereby helps control of coronary heart disease. Oral administration of fractions of peptic digests into spontaneously hypertensive rats at 200 mg/kg of body weight resulted in marked antihypertensive effects (Suetsuna and Chen, 2001). Interest in this field has resulted in collection of cyanobacterial species for production of their secondary metabolites such as dolastatins, originally isolated from the Indian Ocean seashore. Cyanobacterial metabolites have proven to be invaluable as tools in the dissection of signal transduction pathways in mammalian cells and some are currently under clinical evaluation as drug candidates. A plasmid from the marine cyanobacterium, *Synechococcus* spp., whose copy number is dependent on salinity, is being used to develop a stable and controllable gene expression system (Otero and Vincenzini, 2003; Burja and Radianingtylas, 2005). The red microalga of genus *Porphyridium* is a source of biochemicals possessing nutritional and therapeutic value. These compounds include polysaccharides (having anti-inflammatory and antiviral properties), long-chain polyunsaturated fatty acids, carotenoids such as zeaxanthin and fluorescent phycolipoproteins. The red phycoliporoteins, phycoerythrin and the blue phycobiliproteins and phycocyanin, are insoluble in water and can serve as natural colorants in foods, cosmetics and pharmaceuticals. The algal pigments have potential as natural colorants for use in food, cosmetics and pharmaceuticals, as substitutes for synthetic dyes. The *all-trans*-β-carotene and isomers and other minor carotenoids isolated from *Dunaliella salina* have been recognized as good antioxidants. The phycobiliproteins are water soluble pigments consisting of red-colored phycoerythrin and blue-colored phycocyanin and allophycocyanin. *Botryococcus braunii* is a green colonial microalga that is used mainly for the production of carotenoids, hydrocarbons and polysaccharides. Commercial scale production of some microalgal pigments such as β-carotene from *Dunaliella*, astaxanthin from *Haematococcus* and lutein from chlorophycean strains have been initiated. Extracted β-carotene is sold mostly in vegetable oil ranging in concentrations from 1 to 20% to color various food products in soft gels for personal uses. The purified natural β-carotene is generally accompanied by other carotenoids of *Dunaliella*, predominantly lutein, neoxanthin, zeaxanthin, violaxanthin, cryptoxanthin, and α-carotene (Arad and Yaron, 1992; Jeffrey et al., 1997; Shi et al., 2002; Campo et al., 2007). The advantages of carotenoids production by microalgae are given in Table 7.2.

Similar to marine fatty fish such as cod, herring, tuna, and others, microalgae are also sources of polyunsaturated fatty acids (PUFA) (Berge and Barnathan, 2005). Under certain conditions such as algal bloom, the proportion of PUFA of the omega-3 type can reach 50% of total lipids. Because of therapeutic importance of these fatty acids, and paucity of availability of marine fishes as their conventional sources, there has been interest in cultivation of microalgae for large scale production of PUFA. *Chlorella* is an important source of PUFAs (Adams, 1997; Tokusoglu and Unal, 2003). The PUFA, DHA (docosahexaenoic acid), has been produced by heterotrophically grown *Crypthecodinium cohnii*. These algal DHA are used as food supplements such as infant formula. Algal PUFA can be used for nutritional enrichment of food products and also for production of PUFA-enriched fish feeds. Haptophyta biomass is used as feed for bivalve mollusks, crustacean larvae and zooplankton which in turn, is used as feed for fish and crustacean larvae. The polysaccharide of unicellular rhodophyte, *Porphyridium* spp. has potential as a thickening agent and food additive because of its high viscosity over a wide range of pH, temperature and salinity (Jeffrey et al., 1997). Table 7.3 gives some bioactive compounds produced by microalgae.

Table 7.2 Advantages of cultivation of microalgae

Ease of cultivation
High pigment production (3-5% on dry weight basis)
Have both *cis* and *trans* isomers of carotenoids
The algae protein is nutritive and can be used as protein supplements
The algal biomass can be used as poultry feed

Table 7.3 Some bioactive compounds produced by microalgae

Microalgae	Bioactive compounds
Cyanobacteria	Enzyme inhibitors such as tyrosinase inhibitor Plant growth regulators UV-absorbing compounds Sulfated polysaccharides having anti-HIV activity Novel antibiotics Herbicides Anti-malarial compounds Immuno-suppressive agents Compounds having cytotoxic activity Polyunsaturated fatty acids
Rhodophyta	Polysaccharide of rhodophyte *Porphyridium* spp. used as a thickening agent Polysaccharides have hypocholesterolemic agent and antiviral activity against animal viruses including *Herpes simplex* types 1 and 2 and also retroviruses Anti-cancer compounds
Chlorophyta	Anti-inflammatory activity Immuno-suppressive activity
Cryptohyta	Significant content of polyunsaturated fatty acids and hence used in aquaculture feeds
Heterokontophyta	Microalgae belonging to this phylum are good raw materials for production of polyunsaturated fatty acids (EPA and DHA) Diatoms as potential antioxidants and domoic aid Used as feed in aquaculture
Dinophyta	Culturing is difficult and hence difficulties in isolation of bioactive compounds. However DHA is produced by heterotrophically grown *Crypthecodinum cohnii* Sulfated polysaccharides from *Gymnodinum* spp. have antiviral activity
Haptophyta	Used as feed DHA production
Euglenophyta	Culturing is difficult for most organisms. *Euglena gracilis* when cultured photoheterotrophic or photo-autotrophic conditions produces antioxidant vitamins such as β-carotene, and vitamin C and E

Source: Matsunaga et al. (2005); Plaza et al. (2009).

Cultivation of Microalgae

Microalgal biotechnology has the potential to produce a vast array of products from virtually untapped sources (Schweder et al., 2005). Microalgae can be cultivated for specific biochemicals such as pigments and polyunsaturated fatty acids (PUFA). Growing microalgae also facilitates production of hydrogen, methane, ethanol and triglycerides, through biophotolysis, anaerobic digestion (with yeast), and extraction of lipids, respectively. Other potentials are production of fuels (through trans-esterification of lipids) and liquid hydrocarbons (Campo et al., 2007; Liebezeit, 2005; Jeffrey et al., 1997; Arad and Yaron, 1992). Production of these compounds is essentially based on the relatively efficient photosynthetic machinery of these algae and selection of species for the required applications. In culturing marine invertebrates, basically three different approaches can be followed, viz., *in situ* cultivation, bioreactor and cell cultures (Liebezeit, 2005). Microalgae can be cultivated in ponds having large amounts of surface area to be economically viable. The growth rate and maximum biomass yield of microalgae are influenced by cultivation parameters (light, temperature and pH) and nutritional status (CO_2, nitrogen and phosphate concentrations). Stringent nitrogen limitation conditions are required to stimulate the organisms to produce required compounds such as lipids. The open ponds must be maintained at low densities of the organism to allow light to penetrate into the system for photosynthesis. However, this creates processing problems since large volumes of water must be processed to recover the small volumes of biomass. On the other hand, increasing the density of cultures decreases photon availability to individual cells, since light penetration of microalgal cultures is poor, especially at high cell densities, adversely affecting specific growth rates. The open culture systems, however, offer advantages such as low construction cost and ease of operation. Some of the open systems include shallow ponds, which may be unstirred or paddle-wheeled, slopping cascade, tubular reactors (helix, plane or double layered), or laminar reactors. Microalgae such as *Chlorella, Dunaliella, Spirulina, Porphyridium,* and *Haematococcus* spp. have been commercially cultured. *Dunaliella salina* is cultured in large (up to 250 ha, approximately) shallow open-air ponds, with no artificial mixing. Similarly, *Chlorella* are grown outdoors in either paddle-wheel mixed ponds or circular ponds with a rotating mixing arm of up to about 1 ha in area per pond.. The production of microalgae for aquaculture feed is carried out indoors, in 20 to 40 litre carboys or in large plastic bags of up to 1000 litre. In India, initiated by Indo-German cooperation, large scale cultivation of *Spirulina platensis* was conducted with the aim of utilizing the microalga as animal feed and nutritional supplement. (Becker and Venkataraman, 1984) The alga was cultured in large shallow open-air ponds with no artificial mixing or using paddle wheel mixed ponds or circular ponds. Cultivation was carried out in a high salt and bicarbonate/carbonate (alkaline pH of about 10) medium under light irradiation and aeration to help optimal photosynthesis by the alga. Batch tests performed at 25°C in open tanks suggested that inorganic carbon is preferentially assimilated in the form of bicarbonate and that its utilization efficiency depended either on pH. For the rapid growth of cells often CO_2 is injected into the water. The efficiencies of photosynthesis and utilization of CO_2 reach maximum values of 6 and 38%, after 4 and 7 d, respectively. Instead of mineral medium, the possibility of growing the alga on biogas slurry, swine waste water, cattle waste or water from aquaculture ponds with proper supplementation of nutrients including CO_2 and minerals suggests the possibility of recycling waste water for economic benefits. *Spirulina* grows relatively slowly, with an output of about 5 g dry mass/m^2/day, in a medium of solid cattle waste extract in water containing sodium chloride at 10 g/l. The growth rate and output rate were substantially increased when the waste-based medium was fortified with carbon, nitrogen and phosphorus. Aeration of the algal cultures considerably improved the output rate of algal biomass (Mitchell and Richmond, 1998). After cultivation, *Spirulina* and other microalgae can be harvested by simple filtration process through cloth. They are dewatered by centrifugation and dried to a powder. The biomass is usually dehydrated by sun-, drum-, or spray-drying. *Spirulina* are available in powder, flake, capsule, and tablet form (Becker and Venkataraman, 1984; Mitchell and Richmond, 1998).

The development of efficient photo bioreactors significantly extends the number of species that can be cultivated under controlled conditions and the range of extractible products, for example: β-carotene, phycocyanin, phycoerythrin and glycerol (Borowitzka, 1999). Nutrient composition of the biomass of microalgae cultivated in a photo-reactor is highly influenced by residence time in the reactor. The biomass harvested for short residence times is generally richer in protein and unsaturated fatty acids than biomass harvested for high residence time (Rebolloso-Fuentes et al., 2001; Madigan and Martinko, 2005). A helical tubular photo bioreactor system has also been developed which allows these algae to be grown reliably outdoors at high cell densities in semi-continuous culture. Other closed photo bioreactors such as flat panels are also being developed. The high cost of microalgae culture systems relates to the need for light and the relatively slow growth rate of the algae. Developments in photo bioreactors allow successful cultivation of different species of microalgae under controlled conditions for a range of extractible products including β-carotene, phycocyanin, phycoerythrin, glycerol, etc. These organisms range in size from microscopic single cells to giant kelp averaging several meters. The problems of propagation of mixed culture populations and frequent contaminations by bacteria, fungi, protozoa etc. hamper the production affecting the quality of the microalgae. For safety reasons, the microalgae used for human and animal consumption should be grown in good quality water (Borowitzka, 1999; Lorenz and Cysewski, 2000; Matsunaga et al., 2005; Venugopal, 2008).

Biotechnological applications of dinoflagellates are limited because most dinoflagellates cannot be easily cultured. The microalgae *Dunaliella salina* and the diatoms (*Nitzschia* spp.), when grown under controlled conditions of temperature, pH, photon flux density and salinity, can give high total lipid contents up to 28% in the cells on dry weight basis (Renaud et al., 1994). The green microalga, *Chlorella zofingiensis* has been cultivated for astaxanthin. The alga showed excellent growth on glucose-supplemented media in batch culture. In the absence of light, formation of secondary carotenoids was mostly dependent on the initial carbon and nitrogen balance in the medium. Enhanced biosynthesis of astaxanthin is dependent on a high C/N ratio of 180. The light-independent astaxanthin-producing ability of *C. zofingiensis* suggests that the alga might be potentially employed for commercial production of astaxanthin on a large-scale (Ip and Chen, 2005). The influence of culture conditions on yield and carotenoid production has also been shown in the case of *D. salina*. Under stress conditions, such as formation of excessive free radicals, cell division inhibition, nitrogen starvation, high salinity and temperature, this balance is disturbed and the cells generate additional amounts of the carotene. With nitrogen starvation, β-carotene production can be enhanced about seven-fold. Maximum carotene (8.28 pg/cell) could be obtained upon exposure of the cells to high light irradiance (6000 lux) at a temperature of 35°C (Pisal and Lele, 2005) The green microalga *Chlorella protothecoides* has been shown to produce enhanced quantity of lutein under a combination of nitrogen limitation and high-temperature stress when grown in batch mode in a 3.7 litre fermenter containing 40 g/L glucose and 3.6 g/L urea, followed by a relatively reduced supply of nitrogen source to establish a nitrogen-limited culture. This N-limited fed-batch culture was scaled up to 30 l, and a three-step cultivation process was developed for high yield of the carotene. (Shi et al., 2002) The process of isolation of lutein consists of extraction of the crude pigment with dichloromethane from the microalga after saponification, which gives the carotenoid a minimum purity of 90% and yield of 91% (Gouveia et al., 2007). A functional food oil, rich in fatty acids and antioxidants, colored with carotenoids extracted from *Chlorella vulgaris* has been produced. The alga was subjected to supercritical CO_2 extraction at a pressure of 300 bar or using acetone containing vegetable oil at room and high temperatures. The recovery of carotenoids was 100% with oil at room temperature for 17 h. In supercritical extraction the degree of crushing strongly influenced the extraction recovery and higher pigment recoveries were obtained with well-crushed biomass (Gouveia et al., 2007; Tokusoglu and Unal, 2003). Production of β-carotene by *Spirulina* algae is very high, and depends upon light intensity. The natural β-carotene extracted from the algae is a mixture of numerous carotenoids and essential nutrients that are not present in synthetic β-carotene. On an average,

D. salina under ideal conditions can yield 400 mg β-carotene/m^2 of cultivation area. Growth of the microalga is comparable with that of *Spirulina* and requires bicarbonate as a source of carbon and other nutrients such as nitrate, sulfate and phosphate. Initial growth phase requires 12 to 14 d in nitrate rich medium, and 25-30k lux light intensity for vegetative growth. For optimal carotene synthesis the light should be reduced to 15 k lux, besides, nitrate depletion and maintenance of the initial salinity. Harvesting is done by flocculation followed by filtration; the biomass can be directly utilized for food formulation. Most of the pharmaceutical formulations are made by extracting the alga with olive or soybean oil (Dufosse et al., 2005). Although not from marine habitat, *Haematococcus pluvialis* is a green alga that can grow both under autotrophic and heterotrophic conditions. It is known for its ability to synthesize astaxanthin, up to 2.0% on dry weight basis and hence, is one of the potential organisms for commercial production of astaxanthin. Astaxanthin from *H. pluvialis* is obtained by crushing the algae and supercritical carbon dioxide extraction. Several European countries have approved its marketing as a dietary supplement for human consumption. The maximum total recovery of astaxanthin exceeded 97%. Another source of the pigments is *Phaffia rhodoxyma* (*Xanthophyllomyces dendrorhous)*, which requires a large amount of feed for sufficient pigmentation leading to higher ash contents (Dufosse et al., 2005). Some current commercial microalgal products include a U.S. FDA (Food and Drug Administration) approved dietary supplement containing astaxanthin, called *Zanthin*, extracted from the microalga, *H. pluvialis.* Martek DHA^{TM} is a commercial oil product from microalga that contains significant amount of DHA. A carbohydrate extract from the green microalga, *Chlorella pyrenoidosa* is claimed to boost response of the immune system to the flue vaccine (Ohr, 2005).

CORAL REEF AND CORALS

Coral reefs are massive deposits of calcium carbonate in the oceans, which harbor a rich and diverse ecosystem of animals. The reefs are produced primarily by corals with minor additions from calciferous algae and other organisms that secrete calcium carbonate. The coral carbonate is composed of two distinct mineral forms—calcite and aragonite, the latter containing significant amounts of magnesium. The reefs are unique among marine associations in that they are built up entirely by biological activity. Some major physical factors that influence coral reef development are temperature, depth, light, salinity, sedimentation and emergence into air. Coral reefs are also found in the clean coastal waters of the tropics and subtropics, which give optimal conditions such as moderate temperature and good sunlight favoring growth of reef forming organisms. It has been estimated that coral reefs occupy about 600,000 square miles of the earth surface, representing about 0.17% of the total area of the planet. On the continental shelves of northern and western Europe, extensive reefs are found at depths of 60 to 2000 m. The age of the reefs extends to thousands of years; the Great Barrier Reef of Australia is said to be more than 9000 yr old. The reefs, in general, are grouped into one of three categories, namely, atolls, barrier reefs, and fringing reefs. Atolls are usually easily distinguished because they remodified ring-shaped reefs that rise out of very deep water far from land and enclose a lagoon. With few exceptions atolls are found only in the Indo-Pacific area. Barrier reefs and fringing reefs tend to grade into each other and are not readily separable.

Corals, the major organisms that form the basic reef structure, are members of the phylum Cnidaria, class, Anthozoa and order Madreporaria, which include such diverse forms as jellyfish, hydroids, and sea anemones. There is a bewildering array of other organisms associated with reefs such as fish and shellfish. Mollusks of various types including various giant clams, sea urchins, sea cucumbers, starfish and feather stars are present on the reefs, which make significant contributions to $CaCO_3$ deposits of the reef (Sakthivel et al., 2005; www.wri.org/project/coral-reefs, August 30, 2010). Stony corals are the foundation of coral reef ecosystems. The rate of growth of different tiny corals varies widely. For example, members of the genera, *Acropora* (Stag's horn coral) and *Pocillopora* grow rapidly and they represent a considerable proportion of tropical coral reefs. Coralline algae (algae that also secrete calcium carbonate and resembling corals) contribute to the calcification of many reefs. Shallow water corals owe their beautiful colors in part to symbiotic algae, which live inside the coral cells. Sponges are abundant on reefs; but they have little to do on reef construction. Siliceous sponges (Demospongiae) are important in holding coral and rubble together. The important genera of sponges are *Callyspongia, Oceanapia, Haliclona, Axinella,* and *Sigmadocic* spp. (Sakthivel et al., 2005). Marine sponges are the most primitive multi-cellular animals and contain many new metabolites, many of which have shown to possess diverse biological activities. These organisms are difficult to classify due to the few available useful morphological characteristics.

Corals generally form associations with other reef species. Symbioses between photosynthesizing organisms (e.g., cyanobacteria and dino-flagellates, diatoms and algae) and several invertebrate corals including poriferans, cnidarians, ascidians and also mollusks have been reported. These associations are particularly operating in tropical coral reefs. Symbiosis is influenced by physical and environmental factors like depth-dependent light and temperature, and seasonal fluctuations in these parameters (Hooper and van Soest, 2002; Douglas, 2003). The importance of sponge-associated bacteria as a valuable resource for the discovery of novel bioactive molecules has been recognized. In many cases, these associated bacteria have been found to be the actual producers of these compounds. It is likely that bacteria and microorganisms co-evolved with sponges during the long period of time, and thus acquired a complex common metabolism. Many of these associations play a role in maintaining the health and diversity of reef systems (Cerrano et al., 2004).

Coral reefs have been recognized as vulnerable to destruction due to the action of bio-eroding organisms, coastal pollution over-fishing, coral mining, and recreational activities, among others. Climate change poses a serious threat to these complex ecosystems. The release of sewage affects the corals by spreading through the massive sediments. Infectious diseases are recognized as significant contributors to the dramatic loss of corals observed worldwide (Santhanam and Venkataramanjuam, 1996; Voss and Richardson, 2006). In the light of the deteriorating state of coral reefs worldwide, the necessity of restoring corals has been seriously felt, attracting worldwide efforts. These efforts include construction of artificial reefs, mariculture and restocking. Construction of artificial reefs makes use of low-profile structures such as shipwrecks in the seabed to mimic a natural reef. The ability of benthic artificial reef communities, mainly filter feeders like bryozoans, bivalves, sponges and tunicates to resemble those of a natural reef is of great use in rehabilitation and restoration of degraded marine habitats. In spite of efforts in this direction, understanding of the interactions between artificial and natural reefs is poor and doubts exist on the ability of artificial reefs to mimic adjacent natural reef communities, performance of artificial reefs and their possible effects on the natural surroundings (Baine, 2001). Reef Watch Marine Conservation (www. reefwatchindia.org) is one of the leading organizations in India working on marine and coastal conservation issues.

Bioactive Compounds from Corals

Coral reef organisms are known to possess several biologically active chemicals that can be of immense therapeutic applications. The probability of discovering a drug from marine sources, particularly corals, is approximately a thousand times more than that from terrestrial ones. Therefore corals and coral reefs are being targeted for studies world over for their bioactive compounds. These efforts have yielded a number of compounds from these sources, which include analgesic, anti-Alzheimer, anti-angiogenic, anti-asthma agents, anti-cancer drugs, anti-fungal agents, anti-inflammatory compounds, immuno-stimulatory and immuno-suppressive compounds. While some of these compounds are being commercially produced, many are at various levels of clinical trials. For example, zoanthamine, capable of preventing osteoporosis, was isolated in 1984 by Faulkner's group from a *Zoanthus* spp. (Phylum Cnidaria, Class Anthozoa, Order Zoanthides), collected off the Visakhapatnam coast of India (Rao et al., 1984). Some of these compounds also possessed anti-inflammatory, cytotoxic and analgesic activities. Potential of these compounds from dinoflagellate has also been indicated recently (Sudarsanam and Anita Devi, 2004). Hydroxyapatite made from the rigid exoskeletons of marine corals can fill voids caused by fractures or other trauma in the upper, flared-out portions of long bones. The material is similar to the human bone in structure. The U.S. Food and Drug Administration has approved coral-derived implants for applications such as certain types of bone loss. Marine carbonate as highly interconnected microporous materials is receiving attention mainly in medical applications. Three-dimensional microporous skeletons are found in certain species of coral. The carbonate framework may be used as a template for the deposition of metals, ceramics, or polymers, which after removal of the carbonate by mid acid treatment, provides an interconnected porous composite structure for varied applications (White and White, 2003). The Caribbean Sea whip contains compounds called pseudopterosins. These compounds appear to have anti-inflammatory properties and have the potential for treatment for skin irritations resulting from injury or infection (Henkel, 1998).

Sponges are sessile marine filter feeders that have developed efficient defense mechanisms against foreign attackers such as viruses, bacteria, or eukaryotic organisms. Protected by a highly complex immune system as well as by the capacity to produce efficient antiviral, antimicrobial and cytostatic compounds, they have survived in the ocean for several millions of years. Marine sponges are therefore considered as a gold mine with respect to the diversity of their secondary metabolites that are capable of providing future drugs against important diseases, such as cancer, a range of viral diseases, malaria, and inflammations (Muller, 2004; Sipkema et al., 2005). Sponges are also rich in antioxidant enzymes such as superoxide dismutase, catalase, glutathione S-transferases, glutathione reductase, and glutathione peroxidases. The bacteria associated with the sponge *Suberites domuncula* showed angiogenesis inhibitor, antimicrobial and hemolytic activities. The bacterial extracts were also strongly active against *Staphylococcus aureus* and *S. epidermidis*, isolated from hospital patients (Thakur et al., 2005). Marine sponges contain glycolipids and phospholipids in their membranes, and are rich sources of unusual fatty acids containing very long chains (C_{23} to C_{34}), known as demospongic acids, which constitute up to 80% of the total fatty acids. Dragmacidins, a newly discovered class of bis (indole) alkaloids, have been isolated from a variety of marine sponges, which have anti-viral and anti-cancer activities (Capon et al., 1998). Eight new noradrosinane sesquiterpenoids, laevinols A-H, a new neolemnane sesquiterpenoid, levinone A and also other two previously known compounds were isolated from soft coral, *Lemalia laevis* (El-Gamal et al., 2005). Polyaromatic alkaloids, known as lamellarins, have been isolated from the marine ascidian, *Didemnum* spp. and the mollusk, *Lamellaria* spp. New cytotoxic cyclicheptapeptide, mollamide and patellins (cyclic peptides) have also been isolated from ascidians *Didemnum molle* and *Lissoclinum* sp. collected on the Great Barrier Reef (Gross and Kong, 2006) Dercitin, an alkaloid, isolated from the sponge, *Dercitus* spp. has been recognized to have anti-leukemic activity (Ireland et al., 2000; Ranga Rao et al., 2006).

Sponge Culture

Owing to the discovery of many commercially important compounds in sponges, there is increasing interest in production of marine sponge biomass. *In situ* sponge aquacultures, based on conventional methods for producing commercial bath sponge are still the earliest and least expensive way to obtain sponge biomass in bulk. During the last 20 yr, sponges have also been experimentally cultured both in the sea and in tanks on land for their biologically active metabolites. Many studies have shown that sponges grow quickly, often doubling in size every few months. Sea-based culture holds great promise, with several small-scale farming operations producing bath sponges or metabolites. These studies have focused on identification of the optimal environmental conditions that promote production of bath sponges or bioactive metabolites. The ideal farming condition will vary between species and regions, but will generally involve threading sponges on rope or placing them inside a mesh. A better-defined production system has been made for culturing sponges in semi-controlled systems, but these still use unfiltered natural seawater. Other interesting developments include partially harvesting farmed sponges to increase biomass yields, seeding sexually reproduced larvae on farming structures, using sponge farms as large biofilters to control microbial populations, and manipulating culture conditions to promote metabolite biosynthesis. An adequate supply of food seems to be the key to successful sponge culture. Feeding diverse rather than a single type of food had better effect on growth rate of the sponge (Duckworth, 2009; Osinga et al., 1999). Cell cultures of marine sponges might be considered as an alternative to *in situ* culture. The advantage of cell cultures is that they can be completely controlled and easily manipulated for optimal production of the target metabolites. However, the technique is still in its infancy and a continuous cell line has yet to be established. Laboratory scale cultures have been standardized for bryozoan species, because of the interest in bryostatin-1, a potent antineoplastic agent, produced by the bryozoa, *Bugula neritina* (Liebezeit, 2005). As mentioned earlier, some sponge metabolites are, in fact, produced by symbiotic bacteria or algae that live in the sponge tissue. Only a few of these symbionts have been cultivated so far. Another promising way to utilize the bioactive potential of the microorganisms is the cloning and heterologous expression of enzymes involved in secondary metabolism (Thakur et al., 2005).

Several countries have initiated serious work for the discovery of drugs from marine sources, particularly from corals. Australian research on marine natural products from coral organisms of the Great Barrier Reef has yielded a number of compounds. (Volkman, 1999). Cytotoxic diterpenes, sesquiterpenoids and furan derivatives have been isolated from Formosan soft corals in Taiwan. Among the identified compounds, the toxicity property of some has been investigated in various cancer cell lines. Many of these bioactive compounds are ready for further clinical trials to establish their potential use as drugs against diseases. In the U.S., the National Sea Grant College Program, under the National Oceanic and Atmospheric Administration is undertaking active research in the area. Recent interest in the area is also depicted by collection of thousands of marine organism samples and identification of a number of promising potential drugs for treating cancer, Alzheimer's, malaria, AIDS and other ailments by researchers at the University of Miami. The Sea Grant Program has compiled information on marine-based natural products of biomedical significance. Information provided for each product entry includes compound source, bioactivity, and clinical status, as well as structural and chemical attributes and commercial development highlights. Some of them are now at various stages of development (HBOI, 2006; Proksch, 2002). In India, the government has taken up a national coordinated program on development of drugs from the ocean to harness potential bioactive substances, which is being implemented by the Central Drug Research Institute, Lucknow, Centre for Advanced Studies in Marine Biology of the University of Annamalai, Tamil Nadu and other institutes. Over 4000 samples collected from 800 different species of marine flora and fauna were examined for bioactive compounds, giving about 500 isolates exhibiting various biological activities and identification of two compounds, one having anti-diabetic and another, having anti-hyper-lipidemic activity (Sudarsanam and Anita Devi, 2004). Table 7.4 indicates some drugs isolated from sponges and their activities

MARINE BACTERIA

Marine bacteria are involved in nutrient cycling and the degradation of marine organic matter. Generally they form part of a symbiotic association with hosts such as algae. Recent extensive chemical investigations on microbial extracts from marine environments have led to the discovery of a variety of secondary metabolites (Jensen and Fenical, 2005). A survey

Table 7.4 Some drugs isolated from sponges and their activities

Drug	Function
Benzamide A and B	Tumor growth inhibitor
Contignasterol	Anti-asthma agent
Debromohymenialdisine	Anti-Alzheimer agent; treatment against osteoarthritis
Discodermolide	Apart from anticancer properties, possesses immuno-suppressive and cytotoxic activity
Girodazole	Anti-tumor activity, inhibits protein synthesis in eukaryotic target cells
KRN7000 (α-galactosylceramide)	Anti-tumor, immuno-stimulatory activity. Exhibits cytotoxic and anti-tubulin activity
Manoalide	Anti-inflammatory agent. Holds potentials against rheumatoid arthritis
Topsentins	Anti-inflammatory agent. Show promising for the treatment of pancreatic cancer
Dictyostatin	Inhibits the growth of human cancer cells
Lasonolides	Anti-cancer activity
Peloruside A	Antimiotic agent. Directly induces tubulin polymerization
Dercitin	Alkaloid having anti-leukemic and anti-tumor activity
Salicylamides	Potential to treat tumor and osteoporosis

Source: U.S. National Sea Grant Program, Harbor Branch, California; www.marinebiotech.org; Ireland et al. (2000).

of culturable heterotrophic bacteria associated with the marine ark shell *Anadara broughtoni* inhabiting the Sea of Japan led to isolates having potent antimicrobial activities. Several bacterial strains isolated from green and brown marine algae, displayed antibiotic activity. Antibacterial spectra of all the strains include activity against *Staphylococcus, Alcaligenes, Pseudomonas, Vibrio, Pasteurella*, and *Achromobacter* spp. (Lemos, 1985). Salinosporamide A is a potent anticancer agent that recently entered phase I human clinical trial for the treatment of multiple myeloma. The novel natural product is produced by the recently described obligate marine bacterium, *Salinispora tropica*, which is found in ocean sediment (Feling et al., 2003).

A few marine microorganisms have been observed to produce polyunsaturated fatty acids (PUFA). Microbial production of PUFA has been essentially attributed to five well known genera, namely, *Shewanella, Colwella, Photobacterium, Psychromonas* and *Moritella* (Berge and Barnathan, 2005). The chief advantages of biotechnological production of PUFA from microorganisms are that these organisms produce essentially a single type of PUFA, rather than a complex of diverse PUFA types, synthesized by fish or algae, therefore providing better consistency and purity of the fatty acid. *Shewanella putrefaciens* strain ACAM 342 was found to produce the PUFA, $C_{18:2\omega3}$, $C_{18:3\omega3}$ and $C_{20:5\omega3}$ both under aerobic and anaerobic conditions at 15°C and 25°C. High prevalence of PUFA producing bacteria in arctic invertebrates has been observed. More than 100 bacterial strains including *Pseudomonasa* and *Vibrio* spp., capable of producing DHA and EPA (eicosapentaenoic acid), were isolated from arctic and sub-arctic invertebrates and also from some fish species (Jostenseri and Landfaid, 1997). Recently advanced molecular techniques have enhanced the prospects of isolation and identification of PUFA from marine microorganisms (Nichols, 2003). In addition to bacteria, fungi could also provide PUFA.

A growing number of marine fungi are the sources of novel and potentially life-saving bioactive secondary metabolites (Cole and Schweikert, 2003). Filamentous fungi have long been used by the fermentation industry for the production of metabolites including antibiotics and enzymes. Along with developments in molecular genetics, research on bioprocessing technologies may have a competitive advantage for production of enzymes, healthcare products including generic biopharmaceuticals from fungi (Wang et al., 2005b; Bhadury et al., 2006). A process for growing fungi, *Schizochytrium* and *Thraustochytrium* spp., and for production of PUFA has been patented. These fungi were grown in fermentation medium containing non-chloride sodium salts, in particular, sodium sulfate. The process produced fungi with a cell aggregate size that was useful for the production of feeds for use in aquaculture (Barday, 2006).

Cultivation and Genomics of Marine Microorganisms

Most products obtained from microbial growth are formed as secondary metabolites. Therefore, optimization of culture conditions is essential for maximum recovery of these compounds. The majority of marine microbes cannot be cultured under artificial laboratory conditions and hence have posed problems in detailed taxonomical and physiological characterization. Nevertheless, interesting laboratory conditions have been created for cultivation of microorganisms which are adapted to extreme environments (Antranikian et al., 2005). Recent research extending the last two decades has provided genomics, post-genomic cloning, protein expression and other gene techniques for production of these metabolites, opening up new possibilities for the exploration of pathways responsible for the synthesis of metabolites of biotechnological interest. The function of the majority of genes within the sequenced marine genomes is not well understood. In order to assign potential functions to the genes of a genome, functional genome analysis techniques are used. These techniques include the expression profiling of the whole set of genes by using genomic DNA arrays and/or proteomics. These techniques are not only suitable for exploration of the functions of the proteins but also help to find new potential drug targets. Proteome and transcriptome analysis techniques have led to a shift from direct antimicrobial screening programs toward rational target-based strategies. The proteome technique is mainly based on two-dimensional protein gel electrophoresis, used for protein separation, and mass spectrometry, applied for protein identification. Complete genome sequence of the organism of interest is mandatory for the proteome analysis. The proteome of only few marine microorganisms has been investigated so far. Most of these proteomic studies explore how marine bacteria adapt to alterations in their environmental conditions. The complete genes of three strains of the abundant cyanobacteria, *Prochlorococcus* have been sequenced and analyzed. The genomic database for cyanobacteria is available (http://www.kazusa.or.jp/cyano/cyano. html). Heterogeneous expression techniques are commonly deployed when the original microbe is unable to have a sufficient growth and/ or compound production rate. In this case, the genes responsible for the production of the target compound are isolated and inserted into a more favorable host, one having either a faster growth rate or more highly developed expression/production system. There is scope for applying genetics to microorganisms for improved yield and hence, for reduction of cost of production of PUFA (Berge and Barnathan, 2005).

CONCLUSION AND FUTURE TRENDS

In summary, the sea is rich in a multitude of living resources. While the food potential of the oceans has been well realized, there is much scope for exploitation of marine organisms including seaweed, corals, microalgae and other marine microorganisms for nutraceuticals and bioactive compounds. Microalgae have industrial uses as raw material for carotenoids, vitamin and fatty acid supplements and as feed additives for poultry, livestock, fish and crustaceans. Microalgae have the potential to satisfy the growing worldwide demand for carotenoids to a significant extent. Pigments from microalgae such as *Dunaliella* are distributed in markets under different categories, under the names, β-carotene extracts, Dunaliella powder for human use and dried Dunaliella as animal feed. Future trends involve combinatorial engineering and production of niche pigments not found in plants. Microalgae and also other microorganisms as well as corals can provide numerous bioactive compounds including hormones, antibiotics, and pharmaceuticals. Some of them include benzamide A and B, bryostatins, debromohymenialdisine, KRN7000 (α-galactosylceramide), manoalide, dictyostatin, salicylhalamides, etc. There are a few marine-derived drugs at advanced stages of development particularly in the anti-cancer, anti-viral and anti-inflammatory areas.

The scope exists for searching for novel compounds from the marine environment that can serve as models for drug development. Marine biotechnology encompasses those efforts that help in harnessing marine resources of the world. With the current technologies in bio-engineering and synthetic chemistry, pharmaceutical industry can achieve success in processing of marine organisms for the development of drugs and other important compounds for healthcare.

REFERENCES

Adams, M. (1997). Super-foods for optimal health: *Chlorella* and *Spirulina*. In: *Spirulina platensis* (Arthrospira): Physiology, Cell-biology and Biotechnology. A.Vonshak, (ed). Taylor & Francis, London, UK.

Antranikian, G., Vorgias, C.E. and Bertoldo, C. (2005). Extreme environments as a resource for microorganisms and novel biocatalysts. Adv. Biochem. Eng. Biotechnol. 96: 74-79.

Arad, S.M. and Yaron, A. (1992). Natural pigments from red microalgae for use in foods and cosmetics. Trends Food Sci. Technol. 3: 92-97.

Baine, M. (2001). Artificial reefs: a review of their design, application, management and performance. Ocean Coastal Manage. 42: 241-248.

Barday, W.R. (2006). *Schizochytrium* and *Thraustochytrium* strains for producing high concentrations of ω-3 fatty acids. US Patent No. 7 022 512 BZ.

Becker, E.W. and Venkataraman, L.V. (1984). Production and utilization of the brown-green alga, *Spirulina* in India. Biomass 4: 105-107.

Berge, J-P. and Barnathan, G. (2005). Fatty acids from lipids of marine organisms: molecular diversity, role as biomarkers, biologically active compounds and economical aspects. Adv. Biochem. Eng. Biotechnol. 96: 49-53.

Bhadury, P., Mohammed, B.T. and Wright, P.C. (2006). The current status of natural products from marine fungi and their potential as anti-infective agents. J. Ind. Microbiol. Biotechnol. 33: 325-329.

Bhakuni, D.S. and Rawat, D.S. (2005). Bioactive metabolites of marine algae, fungi and bacteria. In: Bioactive Marine Natural Products. D.S. Bhakuni and D.S. Rawat (authors) Springer, The Netherlands pp. 1-25.

Borowitzka, M.A. (1999). Commercial production of microalgae: ponds, tanks, tubes and fermenters. J. Biotechnol. 50: 313-318.

Burja, A. and Radianingtylas, A. (2005). Marine microbial-derived nutraceuticals biotechnology: an update. Food Sci. Technol. 19: 14-19.

Campo, J.A. del., Garcia-Gonzalez, M. and Guerrero, M.G. (2007). Outdoor cultivation of microalgae for carotenoid production: current state and perspectives. Appl. Microbiol. Biotechnol. 74: 1163-1169.

Capon, R.J., Rooney, F., Murray, L.M., Collins, E., Sim, A.T.R., Rostas, J.A.P., Butter, M.S. and Caroll, A.R. (1998). Dragmacidins: New protein phosphatase inhibitors from a southern Australian deep-water marine sponge, *Spongosorites* spp., J. Nat. Prod. 61: 660-662.

Cerrano, C., Calcinai, B., Cucchiari, E., Di Camillo, C., Nigro, M., Regoli, F., Sara, A., Schiapparelli, S., Totti, C. and Bavestrello, G. (2004). Are diatoms a food source for Antarctic sponges? Chem. Ecol. 20: 57-64.

Cole, R.J. and Schweikert, M.A. (2003. Handbook of Fungal Metabolites. Academic Press, London, UK.

Douglas A.E. (2003). Coral bleaching—how and why? Review. Mar. Pollut. Bull. 46: 385-391.

Duckworth, A. (2009). Farming sponges to supply bioactive metabolites and bath sponges: A Review. Mar. Biotechnol.11: 669-679.

Dufosse, L., Murthy, K.N.C. and Ravishankar, G.A. (2005). Microorganisms and microalgae as sources of pigments for food use: a scientific oddity or an industrial reality? Trends Food Sci. Technol. 16: 389-393.

El-Gamal, A.A.H., Chiu, E.P., Li, C-H., Cheng, S-Y., Dai, C-F. and Duh, C-Y. (2005). Sesquiterpenoids and non-sesquiterpenoids from the Formosan soft coral, *Lemnalia laevis*. J. Nat. Prod. 68: 1749-1953.

Fabregas, J. and Herrero, C. (1990). Vitamin content of four marine microalgae: Potential use as source of vitamins in nutrition. J. Ind. Microbiol. Biotechnol. 5: 259-262.

Falkowski, P. G. (2002). The ocean's invisible forest. Sci. Amer., August issue, p. 47.

FAO (2004). The State of World Fisheries and Aquaculture, Part 4, Outlook, Editorial Production and Design Group, Food and Agriculture Organization of the United Nations, Rome, Italy.

Feling, R.H., Buchanan, G.O., Mincer, T.J., Kauffman, C.A., Jensen, P.R., and Fenical, W. (2003). Salinosporamide A: a highly cytotoxic proteasome inhibitor from a novel microbial source, a marine bacterium of the new genus *Salinospora*. Angew. Chem. Int. Ed. Eng. 42: 355-357.

Gouveia, L., Nobre, B.P., Marcelo, F.M., Mrejen, S., Cardoso, M.T., Palavra, A.F. and Mendes, R.L. (2007). Functional food oil colored by pigments extracted from microalgae with supercritical CO_2 Food Chem. 101: 717-721.

Gross, H. and Kong, G.M. (2006). Terpenoids from marine organisms: Unique structures and their pharmacological potential, Phytochem. Rev. 5: 115-119.

Haard, N.F. (1998). Specialty enzymes from marine organisms. Food Technol. 52: 64-68.

HBOI (2006) Harbor Branch Oceanographic Institution, Drug Discovery Team to Explore Newly Discovered Deep-sea Reefs, http://www.fau.edu/hboi/

Henkel, J. (1998). FDA Consumer magazine (January-February issue, US. FDA, Washington D.C, USA.

Herrero, M., Jaime, L., Martín-Álvarez, P. J., Cifuentes, A., and Ibáñez, E. (2006). Optimization of the extraction of antioxidants from *Dunaliella salina* microalga by pressurized liquids, J. Agric. Food Chem. 54: 5597-5603.

Hooper, N.A. and van Soest, R.W.M. (2002). Systema Porifera: A guide to the Classification of Sponges. Kluwer Academic/Plenum Press, New York, USA.

Ip, P.-F., and Chen, F. (2005). Production of astaxanthin by the green microalga *Chlorella zofingiensis* in the dark. Process Biochem. 40: 733-737.

Ireland, C.M., Copp, B.R., Foster, M.P., McDonald, L.A., Radisky, D.C. and Swersey, J.C. (2000). Bioactive compounds from the sea. In: Marine and Freshwater Products Handbook. R.E. Martin, E.P. Carter, G.J. Flick, Jr. and L.M. Davis (eds). Technomic, Lancaster, UK.

Irigoien, X. et al. (2004). Global biodiversity patterns of marine phytoplankton and zooplankton. Nature 429: 863-869.

Jeffrey, S.W., Vesk, M., and Mantoura, R.F.C. (1997). Phytoplankton pigments: windows into the pastures of the sea. Nature and Resources 33: 14-19.

Jensen, P. R. and Fenical, W. (2005). New natural products diversity from marine actinomycetes. In: Natural Products: Drug Discovery and Therapeutic Medicines. L. Zhang and A.L. Demain (eds). Humana Press, New Jersey, USA pp. 315-324.

Jostenseri, J-P. and Landfaid, B. (1997). High prevalence of polyunsaturated fatty acid producing bacteria in arctic invertebrates. FEMS Microbiol Lett. 151: 395-399.

Kim, S-K., Ravichandran, Y.D., Khan, S.B. and Kim, Y.T. (2008). Prospective of the cosmeceuticals derived from marine organisms. Biotechnol. Bioprocess Eng. 13: 511-523.

Kurano, N. and Miyachi, S. (2004). Microalgal studies for the 21st Century. Hydrobiologia 512: 27-31.

Liebezeit, C. (2005). Aquaculture of 'non-food organisms' for natural substance production. Adv. Biochem. Eng. Biotechnol. 97: 1-28.

Lemos, M. L. (1985). Antibiotic activity of epiphytic bacteria isolated from intertidal seaweeds. Earth Environ. Sci. 11: 149-154.

Lorenz, R.T. and Cysewski, G.R. (2000). Commercial potential for *Haematococcus* microalgae as a natural source of astaxanthin. Trends Biotechnol. 18: 160-163.

Madigan, M.T. and Martinko, J.M. (2005). Biology of Microorganisms, 11th edn., Pearson Education Inc., London, UK.

Matsunaga, T. Takeyama, H., Miyashita H. and Yokouchi, H. (2005). Marine microalgae. Adv. Biochem. Eng. Biotechnol. 96: 165-188.

Mitchell, S.A. and Richmond, A. (1998). Optimization of a growth medium for *Spirulina* based on cattle waste. Biol. Wastes 25: 41-44.

Muller, W.E.G. (2004). Sustainable production of bioactive compounds by sponges: Cell culture and gene cluster approach: a review, Mar. Biotechnol. 6: 117-122.

Nichols, D.S. (2003). Prokaryotes and the input of polyunsaturated fatty acids to the marine food web. FEMS Microbiol. Lett. 219: 1-5.

Nybakken, J.W. (1997). Marine Biology: An Ecological Approach, 4th edn., Addison Wesley, Longman, UK.

Ohr, L.M. (2005). Riding the nutraceutical wave. Food Technol. 8: 95-97.

Osinga, R., Tramper, J. and Wijffels, R.H. (1999). Cultivation of marine sponges. Mar. Biotechnol. 1: 509-514.

Otero A, and Vincenzini M. (2003). Extracellular polysaccharide synthesis by *Nostoc* strains as affected by N source and light intensity. J. Biotechnol. 102: 143-148.

Pisal, D.S. and Lele, S.S. (2005). Carotenoid production from microalga, *Dunaliella salina*. Ind. J. Biotechnol. 4: 476-480.

Plaza, M., Herrero, M. Cifuentes, A and Ibaflez, E., (2009). Innovative natural functional ingredients from microalgae. J. Agric. Food Chem. 57: 7159-7170

Prabhudas, P., Srivastaav, P.P. and Mishra, M.N. (2009). Microaglae (*Spirulina*) biomolecules and its food uses—a review. Indian Food Industry, 28: 39-44.

Proksch, P., Edrada, R.A. and Ebel, R. (2002). Drugs from the seas—current status and microbiological implications. Appl. Microbiol. Biotechnol. 59: 125-131.

Ranga Rao, A., Sarada, R., Bhaskaran, V. and Ravishankar, G.A. (2006). Antioxidant activity of *Botryococcus braunii* extract elucidated in vitro models. J. Agric. Food Chem. 54: 4593-4597.

Rao, C.B., Anjaneyula, A.S.R., Sarma, N.S., Venkateswaralu, Y., Rosser, R.M., Faulkner, D.J., Chen, M.H.M. and Clardy, J. (1984). Zoanthamine: a novel alkaloid from a marine Zoanthid. J. Am. Chem. Soc. 106: 7963-7984.

Rebolloso-Fuentes, M.M., Navarro-Pérez, A., García-Camacho, F., Ramos-Miras, J.J. and Guil-Guerrero, J.L. (2001). Biomass nutrient profiles of the micro-alga *Nannochloropsis*, J. Agric. Food Chem. 49: 2966-2972.

Renaud, S. M., Parry, D.L., and Thinh, L-V. (1994). Microalgae for use in tropical aquaculture I: Gross chemical and fatty acid composition of twelve species of microalgae from the Northern Territory, Australia. J. Appl. Phycol. 6: 337-345.

Sakthivel, M., Ramathilagam, G. and Pusharaj, A.(2005). Field study on corals and coral living organisms in Van Tivu, in the Gulf of Mannar. Fishery Technol. 42: 11-17.

Santhanam, K. and Venkataramanjuam. (1996). Impact of industrial pollution and human activities on coral resources of Tuticorin (South India) and methods for conservation. Proc. Int. Coral Reef Symp., Panama City, Panama.

Schweder, T., Lindequist, U. and Lalk, M. (2005). Screening for new metabolites from marine microorganisms. Adv. Biochem. Eng. Biotechnol. 96: 1-9.

Sen, D.C. and Sarkar, S. (2004). *Spirulina*: A classical health food. Beverage Food World (India) 31: 45-49.

Shi, X-M., Jiang, Y. and Chen, F. (2002). High-yield production of lutein by the green microalga, *Chlorella protothecoides* in heterotrophic fed-batch culture, Biotechnol. Progr. 18: 723-727.

Sipkema, D., Franssen, M.C.R., Osinga, R., Tramper, J. and Wijffels, R.H. (2005). Marine sponges as pharmacy. Mar. Biotechnol, 7: 142-148.

Sudarsanam, D. and Anita Devi, C.P. (2004). Marine wealth and human health: A pharmaceutical perspective. Aquaculture Foundation of India, Proc. Int. Conf. Marine Living Resources of India for Food and Medicine, February, 27-29, 2004, Chennai, India.

Suetsuna, K. and Chen, J.-R. (2001). Identification of antihypertensive peptides from peptic digest of two microalgae, *Chlorella vulgaris* and *Spirulina platensis*. Mar. Biotechnol. 3: 305-309.

Thakur, A.N., Thakur, N.L., Indap, M.M., Pandit, A.R., Vrushali V., Datar, V.V. and Müller, W.E.G. (2005). Antiangiogenic, antimicrobial, and cytotoxic potential of sponge-associated bacteria. Mar. Biotechnol. 7: 245-250.

Tokusoglu, O., and Unal, M.K. (2003). Biomass nutrient profiles of three microalgae: *Spirulina platensis*, *Chlorella vulgaris*, and *Isochrisis galbana*. J. Food Sci. 68: 4-9.

Venugopal, V. (2008). Drugs and pharmaceuticals from marine sources. In: Marine Products for Healthcare: Functional and Bioactive Nutraceuticals from the Ocean. V. Venugopal (author) CRC Press, Boca Raton, Fl., USA.

Volkman, J.K. (1999). Australasian research on marine natural products: chemistry, bioactivity and ecology. Mar. Freshwater Res. 50: 761-767.

Voss, J.D. and Richardson, L.L. (2006), Nutrient enrichment enhances black band disease progression in corals. Coral Reefs. 25: 569-573.

Wang, Y., Chang, C.F., Chou, J. Cheng, H.L., Deng, X., Harvey, B.K., Cadet, J.L. and Bickford, P.C. (2005a). Dietary supplementation with blueberries, spinach, or spirulina reduces ischemic brain damage. Exp. Neurol. 193: 75-79.

Wang, L., Ridgway, D., Gu, T. and Moo-Young, M. (2005b). Bioprocessing strategies to improve heterologous protein production in filamentous fungal fermentation, Research review. Biotechnol. Adv. 23: 115-129.

White, R.A. and White, E.W. (2003). Biomedical applications of marine natural products. In: Marine Biotechnology in the 21st Century: Problems, Promises and Products, National Academy Press, Washington, D.C., USA.

8

Microbial Remediation of Fish and Shrimp Culture Systems and their Processing Industry Wastes

Wanchai Worawattanamateekul[1,*] and *Ramesh C. Ray*[2]

INTRODUCTION

Aquaculture is the world's fastest growing food production sector (Montet and Ray, 2009). It was once considered an environmentally sound practice because of its traditional polyculture and integrated systems of farming based on optimum utilization of farm resources, including farm wastes. Increased production is being achieved now by the expansion of land and water under culture and the use of more intensive and modern farming technologies that involve higher usage of inputs such as water, feeds, fertilizers and chemicals. These fed fish farms produce large amounts of wastes, including dissolved inorganic nitrogen and phosphorus. Likewise, marine aquaculture effluents generate substantial amounts of nutrients (nitrogen and phosphorus) and organic matter that exert a biological oxygen demand (BOD) contributing to the deterioration of the quality of receiving water. Effluent water from fish and shrimp farms typically contains dissolved nutrients and suspended particulates which have been implicated as causing significant environmental impacts such as eutrophication (Jones et al., 2001). The cumulative impact of shrimp/fish farm wastewater discharges on coasts as a whole and the discharges of several farms on a single bay is therefore, a serious concern. As the aquaculture industry develops, efficient, cost-effective and environmentally

[1]*Department of Fishery Products, Faculty of Fisheries Kasetsart University, Bangkok 10900, Thailand;
Tel.: 66-2-5790113 press 4088; Fax: 66-2-9428644 press 12; E-mail: ffiswcw@ku.ac.th*
[2]*Central Tuber Crops Research Institute (Regional Centre), Bhubaneswar 751 019, India;
Tel/Fax: 91-674-2470528; E-mail: rc_rayctcri@rediffmail.com*
*Corresponding author

friendly preventive and bioremediation methods of improving effluent water quality prior to discharge into receiving waters of sensitive areas are the priority (Jones et al., 2001). Therefore, regulatory agencies have developed standards and criteria for the aquaculture industry concerning effluent disposal (Kinne et al., 2001). Similarly, fish and shrimp processing industry generate large amounts of wastes that generally account for > 50% of the live weight. Major by-products generated through processing include visceral wastes, scales, waste (wash) water, filling wastes (head, frame bones, skins and fins), air bladders, body/head shell wastes, calcareous shells, etc. These wastes, unless undergoing bioremediation can cause serious environmental problems in term of obnoxious odour and emission of toxic gas like ammonia, hydrogen sulphide, etc. In this context, disease management, productivity, good processing practices and environment protection are the four key factors for successful and sustainable aquaculture. The present chapter presents an overview of the principles and applications of bio-/microbial-remediation in addressing the waste management of fish and shrimp culture systems and processing industry wastes.

BIOREMEDIATION

The current approach to improving water quality in aquaculture as well as detoxifying the fish and shrimp processing industry wastes is by bioremediation. Bioremediation is a general concept that includes all those processes and actions that take place in order to bio-transform an environment, already altered by contaminants, to its original status. Although the processes that can be used in order to achieve the desirable results vary, they still have the same principles; the use of microorganisms or their enzymes, that are either indigenous and are stimulated by the addition of nutrients or optimization of conditions, or are seeded into the medium (Thassitou and Arvanitoyannis, 2001). Bioremediation is recognized as an inexpensive, effective, and environmentally safe technology, which offers new and innovative ways to clean up hazardous wastes.

When microorganisms and/or their products are used as additives to improve water quality, they are referred to as bio(microbial)-remediators or microbial remediating agents (Moriaty, 1998). They result in a lower accumulation of slime or organic matter at the bottom of the pond , better penetration of oxygen into the sediment and a generally better environment for the farmed stock (Rao and Karunasagar, 2000). The isolation and development of indigenous bacteria are required for successful microbial remediation (Austin and Brunt, 2009). A successful microbial remediation process in the aquaculture system involves:

- maximizing carbon mineralization to carbon dioxide to minimize sludge accumulation; maximizing primary productivity that stimulates shrimp production and also secondary crops;
- optimizing nitrification rates to keep low ammonia concentration;
- optimizing denitrification rates to eliminate excess nitrogen from ponds as nitrogen gas;
- maximizing sulphide oxidation to reduce accumulation of hydrogen sulphide; and
- maintaining a diverse and stable pond community where undesirable species do not become dominant.

WASTE PRODUCTION IN FISH AND SHRIMP CULTURE SYSTEMS

The physical, chemical and biological conditions of the culture environment have an influence on the health and productivity of fish and shrimp. Exposure of fish and shrimps to toxins like ammonia, carbon dioxide and hydrogen sulphide lead to stress and ultimately disease. The types of wastes produced in aquaculture farms are basically similar. However, there are differences in quality and quantity of components depending on the species cultured and the culture practices adopted. The wastes in hatcheries or aquaculture farms can be categorized as: (1) residual food and faecal mater; (2) metabolic by-products; (3) residues of pesticides; (4) fertilizer derived wastes; (5) wastes produced during moulting; and (6) collapsing algal blooms (Sharma and Scheeno, 1999).

Microbial Remediation of Organic Detritus

Organic detritus is accumulated in aquaculture ponds because of excess feeds, faeces and dead algae. Aeration by itself is not enough to decompose these organic detritus. Active bacterial populations must be changed to species that are adapted to rapid degradation of complex organic matter. Members of the genus *Bacillus,* i.e., *Bacillus subtilis, B. licheniformis, B. cereus, B. coagulans,* and *B. polymyxa* are good examples of bacteria suitable for microbial remediation of organic detritus. However, these are not normally present in the required amounts in the water column, their natural habitat being the sediment. When certain *Bacillus* strains are added to the water frequently and at high density, they degrade organic matter faster than in situations where only the natural populations are available. As a part of bio-augmentation, the *Bacillus* can be produced, mixed with sand, clay, rice husk and wheat bran or cow dung and broadcasted to be deposited at the bottom of the pond. These bacteria produce a variety of enzymes that break down proteins and starch to small molecules, which are then taken up as energy sources by other organisms. The removal of large organic compounds reduces water turbidity (Haung, 2003). *Bacillus* based products are now commercially available which can degrade organic matter in a aquaculture pond at a faster level (Fig. 8.1) (Moriarty et al., 2005).

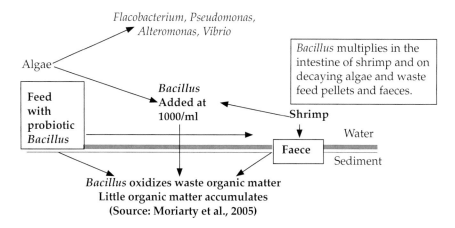

Fig. 8.1 Effect of *Bacillus* at high population density in the pond for organic matter from algae, feed and animals. Specially selected *Bacillus* displaces pathogenic *Vibrio* (Moriarty et al., 2005).

Color image of this figure appears in the color plate section at the end of the book.

Microbial Remediation of Nitrogenous Compounds

Nitrogen applications in excess of pond assimilatory capacity can lead to deterioration of water quality through the accumulation of nitrogenous compounds (e.g., ammonia and nitrite) with toxicity to fish and shrimp (Antony and Philip, 2006). The principal sources of ammonia are fish excretion and sediment flux derived from the mineralization of organic matter and molecular diffusion from reduced sediment, although cyanobacterial nitrogen fixation and atmospheric deposition are occasionally important (Ayyappan and Mishra, 2003). Nitrification proceeds in two steps as follows:

$$NH_4^+ + 1^1/_2 O_2 \longrightarrow NO_2^- + 2H^+ + H_2O$$
$$NO_2^+ 11/_2 \longrightarrow O_2 NO_3^-$$

Bacteriological nitrification is the most practical method for the removal of ammonia from closed aquaculture systems and it is commonly achieved by setting of a sand and gravel bio-filter through which water is allowed to circulate.

The ammonia oxidizers are placed under four genera:

Nitrosomonas. Ellipsoidal or short rods;
Nitrosococcus. Spherical cells;
Nitrosolobus. Lobate and pleomorphic cells; and
Nitrosospira. Spiral-shaped cells.

The nitrite oxidizers belong to only one genus: *Nitrobacter.* Short rods. However, *Nitrosomonas* and *Nitrobacter* are the most common nitrifiers commonly encountered in the aquaculture system.

There are also some heterotrophic nitrifying bacteria (i.e., *Pseudomonas, Bacillus, Alcaligenes,* etc.) that produce only low levels of nitrite and nitrate and often use organic sources of nitrogen rather than ammonia or nitrite (Focht and Verstraete, 1977).

Nitrifers in contaminated cultures have been demonstrated to nitrify more efficiently. Nitrification not only produces nitrate but also alters the pH slightly towards the acidic range, facilitating the availability of soluble materials (Ayyappan and Mishra, 2003). The vast majority of aquaculture ponds accumulate nitrate, as they do not contain a denitrifying filter. Denitrifying filters helps to convert nitrate into nitrogen. It creates an anaerobic region where anaerobic bacteria can grow and reduce nitrate to nitrogen gas.

$$NO_3^- \longrightarrow NO_2^- \longrightarrow NO \longrightarrow N_2O \longrightarrow N_2$$

Nitrogen (N_2) gas diffuses into the air and does not pose a toxicity threat to culture animals. However, ammonia and nitrite released by certain denitrifying bacteria can be toxic to fish, shrimp, and other aquaculture organisms.

At National Centre for Aquatic Animal Health, Cochin University of Science and Technology, Cochin, India, four nitrifying bacterial consortia, which efficiently oxidize ammonia to nitrate, have been developed. Their effectiveness has been demonstrated under different conditions in hatcheries. They have been integrated into two reactor systems, namely, *In-Situ* Stringed Bed Suspended Bioreactors and *Ex-Situ* Packed Bed Bioreactors in immobilized form (Patent No. 828/DEL/2000).

Microbial Remediation of Hydrogen Sulphide

Sulphur is of some interest in aquaculture because of its importance in anoxic sediments. In aerobic conditions, organic sulphur decomposes to sulphide, which in turn gets oxidized to sulphate. Sulphate is highly soluble in water and so gradually disperses from sediments. Sulphide oxidation is mediated by microorganisms in the sediment: the predominant microorganisms are *Thiobacillus*, *Bacillus*, *Arthrobacter*, *Flavobacterium* and *Pseudomonas* (Boyd, 1995). Under anaerobic conditions, sulphate may be used in place of oxygen in microbial metabolism. This process leads to the production of hydrogen sulphide (H_2S) gas. The H_2S is produced by a series of microbially mediated reductions (Boyd, 1995).

$$SO_4^{2-} + 4H_2 + 2H^+ \longrightarrow H_2S + 4H_2O \text{ (Djurle, 2003)}$$

Organic loading can stimulate H_2S production and reduction in the diversity of benthic fauna (Mattson and Linden, 1983). H_2S is soluble in water and has been suggested as the cause of gill damage and other ailments in fish (Beveridge, 1987). Bioassays of several species of fish suggest that any detectable concentration of H_2S should be considered detrimental to fish production (Boyd, 1979). The photosynthetic benthic bacteria that break H_2S at the bottom of the pond have been widely used in aquaculture to maintain a favourable environment (Boyd and Silapajarn, 2006). These bacteria contain bacterio-chlorophyll that absorb light (blue to infrared spectrum, depending on the type of bacterio-chlorophyll) and perform photosynthesis under anaerobic conditions (Haung, 2003). These are purple and green sulphur bacteria that grow at the anaerobic portion of the sediment-water interface. Photosynthetic purple non-sulphur bacteria can decompose organic matter, H_2S, NO_2 and harmful wastes of ponds. The green and purple sulphur bacteria split H_2S to utilize the wavelength of light not absorbed by the overlying phytoplankton. The purple and green sulphur bacteria obtain reducing electrons from H_2S at a lower energy cost than H_2O splitting photoautotrophs and thus require lower light intensities for carrying out photosynthesis. The general equation of this reaction is as follows:

$$CO_2 + 2H_2S \longrightarrow (CH_2O) + H_2O + 2S$$
$$S + CO_2 + 3H_2S \longrightarrow (CH2O) + H_2SO_4$$
$$CO_2 + NaS_2O_3 + 3H_2O \longrightarrow 2(CH_2O) + NaS_2O_4 + H_2SO_4$$

Chromatiaceae and Chlorobiaceae are the two families of photosynthetic sulphur bacteria that favour anaerobic conditions for growth while utilizing solar energy and sulphide. Chromatiaceae contain sulphur particles in cells but Chlorobiaceae precipitate them out. The family Rhodospirillaceae is useful for H_2S removal as they mainly utilize organic material, such as lower fatty acid, as a source of hydrogen. But they can be used as efficient mineralizers at the bottom of the pond as they grow in both aerobic and anaerobic conditions as heterotrophic bacteria even in the dark without utilizing solar energy (Boyd and Silapajarn, 2006). Photosynthetic bacteria of importance in aquaculture belong to the following three families (Haung, 2003):

Rhodospirillaceae: *Rhodospirillum, Rhodopseudomonas, Rhodomicrobium*
Chromatiaceae: *Chromatium, Thiocystis, Thiosarcina,Thiospirillum, Thiocapsa, Lamprocystis, Thiodictyon, Thiopedia, Amoebobacter, Ectothiorhodospira.*
Chlorobiaceae: *Chlorobium,Prosthecochloris,Chloropseudomonas, Pelodictyon, Clathrochloris.*

For bioremediation of H_2S toxicity, the bacterium that belongs to Chromatiaceae and Chlorobiaceae can be mass cultured and can be applied as pond bioremediators. Being autotrophic and photosynthetic, mass culture is less expensive and the cultured organisms can be adsorbed on to the sand grains and applied so that they may reach the bottom of the pond to enrich the hypolimnion and ameliorate H_2S toxicity (Boyd and Silapajarn, 2006).

Microbial Remediators as Disease Controlling Agents

In recent years, there has been growing interest in biocontrol of microbial pathogens in aquaculture using antagonistic microorganisms. A study on the role of antagonistic bacteria, especially the co-existing bacteria, as biocontrol agents appears worthwhile in lieu of the negative impacts of antibiotics (Abraham et al., 2001). Most probiotics proposed as biological control agents in aquaculture belong to the lactic acid bacteria (*Lactobacillus, Carnobacterium, Enterococcus* etc.), *Vibrio* (*Vibrio fluvialis*), *Bacillus*, and *Pseudomonas*. This subject has been elaborately discussed in Chapter 7 of Volume 1 of this 2-volume series by Austin and Brunt (2009).

SCREENING AND MASS MULTIPLICATION OF MICROBES FOR UTILIZATION AS BIOREMEDIATORS

Microorganisms both Gram positive and Gram negative have been tested for their efficacy as microbial remediators in aquaculture by various workers (Table 8.1). *Bacillus* is the most commonly used organism followed by *Aeromonas* and *Pseudomonas*. These microorganisms are mass multiplied in bioreactor by either solid state or submerged fermentation.

Solid State Fermentation

Solid state fermentation refers to the process where growth and production of bio-remediating microorganisms occur on the surface of solid materials. This process occurs in the absence of "free" water, where the moisture is absorbed to the solid matrix such as rice husk, wheat bran, dried cow dung, etc. Solid state fermentation is usually carried out in an uncontrolled environment with no provision for maintaining pH, temperature, aeration, etc., and has a series of advantages over submerged fermentation including lower cost, improved product characteristics, higher product yield, easiest product recovery and reduced energy requirement (Ray et al., 2008). However, chance of outside contamination of undesired microorganisms cannot be ruled out in this system.

Table 8.1 Representative microorganisms used as bioremediating agents

Bioremediator Identity	Source	Used with	Method of application
Gram-positive bacteria			
Bacillus sp. 48	Common snook	*Centropomus undecimalis*	Added to water; reduced salinity
Bacillus sp.	Commercial product	Penaeids	Water
Bacillus sp.	Commercial product	Channel catfish	Spread in pond water
Bacillus S11	Gastro-intestinal tract of Penaeids	Shrimp, *Penaeus monodon*	Premix with feed
Bacillus (DMS series)	Commercial product	Shrimp, *P. monodon*	Added to water
Bacillus (BIOSTART)	Commercial product	Catfish, *Ictalurus punctatus*	Added to water
Mixed culture, mostly *Bacillus* sp.	Commercial product	*Brachionus plicatilis*	Mixed with water
Enterococcus faecium SF68	Commercial product	*Anguilla anguilla*	Oral administration
Gram-negative bacteria			
Aeromonas media	Unknown	*Crassostrea gigas*	Mixed with water
Aeromonas media	Unknown	*Crassostrea gigas*	Mixed with water
Lactobacillus plantarum	Turbot larvae	*Scophthalmus maximus*	Indirectly *via* rotifers
Lactobacillus plantarum (906)	Human faeces	Sea bream, *Sparus aurata*	Dry feed as vector
Lactobacillus rhamnosus ATCC 53103	Culture collection	Rainbow trout, *Oncorhynchus mykiss*	Mixed with feed
Photorhodobacterium sp.	Unknown	*Penaeus chinensis*	Mixed with water
Pseudomonas fluorescence	*Onchorhynchus mykiss*	*Onchorhynchus mykiss*	Mixed with water to 10^5 or 10^6 cells ml^{-1}
Pseudomonas sp.	*Onchorhynchus mykiss*	*Onchorhynchus mykiss*	Mixed in water
Roseobacter sp. BS 107	Unknown	Scallop larvae	Mixed in water

Source: Antony and Philip (2006), Austin and Brunt (2009).

Submerged Fermentation

In contrast to solid state fermentation, submerged fermentation is the process in which the growth and mass multiplication of microorganisms is carried out in well-defined liquid medium enriched with carbohydrate, nitrogen and other essential nutrients in sterilized bioreactor designed to maintain optimum pH, temperature, aeration, etc. for quality bio-inoculants production and the product (bio-inoculants) is totally free from outside contamination in such system (Ray et al., 2008).

COMMERCIAL PRODUCTS

Microbial remediators commercially available in the market mainly include organic detritus decomposers, nitrifiers, sulphur bacteria; *Bacillus* spp. and *Pseudomonas* spp. (Table 8.2) as well as probiotics (lactic acid bacteria). The bacteria used in these formulations are selected specifically for their ability to degrade organic materials, nitrifying ammonia and/or oxidizing hydrogen sulphide. In brief, they perform the following functions:

- Improve water in aquaculture systems
- Reduce ammonia levels, nitrates, algae and sludge buildup
- Reduce risk of infection and improves survival rates
- Reduce stock losses due to toxic ammonia
- Reduce or eliminates time required to oxidize the bottom of ponds between growing cycle.

Probiotics are mostly used to control various bacterial and viral diseases. Initial product applications may temporarily increase BOD level; it is therefore advisable to provide sub-surface aeration to expedite the establishment of the beneficial microbial species. A minimum dissolved oxygen level of 3% is required for their effective utilization.

NEW TECHNOLOGIES FOR MICROBIAL REMEDIATION

Microbial Mats

Microbial mats occur in nature as laminated heterotrophic and autotrophic vertically stratified communities typically dominated by cyanobacteria, eukaryotic microalgae like diatoms, anoxygenic phototrophic bacteria and sulphate reducing bacteria (D'amelio et al., 1989), but they can be cultured on a large-scale and manipulated for a variety of functions. Their physiological flexibility includes anoxygenic and oxygenic photosynthesis, inclusion of aerobes and anaerobes within the same matrix simultaneously (Caumette et al., 1994). The functional uses of mats broadly cover the areas of aquaculture and bioremediation. Regarding aquaculture, mats were shown to produce protein, *via* nitrogen fixation, and are capable of supplying nutrition to fish and shrimps. Most research on microbial mats has addressed bioremediation, within which two major categories of contaminants are examined: metals and radionuclides, and organic contaminants. Mats sequester or precipitate metals/radionuclide in a small volume. Organic contaminants are degraded and may be completely mineralized (Bender and Phillips, 2004).

Table 8.2 List of commercially available microbial remediators for aquaculture applications

Product	Microbial content	Company / firm
ABIL nitrifying package	Nitrifiers	Tropical marine centre, London, UK.
Alken Clear-flo 1002	*Bacillus* sp.	Alken Murray Corp., New York, USA.
Alken Clear-flo 1100	Nitrifying bacteria	Alken Murray Corp., New York, USA.
Alken Clear-flo 1400	Three species of *Bacillus* + two species of Nitrifying bacteria	Alken Murray Corp., New York, USA.
Ammonix	Nitrifying bacteria	Prowins Bio-Tech Pvt. Ltd., India.
Bactaclean	Nitrifiers	Enviro-Comp. Services, Inc., Dover,USA.
Biogreen	*Bacillus subtilis*	Activa Biogreen Inc., Wood Dale, USA.
Biostart	*Bacillus* sp.	Bio-CAT, Inc., Virginia, USA.
BRF-13A	*Nitrobacter, Nitrosomonas*	Enviro-reps., Ventura, CA, USA.
BRF-1A	Nitrifying bacteria	Enviro-Reps., Ventura, CA, USA.
BRF-4		Enviro-reps., Ventura, CA, USA.
BRF-4	Nitrifying bacteria	Enviro-reps., Ventura, CA, USA.
BZT ® Aquaculture	Nitrifiers	United-Tech, Inc., Indiana, USA.
Detro Digest	*Bacillus* sp.	NCAAH, CUSAT, India.
Eutro Clear	Nitrifying bacteria	Bioremediate. Com, LLC, Atlanta, USA.
Nitro Clear	*Nitrobacter, Nitrosomonas*	Bioremediate. Com, LLC, Atlanta, USA.
PBL - 44	Nitrifying bacteria / *Bacillus* sp.	Enviro-reps. Ventura, CA, USA.
Probac BC	*Bacillus* sp.	Synergy Biotechnologies, India.
Pronto	*Bacillus* sp.	Hort-Max ltd., New Zealand.
Ps-1	*Pseudomonas* sp.	NCAAH, CUSAT, India.
Remus	Nitrifying bacteria	Avecom, Belgium.
Super PS	Sulphur bacteria	CP aquaculture Pvt. Ltd., India.

Source: Antony and Philip (2006), modified.

A study was undertaken to investigate the biotechnological feasibility of using indigenous microbial consortia for the construction of microbial mats for the bioremediation of shrimp culture effluents. The treatment concept relied on the immobilization of natural marine microbial consortium on glass wool to mitigate the levels of dissolved nitrogen from a shrimp culture effluent. The results indicated that average efficiencies of ammonia nitrogen removal from shrimp (*Litopenaeus vannamei*) effluent was 97% and 95% for nitrate nitrogen, over a 20 d period of treatment. This treatment *via* constructed microbial mats was a technically feasible method for simultaneously reducing effluent nutrient loading (especially nitrate and ammonia) and for reducing organic loading (especially BOD_5) of shrimp culture effluents. Throughout the experiment, high values in oxygen concentrations (8 mg/l) were achieved. This was probably due to photosynthetic oxygen production by microalgae and cyanobacteria of the mat. The photoautotrophic (top of the mat) and heterotrophic community was dominated by diatoms (*Nitzchia* sp. and *Navicula* sp.) and filamentous forms of cyanobacteria (*Microcoleus chthonoplastes*, *Spirulina* sp., *Oscillatoria* sp., *Schizothrix* sp., *Calothrix* sp. and *Phormidium* sp.) as well as green algae as comprised by *Chlorella* sp. and *Dunaliella* sp. Mixed microbial mats containing filamentous cyanobacteria (*Oscillatoria* sp.) as the dominant species have been shown to remove nitrogenous compounds and other toxic chemicals from polluted sites (Bender and Phillips, 2004; Zachleder et al., 2002). In summary, marine bioremediation by constructed microbial mats systems could represent an efficient and environmentally sound alternative for the improvement of the environmental conditions of coastal sensitive regions and overall effluent water discharges (Paniagua-Michel and Garcia, 2003).

Aerobic and Anaerobic Sludge Treatments

Most smolt hatcheries are flow-through plants with water and energy demands of up to 700,000 m^3 and 105-245 MW h per 100,000 smolts produced. The anaerobic treatment of sludge from salmon smolt hatchery in a continuous stirred tank reactor at mesophilic temperature (35°C) and 55-60 d HRT (hydraulic retention time) was investigated (Gebauer and Eikebrokk, 2006). The main components of treated sludge with 1.5-3.3% total (dry) solids were 32% nitrogen, 8.5% phosphorous, 1.2% potassium, acceptable concentration of heavy metal apart from zinc and high levels of volatile fatty acids (VFA). The treated sludge was in liquid form and could be used as a liquid fertilizer on cultivated land and meadows; however, requirements for special means of application were needed. Furthermore, the methane content in biogas was stabilized at 59.4-60.5% vol., methane yield was 0.14-0.15 l/g COD, nitrogen mineralization increased to 70%, and 44.8-53.5% COD removed. The potential of the sludge for energy production was also exploited. The net energy production from the biogas was 43-47 MW h/year and could cover 2-4% of the energy demand in flow-through hatcheries, and at least twice as much in recirculation hatcheries. (Arvanitoyannis and Kassaveti, 2008).

Lanari and Franci (1998) examined the potential of biogas production by fish farm effluents in a small-scale close system with partially recirculated water. The system consisted of two fish tanks with a recirculation rate of 60% and a rainbow trout daily feeding allowance of 1%, 1.5% and 2% of live weight, an up-flow anaerobic digester connected with a sedimentation column and equipped with an aerobic filter run at psychrophilic conditions (24-25°C) and with hydraulic retention time (HRT) of 22-38 d, a zeolite column for final treatment of effluents, a gas flow meter and a methane analyzer. Biogas and methane production amounted to 49.8-144.2 l/day and 39.8-115.4 l/d, respectively. The highest biogas and methane production was reported at the highest feeding allowance, while the biogas methane content at 2% feeding allowance was higher than 80%. A remarkable reduction of volatile solids (92-97%), suspended solids (96-99%) and total ammonia nitrogen content (59-70%) in the anaerobic digester was reported; while the zeolite ion-exchange column improved water quality of effluent produced by the digester, as the chemical oxygen demand (COD) was reduced up to 45%. The produced biogas can be used directly in a burner to produce thermal energy or, following depuration, can be employed as fuel in a cogeneration plant to produce thermal and electrical or mechanical energy.

Marine Underwater Depuration System (MUDS)

An underwater device, able to favour the sea auto-cleaning capacities, is described herein. This system, called MUDS, consists of a percolating filter and is placed at sea over effluent water from fish farms outflow of a submarine pipeline. Due to the density difference, the water effluent flows through the percolating filter: this favours the mixing and a prompt recycling of organic matter, activating a marine trophic web. Rich microbenthic communities develop on the MUDS, both interstitially, inside the filter, and on the structure. The community mainly consists of ciliates, nematods, harpacticoid copepods and polychaetes, all of which being organisms that increase the depuration efficiency by consumption of organic matter. This structure acts also as a deterrent for the illegal trawling activity in the area, and attracts large numbers of several finfish species, thus working as a fish aggregating device (FAD) (Cattaneo-Vietti et al., 2003).

Foam Fractionation, Micro-algae Production and Oyster Filtration

Wastewater retention ponds in commercial farms at Atlantic coasts of France are efficient in removing up to 1 metric ton of particulate material (dry weight)/ha/d (faeces and unconsumed feed), but are inefficient in reducing dissolved wastes, both organic (urea, amino acids, protein) and inorganic (total ammonia nitrogen, phosphates). Forthcoming outdoor technologies to treat these forms of waste were examined by trials at different sites: treatment by foam fractionation in extensive systems (Italian fish pond culture), treatment by micro-algae production (*Skeletonema costatum*) and oyster filtration (*Crassostrea gigas*) in intensive systems (sea bass farm, *Dicentrarchus labrax*). It can be concluded that foam fractionation coupled with aeration and water circulation was a good way to treat and recirculate wastewaters in extensive systems, but that a multiple treatment combining a retention pond, foam fractionation and micro-algae-bivalve filtration, is the best solution to treat all these forms of wastes from intensive systems (Hussenot et al., 1998).

Cultivation of Macro-algae

To exploit fish-farm nutrients as a resource input, and at the same time to reduce the risk of eutrophication, the high-temperature adapted red alga *Gracilaria lemaneiformis* (Bory) Dawson from South China was co-cultured with the fish *Sebastodes fuscescens* in North China in warm seasons. Growth and nutrient removal from fish culture water were investigated in laboratory conditions in order to evaluate the nutrient bioremediation capability of *G. lemaneiformis*. Feasibility of integrating the seaweed cultivation with the fed fish-cage aquaculture in coastal waters of North China was also investigated in field conditions. Laboratory seaweed/fish co-culture experiments showed that the seaweed was an efficient nutrient pump and could remove most nutrients from the system. Field cultivation trials showed that *G. lemaneiformis* grew very well in fish farming areas, at maximum growth rate of 11.03%/ day. Mean C, N, and P contents in dry thalli cultured in Jiaozhou Bay were 28.9 ± 1.1%, 4.17 ± 0.11% and 0.33 ± 0.01%, respectively. Mean N and P uptake rates of the thalli were estimated at 10.64 and 0.38 μmol/g dry weigh/t h, respectively. An extrapolation of the results showed that a 1-ha cultivation of the seaweed in coastal fish farming waters would give an annual harvest of more than 70 t of fresh *G. lemaneiformis*, or 9 t dry materials; 2.5 t C would be produced, and simultaneously 0.22 t N and 0.03t P would be sequestered from seawater by seaweeds. Results indicated that the seaweeds could be suitable as a good candidate for seaweed/fish integrated mariculture for bioremediation and economic diversification (Zhou et al., 2006).

WASTE PRODUCTION IN FISH AND SHRIMP PROCESSING SYSTEMS

Fishery industry waste management has been one of the problems that have the greatest impact on the environment (Arvanitoyannis and Kassaveti, 2008). Fish and fishery products are highly traded, with more than 37% of total production entering international trade as different food and feed products. Export of fish and fish products has increased by 32% in the period 2000-2006. In 2006, more than 110 million tonnes (76.8%) of world fish production was used for direct human consumption (Table 8.3); 37.3% was in live and fresh form. Freezing is the main method of processing fish for food use, accounting for 19.9%, followed by prepared and preserved (11.3%) and cured fish (8.3%).

Table 8.3 World fisheries and aquaculture production and utilization

	FAO (*The State of World Fisheries and Aquaculture*); million tonnes									
	1997	1998	1999	2000	2001	2002	2003	2004	2005	2006
Total capture	93.6 (76.48%)	87.7 (74.20%)	93.8 (73.74%)	95.6 (72.92%)	92.9 (71.08%)	93.3 (69.78%)	90.5 (67.94%)	95.0 (67.62%)	94.2 (66.01%)	92.0 (64.07%)
Total aquaculture	28.8 (23.52%)	30.6 (25.80%)	33.4 (26.26%)	35.5 (27.08%)	37.8 (28.92%)	40.4 (30.22%)	42.7 (32.06%)	45.5 (32.38%)	48.5 (33.99%)	51.7 (35.93%)
Total world fisheries	122.4 (100%)	118.2 (100%)	127.2 (100%)	131.1 (100%)	130.7 (100%)	133.7 (100%)	133.2 (100%)	140.5 (100%)	142.7 (100%)	143.7 (100%)
Capture aquatic plants	1.4 (16.28%)	1.1 (11.34%)	1.2 (11.11%)	1.2 (10.53%)	1.2 (10.17%)	1.3 (10.08%)	1.3 (9.42%)	1.4 (9.15%)	1.3 (8.07%)	1.1 (6.79%)
Aquaculture aquatic plants	7.2 (83.72%)	8.6 (88.66%)	9.6 (88.89%)	10.2 (89.47%)	10.6 (89.83%)	11.6 (89.92%)	12.5 (90.58%)	13.9 (90.58%)	14.8 (91.93%)	15.1 (93.21%)
Total Aquatic plants	8.6 (100%)	9.7 (100%)	10.8 (100%)	11.4 (100%)	11.8 (100%)	12.9 (100%)	13.8 (100%)	15.3 (100%)	16.1 (100%)	16.2 (100%)
Utilization										
Human consumption	91.8 (74.8%)	93.6 (79.2%)	95.4 (75.2%)	96.8 (74.1%)	99.5 (76.4%)	100.7 (75.4%)	103.4 (77.7%)	104.5 (74.4%)	107.1 (75.0%)	110.4 (76.8%)
- Marketing Fresh	45.5 (37.1%)	46.9 (39.7%)	48.2 (37.9%)	48.6 (37.0%)	49.5 (37.8%)	50.6 (37.8%)	51.2 (38.5%)	51.4 (36.6%)	51.3 (36.0%)	53.6 (37.3%)
- processing (A+B+C)	46.3 (37.7%)	46.6 (39.5%)	47.3 (37.3%)	48.5 (37.1%)	50.6 (38.6%)	50.1 (37.6%)	52.2 (39.2%)	53.1 (37.9%)	55.7 (39.1%)	56.8 (39.5%)
(A) - Freezing	24.8 (20.2%)	24.7 (20.9%)	24.9 (19.6%)	25.3 (19.4%)	26.4 (20.1%)	26.3 (19.7%)	26.6 (20.0%)	26.8 (19.1%)	28.1 (19.7%)	28.6 (19.9%)
(B) - Curing	9.4 (7.6%)	10.1 (8.6%)	10.5 (8.3%)	10.9 (8.3%)	11.5 (8.8%)	10.8 (8.1%)	11.5 (8.6%)	11.5 (8.2%)	11.8 (8.3%)	11.9 (8.3%)
(C) - Canning	12.1 (9.9%)	11.8 (10.0%)	11.9 (9.4%)	12.3 (9.4%)	12.7 (9.7%)	13.0 (9.8%)	14.1 (10.6%)	14.8 (10.6%)	15.8 (11.1%)	16.3 (11.3%)
Non food uses	31.0 (25.2%)	24.6 (20.8%)	31.8 (24.8%)	34.2 (25.9%)	31.3 (23.6%)	32.9 (24.6%)	29.8 (22.3%)	36.0 (25.6%)	35.6 (25.0%)	33.3 (23.2%)
- Reduction	26.0 (21.2%)	19.8 (16.7%)	25.7 (20.2%)	27.9 (21.3%)	23.9 (18.3%)	24.9 (18.7%)	20.8 (15.6%)	25.1 (17.9%)	23.5 (16.5%)	20.1 (14.0%)
- miscellaneous pur poses	4.9 (4.0%)	4.8 (4.1%)	5.8 (4.6%)	6.1 (4.6%)	6.9 (5.3%)	8.0 (6.0%)	9.0 (6.7%)	10.9 (7.8%)	12.1 (8.5%)	13.1 (9.1%)

Source: FAO, 2009.

Almost all of the remaining world fish production of 33.3 million tonnes (23.2%) was destined for non-food products mainly for the manufacture of fishmeal and fish oil (Table 8.3).

BIOREMEDIATION OF FISH AND SHRIMP PROCESSING WASTES

The fish processing industry is very widespread and quite varied in terms of types of operation, scales of production and outputs. The species of fish processed include cod, tuna, herring, mackerel, pollock, hake, haddock, salmon, anchovy, pilchards and tropical fish.

Fish Cannery Waste Water

Water is used for treating and transporting fish, for cleaning equipment and work areas, and for fluming offal and blood during the canning process. Waste water is most often not utilized and is dumped directly or after being clarified through activated sludge, into the external environment (Fig. 8.2). The discharge of fish drainage water into rivers or coastal waters often causes water pollution and lead to problems such as red tide. However, the use of microorganisms in the activated sludge for drainage clarification produces a secondary waste of cell biomass that must be buried in landfills. Therefore, drainage remediation should utilize microorganisms carefully selected so that the resulting biomass can be used for other industrial objectives such as feeds in aquaculture and an efficient recycling system can be established.

A novel technique for managing fish processing wastewater by cultivating proteolytic yeast, *Candida rugopelliculosa*, as a possible diet of the rotifer, *Brachionus plicatilis* was studied (Lim et al., 2003). It was feasible to use Alaska Pollack processing wastewater as a growth medium for *C. rugopelliculosa*, which was stimulatory for growth of the rotifer by 18.3% over the commercial diet of *Saccharomyces cerevisiae*. Maximum growth of *C. rugopelliculosa* and reduction of influent soluble chemical oxygen demand (SCOD) concentration were $6.09 \pm 0.04 \times 10^6$ cells/l and 70.0%, respectively at 6.3 h hydraulic retention time (HRT). The maximum microbial growth rates (Mmax); half saturation coefficient (Ks); microbial yield coefficient (Y), and microbial decay rate coefficient (kd) were determined to be 0.82 ± 0.22/h, 690 ± 220mg SCOD/l, $(1.39 \pm 0.22) \times 10^4$ cells/mg SCOD and 0.06 ± 0.01/h, respectively.

Several yeasts were also isolated from a drainage canal in a Japanese fish processing factory. They were characterized by the decomposition of organics such as protein and reducing sugars, their growth in the wastewater, the decrease in total organic carbon (TOC), and taxonomy.

Two strains of yeast dominated the sample: *Debaryomyces occidentalis* (P1) and *Trichosporon ovoides* (P19), Strain P19 had the highest TOC-decreasing activity and was immobilized onto chitosan beads. The immobilized yeast reduced the TOC from 1.2×10^3 to 3.0×10^2 mg of C/L per day in the fish cannery wastewater (Urano et al., 2002).

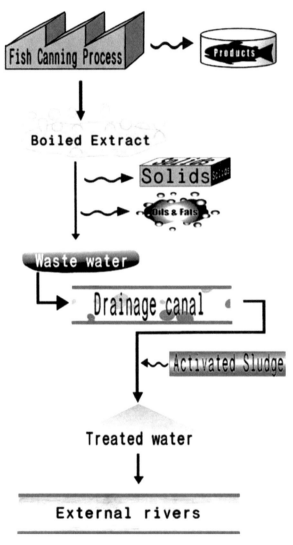

Fig. 8.2 Drainage treatment system in a fish food processing factory. Organisms used as bioremediators (Urano et al., 2002).

Fishmeal and Fish Oil Waste Water

Fishmeal wastewater (FWW), one of the main wastes of the fish processing industry, contains soluble protein, high concentrations of suspended solids with little amount of Ca, Mg, P, Na and S. Figure 8.3 shows the process for fish meal and fish oil production including approximate figures for quantities of inputs and outputs. Anaerobic nitrification and denitrification can result in 20-70% removal of the initial chemical oxygen demand (COD) especially due to the presence of fats and proteins (Huang et al., 2008). Physical-chemical methods such as coagulation-flocculation, centrifugation or nano-filtration can be used to recover solids (especially insoluble proteins) to be recycled to the fish-meal production (Garrido et al., 1998).

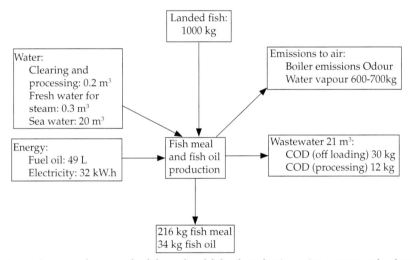

Fig. 8.3 Inputs and outputs for fish meal and fish oil production using average technology.

Once the solids are separated, most of the soluble COD of the waste waters can be efficiently removed by anaerobic digestion. In spite of the high efficiency reached in anaerobic treatment, an organic load accounting for about 20% of the initial COD remains in the waste water, thereby necessitating a further treatment for COD and nitrogen removal. The particular characteristics of these waste waters are their high nitrogen content as ammonia or organic nitrogen, their salinity and quite a high COD, which can complicate the nitrification step. After their treatment by anaerobic digestion, effluents with high ammonia and organic nitrogen contents and a fraction of residual COD are produced. A biological nitrification/denitrification process is proposed to reduce both the residual carbon and the nitrogen. The aerobic stage has the double objective of reducing COD and nitrifying the ammonia and organic nitrogen for its further removal at the denitrification stage. Under operating conditions, ammonia accumulation in the aerobic reactor occurs, inhibiting nitrite-oxidizing bacteria. Specific removal rates of COD (up to 1.5 g COD/g VSS (volatile soluble solids)/d) and TKN (total kjeldahl nitrogen) (up to 0.25 g N/g VSS/d) were obtained. The effect of nitrogen loading rate (NLR) on ammonia conversion was studied. A linear relationship was found between NLR and ammonia oxidation rate up to NLR of 0.6 g-TKN/L/d. Percentages of COD removal ranged from 95% to 60% and the nitrification percentages from 65% to 20% (Garrido et al., 1998).

Fishmeal waste water (FWW) and starch waste water (SWW), without amendment of nutrients, were investigated for lactic acid production using two strains of *Rhizopus oryzae*. *R. oryaze* AS 3.254 could use FWW and SWW simultaneously for lactic acid production and fungal growth while *R. oryzae* 3.41 only utilized this complex for fungal growth. For *R. oryzae* AS 3.254, the optimal C/N for lactic acid production was 100:1. *R. oryzae* 3.41 could produce lactic acid by use of glucose and FWW and the optimal C/N ratio was 124. Both fungal strains showed higher biomass production at lower C/N ratio. The direct lactic acid production from SWW and FWW by *R. orazae* AS 3.254 was characterized by starch hydrolysis, accumulation of reducing sugars, COD and nitrogen removal, and production of lactic acid and fungal biomass. Starch hydrolysis could arrive at 1.23 g/l/h while COD and nitrogen removal were 96.2% and 89.4%, respectively at fermentation time 88 h. The highest lactic acid yield was 0.674 g COD/g COD based on the equivalent COD of SWW and FWW associated with 2.65 g/l fungal biomass at fermentation time 72 h while the maximum productivity 0.783 g/l/h was obtained at fermentation time 40 h (Huang et al., 2008).

CONCLUSIONS

Minimization and enforced reduction of pollutant loads from aquatic animal production and processing facilities are increasingly demanded by various segments of the public and the regulatory community. Environmental rules and laws will continue expanding at all levels of government, requiring increased public and private expenditures. Among the challenges facing the fishery industries are new restrictions on the use of cleaners such as chlorine, the disposal of solid waste, the use of packaging materials, and the discharge of wastewaters (Martin, 1999).

The Environmental Protection Agency estimates that aquaculture industry cost will double in the next 10 yr as a result of strict enforcement of bioremediation of these pollutants. The uses of microorganisms as bioremediating agents in pond culture and in other devices such as constructed microbial mats, MUDS, seaweed/fish integrated mariculture, are some of the approaches for bioremediation and economic diversification in a sustainable manner for aquaculture production and managing processing industry wastes.

ABBREVIATIONS

BOD	Biological oxygen demand
COD	Chemical oxygen demand
EPA	Environmental Protection Agency
FWW	Fishmeal waste water
HRT	Hydraulic retention time
H_2S	Hydrogen sulphide
MUDS	Marine underwater depuration system
NLR	Nitrogen loading rate
SCOD	Soluble chemical oxygen demand
SWW	Starch waste water
TKN	Total kjeldahl nitrogen
TOC	Total organic carbon
VFA	Volatile fatty acids
VSS	Volatile soluble solids

REFERENCES

Abraham, J.T., Shanmugham, S.A., Uma, A., Palaniappan, R. and Dhevendaran, K. (2001). Biocontrol of shrimp bacterial pathogens using penaeid larvae associated bacterium, *Alteromonas* sp. J. Aquacult. Tropics 16(1): 11-22.

Antony, S.P. and Philip, R. (2006). Bioremediation in shrimp culture systems. NAGA, WorldFish Center Quarterly 29: 62-66.

Arvanitoyannis, I.S. and Kassaveti, A. (2008). Fish industry waste: treatments, environmental impacts, current and potential uses. Int. J. Food Sci. Technol. 43: 726-745.

Austin, B. and Brunt, J.W. (2009). The use of probiotics in aquaculture. In: Aquaculture and Microbiology and Biotechnology, Volume 1. D. Montet and R.C. Ray (eds). Science Publishers Inc., Enfield, New Hampshire, USA pp. 185-207.

Ayyappan, S., and Mishra, S. (2003). Bioamelioration in aquaculture with a special reference to nitrifying bacteria. In: Aquaculture Medicine. I.S.B. Singh, S.S. Pai, R. Philip and A. Mohandas (eds). Cochin University of Science and Technology, Cochin, India, pp. 89-107.

Bender, J. and Phillips, P. (2004). Microbial mats for multiple applications in aquaculture and bioremediation. Bioresour. Technol. 94: 229-238.

Beveridge, M.C.M. (1987). Cage Aquaculture. Fishing News Books Ltd., Farnham, Surrey, UK.

Boyd, C.E. (1979). Water Quality in Warm Water Fish Ponds. Agricultural Experiment Station, Auburn Univ. Alabama, USA.

Boyd, C.E. (1995). Bottom Soil Sediment and Pond Aquaculture. Chapman and Hall, New York, USA.

Boyd, C.E. and Silapajarn, O. (2006). Influence of microorganisms/microbial products on water and sediment quality in aquaculture ponds. In: Microbial Biotechnology in

Agriculture and Aquaculture, Volume 2. R.C. Ray (ed). Science Publishers Inc., Enfield, New Hampshire, USA pp. 225-260.

Cattaneo-Vietti, R., Benatti, U., Cerrano, C., Giovine, M.,Tazioli, S. and Bavestrello, G. (2003). A marine biological underwater depuration system (MUDS) to process waste waters. Biomol. Eng. 20: 291-298.

Caumette, P., Matheron, P., Raymond, U. and Relexans, J.C. (1994). Microbial mats in the hyper-saline ponds of Mediterranean salterns. FEMS Microbiol. Ecol. 13: 273-286.

D'amelio, E.D., Cohen, Y. and Des Marais, D. (1989). Comparative functional ultra structure of two hyper saline submerged cyanobacterial mats. In: Microbial Mats. Y. Cohen and E. Rosemberg (eds). American Society for Microbiology, Washington, USA pp. 97-/113.

Djurle, C. (2003). Development of a model for simulation of biological sulphate reduction with hydrogen as energy source, modeling of bacterial competition using AQUASIM. http://www. aquasim. eawag. ch/

FAO Fisheries and Aquaculture Department. (2009). The State of World Fisheries and Aquaculture 2008, FAO, Rome, Italy.

Focht, D.D. and Verstraete, W. (1977). Biochemical ecology of nitrification and denitrification. Adv. Microb. Ecol. 1: 135-211.

Garrido, J.M., Guerrero, L., Méndez, R. and Lema, J.M. (1998). Nitrification of waste waters from fish-meal factories. Water SA. 24: 245-250.

Gebauer, R. and Eikebrokk, B. (2006). Mesophilic anaerobic treatment of sludge from salmon smolt hatching Bioresour. Technol. 97: 2389-2401.

Haung, H.J. (2003). Important tools to the success of shrimp aquaculture—Aeration and the applications of tea seed cake and probiotics. Aquacult. Int. February 13-16.

Huang, L., Sheng, J., Chen, J. and Li, N. (2008). Direct fermentation of fishmeal wastewater and starch wastewater to lactic acid by *Rhizopus oryzae*. Second International Conference on Bioinformatics and Biomedical Engineering, ICBBE 2008, Article no. 4536014, pp. 3219-3222

Hussenot, J., Lefebvre, S. and Brossard, N. (1998). Open-air treatment of wastewater from land-based marine fish farms in extensive and intensive systems: Current technology and future perspectives. Aquat. Living Resour. 11: 297-304.

Jones, A.B., Dennison, W.C. and Preston, N.P. (2001). Integrated treatment of shrimp effluent by sedimentation, oyster filtration and macroalgal absorption: a laboratory scale study. Aquaculture 193: 155-178.

Kinne, P.N., Samocha, T.M., Jones, E.R. and Browdy, C.L.(2001). Characterization of intensive shrimp pond effluent and preliminary studies on biofiltration. N. Am. J. Aquacult. 63: 25-33.

Lanari, D. and Franci, C. (1998). Biogas production from solid wastes removed from fish farm effluents. Aquat. Living Resour. 11: 289-295.

Lim, J., Kim, T. and Hwang, S. (2003). Treatment of fish-processing wastewater by co-culture of *Candida rugopelliculosa* and *Brachionus plicatilis*. Water Res. 37: 2228-2232.

Martin, A.M. (1999). A low-energy process for the conversion of fisheries waste biomass. Renewable Energy 16: 1102-1105.

Mattson, J., and Linden, O. (1983). Benthic microfauna succession under mussels, *Mytilis edulis*, cultured on hanging long lines. Sarsia 68: 97-102.

Montet, D. and Ray, R.C. (2009). Aquaculture and Microbiology and Biotechnology, Volume 1. D. Montet and R.C. Ray (eds). Science Publishers Inc., Enfield, New Hampshire, USA.

Moriarty, D.J.W. (1998). Control of luminous *Vibrio* species in penaeid aquaculture ponds. Aquaculture 164: 351-358.

Moriarty, D.J.W., Decamp, O. and Lavens, P. (2005). Probiotics in aquaculture. AQUA Culture Asia Pacific Magazine Sept/Oct pp. 14-16.

Paniagua-Michel, J. and Garcia, O. (2003). *Ex-situ* bioremediation of shrimp culture effluent using constructed microbial mats. Aqua. Eng. 28: 131-139.

Patent No. 828/DEL/2000 (2000). National Centre for Aquatic Animal Health, Cochin University of Science and Technology, Cochin, India.

Rao, S.P.S. and Karunasagar, I. (2000). Incidence of bacteria involved in nitrogen and sulphur cycles in tropical shrimp culture ponds. Aquacult. Int. 8: 463-472.

Ray, R.C., Shetty, K. and Ward, O.P. (2008). Solid-state fermentation and value-added utilization of horticultural processing wastes. In: Microbial Biotechnology in Horticulture, Volume 3. R.C. Ray and O.P. Ward (eds). Science Publishers Inc., Enfield, New Hampshire, USA pp. 231-272.

Sharma, R., and Scheeno, T.P. (1999). Aquaculture wastes and its management. Fisheries World. April pp. 22-24.

Thassitou, P.K. and Arvanitoyannis, I.S. (2001). Bioremediation: a novel approach to food waste management. Trends Food Sci. Technol. 12: 185-196.

Urano N., Sasaki, E., Ueno, R., Namba, H. and Shida, Y. (2002). Bioremediation of fish cannery wastewater with yeasts isolated from a drainage canal. Mar. Biotechnol. 4: 559-564.

Zachleder, V., Hendeychova, J., Bisova, B. and Kubin, S. (2002). Viability of dried filaments, survival and reproduction under water stress and survivability following heat and UV exposure in *Lyngbya martesiana*, *Oscillatoria agardhii*, *Nostoc calcicola*, *Spirogyra* sp. and *Vaucheria gemiata*. Folia Microbiol. 47: 61-68.

Zhou, Y., Yang, H., Hu, H., Liu, Y., Mao, Y., Zhou, H., Xu, X. and Zhang, F. (2006). Bioremediation potential of the macroalga *Gracilaria lemaneiformis* (Rhodophyta) integrated into fed fish culture in coastal waters of North China. Aquaculture 252: 264-276.

9

Fish Waste Management: Treatment Methods Based on the Use of Microorganisms

Ioannis S. Arvanitoyannis,[1,]* Aikaterini Kassaveti[1] and
Theodoros Varzakas[2]

INTRODUCTION

As in many parts of the world, aquaculture production in the Mediterranean has been expanding rapidly over recent years (Basurco and Lovatelli, 2003). Mediterranean aquaculture is based on the production of a few species, the most important being sea bass (*Dicentrarchus labrax*), gilthead sea bream (*Sparus aurata*), mullet (*Mugil* spp.), mussels (*Mytilus* spp.), clams (*Tapes* spp.), as well as rainbow trout (*Oncorhynchus mykiss*) in inland waters. Recent research and technological development has given rise to at least 33 new Mediterranean species being reported as suitable to "leave the laboratories" and be produced on a commercial scale. However, not all of these species are being taken up by the sector (Saroglia and Basurco, 2006). To meet future demands for foodfish supplies, aquaculture production needs to increase by 50 million MT by the year 2050 (Tacon and Forster, 2001). However, the increasing number of fish culture industries in Mediterranean areas has begun to create serious environmental problems due to the impact caused by fish farming waste (FAO, 2002).

For decades, wastes from the fisheries industry have been dumped into the sea or in landfills. These practices were not ecologically viable nor did they utilise the energy potential of these wastes. With the prohibition of

---
[1]*Department of Agriculture Icthyology and Aquatic Environment School of Agricultural Sciences, University of Thessaly, Fytokou Street, 38446 Nea Ionia Magnesias, Volos, Hellas, Greece*
[2] *School of Plant Sciences, TEI Kalamatas, Greece*
Corresponding author: E-mail: parmenion@uth.gr

marine dumping and the move towards reuse and recycle, other treatment options have to be explored (Reddignton, 2003). Fish farm waste affects not only the area surrounding it and directly influenced by the effluent, but can alter a wider coastal zone at different ecosystem levels, reducing the biomass, density and diversity of the benthos, plankton and nekton, and modifying natural food webs (Gowen, 1991; Pillay, 1991). Most industrial facilities discharge their wastewater to a local treatment facility. Due to federal regulations regarding pretreatment and fees charged by local wastewater treatment facilities, many industrial facilities provide pretreatment. Federal regulations prohibit the discharge of wastes incompatible with: a) the conveyance of the wastewater to the local facility (such as highly odorous compounds), b) the processes at the local facility (such as chemicals that would inhibit biological treatment) or c) the use or disposal of the treated wastewater or resulting sludge (such as certain pesticides) (http://www.teicrete.gr/LEI/LAB/downloads/industrial_waste.pdf).

The environmental impact of marine aquaculture is managed in the European Union (EU) through the implementation of legislative and regulatory measures and Codes of Conduct and Codes of Practice. In practice, compliance with these measures and codes requires the adoption of Best Practice and Best Available Technology in relation to matters such as: site selection; management practices that minimise food waste and chemical usage and synchronised production, fallowing and disease control (Cho et al., 1994; Ervik et al., 1994). The objectives and principles of the FAO's Code of Conduct for Responsible Fisheries are the basis for an international commitment to responsible fishing: it is now accepted that the right to fish carries with it the obligation to do so in a responsible manner. This has become widely recognised as the demand for fisheries products continues to grow whilst the resource remains limited. Like other countries, the European Union is still in the process of following up all the practical implications of the Code of Conduct but it is clear that post-harvesting practices are an essential part in achieving its objectives (OECD, 2000).

Microbial consortia involved in wastewater treatment have been a major subject of microbial ecology, and many papers have been published in which molecular tools were used for community analyses (Whiteley and Bailey, 2000). The principal sources of aquaculture wastes are uneaten feed and excreta. The bulk of this waste is in the particulate form and in recirculating systems this is often removed in a concentrated form by gravitational or mechanical methods (Chen et al., 1994). Aquaculture wastes, generated by fish and shellfish, can contain toxic levels of ammonia in addition to large volumes of solid wastes. In many cases, these wastes are flushed into nearby estuaries from coastal pond systems, causing water quality problems (Ackefors and Enell, 1990). The world of microorganisms is made of bacteria, fungi, algae, protozoa, and viruses (microorrganisms are important in industrial and aquaculture.html). Bacteria can play an important role in mass transfer and utilisation of substrate in bioremediation of aquaculture (Van Loosdrecht et al., 1987a, b). Microorganisms reside in the sediment and other substrates, and in the water of aquaculture facilities, as well as in and on the cultured species (http://www.microtack.com/html/microorganism.htm).

BIOREMEDIATION

Bioremediation is a general concept that includes all those processes and actions that take place in order to biotransform an environment, already altered by contaminants, to its original status (Thassitou and Arvanitoyannis, 2001). Microorganisms can be isolated from almost any environmental conditions. Microbes will adapt and grow at sub-zero temperatures, as well as extreme heat, desert conditions, in water, with an excess of oxygen, and in anaerobic conditions, with the presence of hazardous compounds or on any waste stream. The main requirements are an energy source and a carbon source. Because of the adaptability of microbes and other biological systems, these can be used to degrade or remediate environmental hazards (Vidali, 2001). The currently recognised probiotics that may influence fish immunity, disease resistance and other performance indices include: *Bacillus subtillis, Bacillus licheniformis, Bacillus circulans, Lactobacillus rhamnosus, Lactobacillus delbrueckii, Carnobacterium maltaromaticum, Carnobacterium divergens, Carnobacterium inhibens, Enterococcus faecium, Saccharomyces cerevisiae* and *Candida sake* (Gatlin et al., 2006). As the aquaculture industry develops, effective, cost-effective and environmentally friendly preventive and bioremediation methods of improving effluent water quality prior to discharge into receiving waters of sensitive areas will be necessary (Jones et al., 2001).

Bioremediation, although considered a boon in the midst of present day environmental situations, can also be considered problematic because, while additives are added to enhance the functioning of one particular bacterium, fungi or any other microorganisms, it may be disruptive to other organisms inhabiting that same environment when done *in situ* (Sasikumar and Papinazath, 2003). A successful bioremediation involves: optimising nitrification rates to keep low ammonia concentration; optimising denitrification rates to eliminate excess nitrogen from ponds as nitrogen gas; maximising sulphide oxidation to reduce accumulation of hydrogen sulphide; maximising carbon mineralisation to carbon dioxide to minimise sludge accumulation; maximising primary productivity that stimulates shrimp production and also secondary crops; and maintaining a diverse and stable pond community where undesirable species do not become dominant (Bratvold et al., 1997).

MICROBIAL MATS

Microbial mats are laminated, cohesive microbial communities, composed of a consortium of bacteria dominated by photoautotrophic cyanobacteria (also referred to blue–green algae) (Nisbet and Fowler, 1999). Cyanobacteria are organisms which thrive in environments where extreme fluctuations in conditions occur (http://www.bio.vu.nl/thb/users/vdberg/berg-mat. html). Although microbial mats are trying to grow almost everywhere almost all of the time, it is only in special places the microbes have the opportunity to grow the large amounts of biomass comparable to what was found on early Earth (http://nai.arc.nasa.gov/students/this_ month/index.cfm). Most mats stabilise unconsolidated sediments and grow actively; they can be several millimeters to a few centimeters thick, and develop along a variety of microgradients established between water and sediments. The formation of microbial mats is an extremely ancient biological phenomenon, as communities of different types of microbes covered the early Earth (Guerrero et al., 2002).

TREATMENT OF FISH WASTE WITH THE USE OF MICROORGANISMS

Aspergillus sp.

Wheat bran, minced fresh sardines and *Aspergillus awamori* spores were mixed and kept in a ventilated incubator for 5 d at 40°C. A decrease in crude protein, crude fat and nitrogen-free extracts up to 2.6, 22.2 and 25.7%, respectively, were recorded. On the contrary, an increase up to 16.1% in crude fibre content was reported; thus indicating that crude fibre may be synthesised by *Aspergillus awamori*. During fermentation glucoamylase (9.71 U), α-glucosidase (0.21 U), α-amylase (21.69 U) and acidic protease (17778 U), were formed. In addition, an increase in dry matter digestibility was mentioned regardless of the levels of fishmeal, and it was significant at 5 and 10% (Yamamoto et al., 2005).

Different combinations of the bonito's body parts (flesh body, head, fins, and viscera) were prepared with or without the addition of enzymes, such as soybean koji (*Aspergilus oryzae*), ang-khak (*Monacus purpureus*) or viscera in order to produce fish sauce. The results indicated that treated bonitos (whole fish and waste) with soybean koji and ang-khak had higher total sugar content and amylase activities values ranging from 10.8-17.4 mg/l and 5.0-100 U/ml, respectively. On the contrary, total sugar content of treated bonitos without the addition of enzymes ranged from 3.6 to

5.9 mg/L, whereas no amylase activity was detected. In addition, protein and amino acids values of treated whole bonitos were slightly higher than treated fish waste. Volatile basic nitrogen (VBN) and trimethylamine (TMA) were found to exceed the thresholds of 6 mg/100 g and 50 mg/100 g, respectively (Shih et al., 2003).

A biofilter reactor was operated to remove nitrogenous wastes, such as ammonium, nitrite and protein from uneaten feed and cell debris, and glucose in a recirculating aquaculture system. In a continuous fixed-slab reactor, *Aspergillus niger* NBG5 was used for the remediation of the above nitrogenous wastes. The results showed that *A. niger* NBG5 consumed the nitrogenous wastes simultaneously, removing more than 95% of the wastes in 50 h, while ammonium and protein were metabolized prior to the nitrite. In addition, the fungus metabolized nitrogen at low temperatures (22°C), whereas carbon was metabolised at high temperatures (35°C) (Fig. 9.1). Furthermore, glucose consumption (specific rates: 2-2.5 g-C/g-cell/day) by *A. niger* NBG5 was observed at 35°C in the presence of ammonium and nitrite (Hwang et al., 2004).

Hwang et al. (2007) investigated the potential removal of nitrogenous substances by *Aspergillus niger* NBG5 in a continuous stirred tank reactor (CSTR) system. In the CSTR system, glucose and aquacultural wastewater were exchanged in a mixer every 24 h. Nitrite was removed during the pretreatment of wastes using ozone. The nitrogenous wastes passed CSTR with a hydraulic retention time of 3.3 h, an operating temperature of

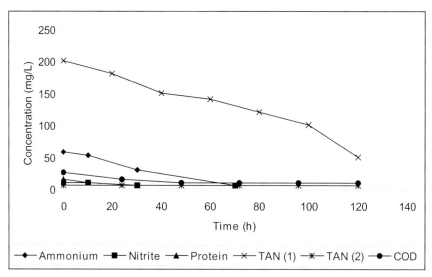

Fig. 9.1 Concentration of ammonium, nitrite, protein and TAN (1) after treatment with *A. niger* at 22°C, and COD and TAN (2) after treatement with *Bacillus subtilis* (Hwang et al., 2004, 2007; Liu and Han, 2004).

Table 9.1 Treatment methods based on the use of *Aspergillus* for aquaculture waste remediation (parameters, quality control and results).

No	Kind of waste	Treatment	Parameters	Methodology	Quality control methods	Results	References
1	Fish waste	Fermentation process using *Aspergillus awamori*	Dry matter (DM), crude protein (CP), lipid, ash, fiber content, glucoamylase, acidic protease activity, released tyrosine	Wheat bran (700 g), minced fresh sardines (2638 g) and *A. awamori* spores (0.7 g) mixed together and kept in a ventilated incubator for 5 days at 40°C	1. DM, CP, lipid, ash and fiber content: proximate chemical analysis (AOAC, 1984) 2. Glucoamylase: using kits 3. Acidic protease activity: using casein as a substrate 4. Released tyrosine: photometrically at 660 nm	1. Decrease of crude protein, fat, and nitrogen-free extracts up to 2.6, 22.2 and 25.7%, respectively 2. 16.1% increase of crude fibre 3. Production of glucoamylase (9.71 U), α-glucosidase (0.21U), α-amylase (2169Y) and acidic protease (17.778U) 4. Increase in DM digestibility	Yamamoto et al., 2005
2	Bonito waste (flesh body, head, fins, and viscera)	Fermentation using *Aspergillus oryzae* and other enzymes (ang-khak or viscera)	pH, protein, reducing sugar, total sugar, trimethylamine (TMA), volatile basic nitrogen (VBN), free amino acids, color, volatile compounds	Different combinations of fish body parts with or without the addition of enzymes (viscera soybean koji or ang-khak) were inoculated into a porcelain vase for fermentation	1. pH and protein: standard methods (AOAC, 1984) 2. Reducing sugar and total sugar: Miller's method (Miller, 1959) and phenol-sulfuric method (Dubois et al., 1956), respectively 3. Trimethylamine (TMA) and volatile basic nitrogen (VBN): Castell et al's method (Castell et al., 1974), and Cobb *et al's* method (Cobb et al., 1973), respectively 4. Free amino acids: amino acid analyzer 5. Colour: colour measuring system 6. Volatile compounds: GC	1. Higher total sugar content (10.8-17.4 mg/L) and amylase activities (5.0-100 U/mL) of soybean koji and ang-khak treated bonitos 2. Lower total sugar content (3.6-5.9 mg/L), and no detection of amylase activities of treated bonitos without the addition of enzymes 3. Volatile basic nitrogen (VBN) and trimethylamine (TMA) exceed thresholds of 6 mg/100 g and 50 mg/100 g	Shih et al., 2003
3	Nitrogenous wastes from uneaten feed and cell debris in a recirculating aquaculture system	Remediation using *Aspergillus niger* NBG5	Ammonium, nitrite, protein, glucose	Under aseptic operating conditions in a continuous fixed-slab reactor (CFSR)	1. Ammonium and nitrite: phenolhypochlorite and the Wood–Armstrong–Richard methods, respectively (APHA, 1995) 2. Protein: Bradfold method (#500, Bio-Rad Lab, CA) 3. Glucose: enzyme method (#510DA, SIGMA, USA)	1. Removing more than 95% of the wastes in 50h 2. Ammonium and protein were metabolized prior to the nitrite 3. Fungus metabolized nitrogen at low temperature (22°C), whereas carbon was metabolized at high temperatures (35°C) 4. Glucose consumption at 35°C (2-2.5 g-C/g-cell/day)	Hwang et al., 2004
4	Aquacultural wastewater	Remediation using *Aspergillus niger* NBG5	Ammonium, nitrite	In a continuous stirred tank reactor (CSTR) system, glucose and aquacultural wastewater were exchanged in a mixer every 24h (HRT 3.3h, 26±2°C, pH 6.5±0.5)	Ammonium and nitrite: phenolhypochlorite (with 640 nm) and colorimetric method (with 543 nm), respectively (APHA, 1992)	1. TAN concentration decreased to 0.35 mg/L 2. Nitrite nitrogen concentration decreased to 0.12 mg/L	Hwang et al., 2007

26 ± 2°C and a pH of 6.5 ± 0.5. Total ammonia nitrogen (TAN) and nitrite nitrogen concentration decreased to 0.35 mg/l and 0.12 mg/l, respectively (Fig. 9.1).

The treatment methods based on the use *Aspergillus* sp. for aquaculture waste remediation (parameters, quality control and results) are presented in Table 9.1.

Bacillus sp.

Prawn nursery wastewaters were first filtered by sand filter, sterilised by ultraviolet reactor, and spores of *Bacillus subtilis* were added to the effluents under conditions of continuous aeration. The remediated water was then filtered through sand filter. Three experiments were conducted; in the first experiment the degradation of dissolved organic matter (DOM) and TAN without nutrient addition was investigated; in the second experiment the effects of four nutrients (glucose, potassium dihydrogen phosphate, micro-elements solution, and vitamin mixture) on bioremediation performance; and finally in the third experiment the detection of an optimal rationale of nutrients addition for removing ammonium and avoiding second contamination during bioremediation. After 48 h, Chemical Oxygen Demand (COD) removal was up to $57.5 \pm 5.5\%$, and DOM was degraded directly. From d 2 to d 5 TAN removal was decreased, whereas the addition of glucose and/or phosphate supported TAN removal (> 85%) (Fig. 9.1). The addition of vitamin mixture and/or micro-elements solution did not affect the TAN removal. The optimal rates of C/N or N/P for wastewater remediation were 5.4/1 or 5-7/1, respectively. After 5 d of microbial remediation the wastewater became suitable for reuse (Liu and Han, 2004).

The role of probiotic cell wall hydrophobicity in bioremediation of aquaculture was examined by Wang and Han (2007). *Bacillus* sp. YB-030518 and *Bacillus* sp. YB-034325 (1% inoculation) with different levels of cell wall hydrophobicity were added to basal shrimp feed at 30°C for 96 h. The results revealed that there was no significant difference in cell wall hydrophobicity between the two strains of *Bacillus* sp. at pH 5.5, 7.0 and 8.5, at 5°C and 20°C. Bioremediation capability (%) at 48 h and 96 h was significantly higher for YB-034325 compared to YB-030518 (Table 9.2).

Table 9.2 Bioremediation capability and cell hydrophobicity of *Bacillus* sp. YB-030518 and -034325 (Wang and Han, 2007)

Bacillus spp.	Bioremediation capability (%)		Cell hydrophobicity (%)	
	48 h	96 h	18 h	96 h
Bacillus sp. YB-030518	55	60	63.3	34.3
Bacillus sp. YB-034325	65	70	71.3	43.3

Two bacteria, *Bacillus subtilis* and *Bacillus megaterium,* were added to the recirculating-water systems of the grow-out facilities of red-parrot fish (male *Cichlasoma citrinellum* × female *Cichlasoma synspilum*). The results indicated that total ammonia nitrogen concentrations were held below 5.4 mg/l, while COD remained below 40.8 mg/l or less, the transparency of the recirculated aquarium water remained at 30-50 cm, and the survival rate and quality of fish were significantly improved (Chen and Chen, 2001).

In another study, a procedure for the extraction of protein and production of peptides by enzymic hydrolysis from bone and skin wastes containing collagen was fully developed. Fat and inorganic components were first removed in a pretreatment step and a high molecular weight protein extracted under acidic conditions (pH 3) using a 1 h reaction time at 60°C. The extract was shown to have a high water retention capacity, was beneficial for repair of rough skin, had no odour problem and was demonstrated to be safe in skin patch tests. It was thereby considered acceptable for use in cosmetic materials. Pretreated fish bone and pig skin were hydrolysed with a commercial enzyme. The hydrolysates had a high anti-radical activity (IPO X 50, 0.18 and 0.45 mg/ml) and a high potential for decreasing blood pressure (IC50, 0.16 and 0.41 mg/ml), proposing the hydrolysates could be potentially useful as additives in food materials (Morimura et al., 2002).

In recent years, there has been growing interest in biocontrol of microbial pathogens in aquaculture using antagonistic microorganisms (Westerdahl et al., 1991; Maeda, 1994; Chapter 1 and 2 in this Volume). The microbial species composition in hatchery tanks or large aquaculture ponds can be changed by adding selected bacterial species to displace deleterious normal bacteria. Abundance of luminous *Vibrio* strains decreased in shrimp ponds and hatchery tanks waters where specially selected, probiotic strains of *Bacillus* species were added. When the probiotic bacteria were used, no disease was experienced and indeed shrimp survival was extremely high (80-100%), even in the presence of luminous *Vibrio* species in all ponds treated with probiotics (Moriarty, 1999).

The treatment methods based on the use of *Bacillus* sp. for fish waste remediation (parameters, quality control and results) are shown in Table 9.3.

Lactobacillus sp.

Fish wastes were mixed with 15% molasses, inoculated with 5% starter culture of *Lactobacillus plantarum*, incubated at $22 \pm 2°C$ for 20 d, and then stored for 20 d. The obtained product was incorporated with bran and ground barley to make different formulas and then fed to broilers. A significant pH decrease (initial value 6.13-6.75) in the fermenting product and then remained constant at 4.2 and 4.5, and nitrogen decrease (5.3-5.7%) was recorded; while non-protein nitrogen(N) increase (220-262%) was reported. In addition, a net increase in the broiler weight relative to the control diet was also noticed (Hammoumi et al., 1998).

Three different experiments were performed by Ayangbile et al. (1997) in order to examine the effect of chemicals on preservation of crab-processing waste and the fermentation characteristics of the waste-straw mixture. In "Experiment 1" 1.5% propionic/formic acid (1:1) and crab-processing waste were mixed with wheat straw, sugarcane molasses and water (32:32:16:20, w/w) with or without 0.1% microbial inoculant (*Streptococcus faecium* and *Lactobacillus plantarum*). In "Experiment 2" a mixture of 0.2% sodium hypochlorite (NaOCl) or 0.4% hydrogen peroxide (H$_2$O$_2$) were added to crab waste, and the mixture was ensiled just like "Experiment 1 "Experiment 3" involved the addition of no chemicals, or 0.2% NaOCl, or 0.2% or 0.4% NaOCl/calcium hypochlorite [Ca (OCl)$_2$] or 1% NaNO$_2$ to crab waste, and the mixture was ensiled as described in "Experiment 1". In "Experiment 1" the addition of 1.5% propionic/formic acid (1:1) resulted in waste degradation for up to 14 d, while 16% pH and 48.9% water soluble carbohydrates were decreased, and an increase in lactic acid concentration up to 252.1% were achieved with or without the presence of microorganisms in ensiled waste. In addition, in "Experiment 2" waste deterioration was prevented for 7 d, whereas trimethylamine (TMA) content was increased up to 246.5% and 1.87% when crab waste was treated with NaOCl and H$_2$O$_2$, respectively. Finally, in "Experiment 3" the addition of 1% NaNO$_2$ and a combination of 0.2% NaOCl/0.2% calcium hypochlorite [Ca(OCl)2] (1:1, w/w) or 0.2% of each alone resulted in waste preservation for up to 10 d, while TMA concentration was increased significantly.

A mixture of fish waste, heads and viscera of different fish, but mainly sardines (*Sardinia ilchardus*), and molasses was inoculated with 32 strains of yeasts *Saccharomyces cerevisiae* and *Candida* sp. and 14 strains of lactic acid bacteria *Lactobacillus plantarum and Pediococcus acidi-lactici* at 28°C for 10 d and fermented at different values of pH (4.0-5.8), temperature (20-30°C), and molasses proportions (20-30%, w/w). The results showed that the final product was free of coliforms and *Salmonella*. In addition, a decrease in pH (17.1%), dry matter (10.5%), reducing sugars (28.4%), protein (10.3%), total nitrogen (12.0%) and TMA (69.2%) in the final product was reported. On the other hand, a considerable increase in ash (7%), fat (10.9%), non-protein-nitrogen (NPN) (130.8%), and total volatile nitrogen (TVN) (49.6%) was recorded (Faid et al., 1994).

Saltwater (SW), freshwater fish (FW), and tilapia filleting residue (TR) were separately mixed with 15% sugar cane molasses, 5% *Lactobacillus plantarum* culture media and 0.25% sorbic acid (w/w). The addition of

Table 9.3 Treatment methods based on the use of *Bacillus* sp. for fish waste remediation (parameters, quality control and results)

No.	Kind of waste	Treatment	Parameters	Methodology	Quality control methods	Results	References
1.	Prawn nursery wastewater	Microbial remediation using *Bacillus subtilis*	pH, water temperature, disswolved oxygen, water salinity, TAN, nitrite, nitrate, phosphate, COD	Filtration by sand filter, sterilization by ultraviolet reactor, addition of *Bacillus subtilis* under conditions of continuous aeration, filtation through sand filter	1. pH, water temperature, dissolved oxygen: by handheld meters model pH 315i or Oxi 315i 2. Water salinity: using optical refractometer 3. TAN, nitrite, nitrate, phosphate: spectrophotometrically 4. COD: by oxidization of organic matter with potassium permaganate according to standard methods (Wu et al.,1980)	5. COD removal up to $57.5 \pm 5.5\%$, and DOM degradation after 48th 6. TAN removal decrease from day 2 to day 5 7. Glucose and/or phosphate addition resulted in TAN removal (>85%) 8. Vitamin mixture and/or micro-elements solution addition did not affect TAN removal 9. Optimal rates of C/N or N/P for wastewater remediation 5.4/1 or 5-7/1, respectively	Liu and Han, 2004
2.	Basal shrimp feed	Bioremediation with *Bacillus sp.* YB-030518 and *Bacillus sp.* YB-034325	Probiotic cell surface hydrophobicity	Two strains of *Bacillus sp.* (1% inoculation) were added to basal shrimp feed at 30°C for 96h	Probiotic cell surface hydrophobicity: by bacterial adherence to hydrocarbons assay, which is based on the partitioning of cells in a two-phase system (Rosenberg et al., 1980; Zhang and Miller, 1994)	1. No significant differences in cell wall hydrophobicity between the two strains of *Bacillus sp.* (pH 5.5,7.0 and 8.5, at 5°C and 20°C) 2. *Bacillus* YB-034325 had higher bioremediation capability (%) at 48h and 96h	Wang and Han, 2007
3.	Recirculating-water	Aquaculture water remediation with two species of *Bacillus*	-	Addition of *Bacillus subtilis* and *Bacillus megaterium* to the recirculating-water system of the grow-out facilities of redparrot fish	-	1. Total ammonia nitrogen concentrations 5.4mg/L 2.COD < 40.8mg/L 3. Recirculated aquarium water transparency remained at 30-50 cm 4. Improvement of survival rate and quality of fish	Chen and Chen, 2001

No.	Substrate	Treatment	Parameters	Process	Methods	Results	Reference
4.	Fish bone wastes	Treatment with enzymes L and K originated from *Bacillus* species	Organic matter, fat content, protein content molecular weight, water retention capacity, water vaporization potential, extract odor, skin patch test, peptidee molecular weight, amino acid composition, physiological activities	Pre-treated (fat removal, deshing), treated enzymatically, agitated (200rpm, 60min, 60°C, Ph8.0), centrifuged (800xg,10min)	1. Organic matter the amount of a sample: vaporized during incineration at 600°Cfor 30 min 2. Fat content: by Soxhlet extractor 3. Protein content: by Lowry-Folin method (Lowry et al., 1951) 4. Molecular weight: by Laemmli method (Laemmli,1970) 5. Water retention capacity: by direct application of human skin and using hygrometer 6. Water vaporozation potential: by direct application of human skin and using Tewameter 7. Extract odor: sample dilution with pH adjustments 8. Skin patch test: by the Association of Industrial Information's method (Hayakawa et al., 1975) 9. Peptide molecular weight: mean degree of polymerization and amino acid composition 10. Amino acid composition: hydrolysis of a sample under both acidic and alkiline conditions 11. Physiological activities: antioxidant activity (by luminol-enhanced chemi-luminescence). ACE inhibitory activity (by a modified method of Cushman and Cheung,1971)	1. Protein recovery, <53% 2. High collagen degradation efficiency by enzyme L (85.5%) compared to enzyme K (69.2%)	Morimura et al., 2002
5.	Shrimp ponds and hatchery tanks waters	Antagonism between *Bacillus* and *Vibrio* speceis	-	Addition of probiotic strains of *Bacillus* species	-	1. No disease was experienced 2. High survival (80-100%) of shrimps even in the presence of *Vibrio* species	Moriarty, 1999

2% formic acid (w/v) and 2% sulphuric acid (w/v) to the above mixtures caused the production of acid silages (Fig. 9.2). In addition, acid silages had higher protein content 69.91% (SW), 44.38% (FW) and 39.59% (TR), compared to fermented silages, 59.61% (SW), 42.09% (FW) and 35.84% (TR), respectively. Furthermore, the levels of histidine, threonine and serine were increased, whereas the levels of valine, isoleucine, and leucine were decreased in all products (Vidotti et al., 2003).

Table 9.4 provides a synoptical presentation of treatment methods based on the use of *Lactobacillus* sp. for fish waste remediation (parameters, quality control and results).

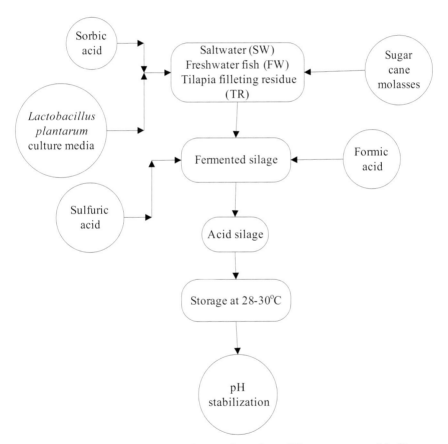

Fig. 9.2 Flow diagram of silages production from three different raw materials (Source: Vidotti et al., 2003; Arvanitoyannis and Kassaveti, 2008).

Other Microorganisms

A mixture of fish waste and peat was composted and extracts from this compost were sterilised and inoculated at a ratio of 5% (v/v) before use as a substrate for the acid-resistant microorganism *Scytalidium acidophilum* ATCC 26774 (150 rpm, 25.0°C, pH 2.0 ± 0.1, 8-10 d) (Fig. 9.3). The results showed that the acid extracts of compost had higher concentrations of nutrients (lipids, carbohydrates, and protein) than the compost. Compost extracts can be effectively utilised as a substrate for the growth of the microorganism *S. acidophilum*, for the biofiltration of liquid and gas effluents, and in the growth of mushrooms (Martin, 1999).

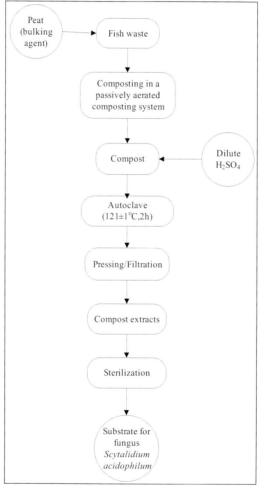

Fig. 9.3 A fundamental process for conversion of fisheries processing waste to valuable compost (Source; Martin, 1999; Arvanitoyannis and Kassaveti, 2008).

Table 9.4 Treatment methods based on the use of *Lactobacillus* sp. for fish waste remediation (parameters, quality control and results)

No.	Kind of waste	Treatment	Parameters	Methodology	Quality control methods	Results	Refferences
1.	Sardine fish waste	Fermentation	pH, dry matter, fat content, total nitrogen, non-protein nitrogen	Chopped, mixed with 15% molasses, inoculated with Lactobacillus plantarum, and incubated at 22°C for 20 days. Incorporation with bran and barley	1. pH: pH-meter 2. Dry meter: oven drying at 105°C 3. Fat content: Soxhlet method using hexane as solvent 4. Total nitogen: Kjeldhal method 5. Non-protein nitrogen: Kjeldhal method	1. pH decrease in the fermenting product (then remained constant at 4.2-4.5) 2. Slight total nitrogen decrease (5.3-5.7%) 3. Non-protein nitrogen increase (220-262%)	Hammoumi et al., 1998
2.	Crab-processing waste	Fermentation	Trimethylamine (TMA), nitrogen, dry matter (DM), ash, pH, lactic acid, water soluble carbohydrates (WSC)	*Experiment 1:* Mixture of 1.5% propionic/ formic acid (1:1) and crab-processing waste ensiled with wheat straw, sugarcane molasses, and water (32:32:16:20, wet basis) with or without 0.1% microbial inoculant (*Streptococcus faecium* and *Lactobacillus plantarum*) *Experiment 2:* Application of either 0.2% sodium hypochlorite (NaOCl) or 0.4% hydrogen peroxide (H₂O₂) to crab waste. Ensiling of the mixture as described in Treatment A *Experiment 3:* Treatment of crab-waste with 1% sodium nitrate (NaNO₂) or 0.2% and 0.4% NaOCl/Calcium hypochlorite [Ca(OCl)₂] or 1% NaNO₂ to crab waste. Ensiling of the mixture as described in Treatment A.	1. Trimethylamine (TMA): colorimetric procedure (Dyer, 1959) after extraction with 7.5% trichloroacetic acid 2. Nitrogen: Kjeldahl method (AOAC, 1984) 3. Dry matter (DM): drying in a forced draft oven at 60°C for 48h 4. Ash: AOAC (1984) 5. pH: electrometrically 6. Latic acid: method of Banker and Summerson (1941) (modified by Pennington and Sutherland,1956) 7. Water soluble carbohydrates (WSC): of Dubois et al. (1956) (adapted to corn plants by Johnson et al., 1966)	*Experiment 1:* 1. Prevention of waste degradation for up to 14days 2. 16% pH and 48.9% watersoluble carbohydrates (WSC) decrease 3. 252.1% lactic acid increase *Experiment 2:* 1. Prevention of deterioration for up to 7 days 2. 246.5% trimethylamine (TMA) concentration increase for NaOCl treated waste 3. 1.87% TMA concentration increase when treated with H₂O₂ *Experiment 3:* 1. Preservation of waste for a minimum of 10 days 2. TMA content increase	Ayangbile et al, 1997

#	Material	Process	Parameters	Treatment	Methods	Results	Reference
3.	Fish waste	Valorisation	pH, dry matter, ash, fat, total nitrogen (TN), non-protein-nitrogen (NPN), total volatile nitrogen (TVN) trimethylanune (TMA), reducing sugars	Fish waste-molasses mixture was inoculated at 28%°C for 10 days with strains of yeasts (32 strains of yeasts *Saccharomyces cerevisiae* and *Candida* sp., and 14 strains of Lactic acid bacteria *Lactobacillus plantorum* and *Pediococcus acidilactictici*) and fermented at different values of pH (4.0-5.8), temprature (20-30°C), and molasses proportions (20-30%,w/w)	1. pH: pH-meter 2. Dry matter: oven drying at 105°C 3. Ash: ignition at 550°C 4. Fat: Soxhlet extraction with hexane 5. Total nitrogen (TN): Kjeldhal method (APHA, 1989) 6. Non-protein-nitrogen (NPH): Kjeldhal method after precipitation with a 2% trichloracetic acid solution 7. Total volatile nitrogen (TVN), method decribed by Conway (1974) 8. Trimethylamine (TMA): method described by Murray and Gibson (1972) 9. Reducing sugars: Bertrand method	1. Final product free of coliforms and *Salmonella* 2. Decrease of pH (17.1%), dry matter (10.5%), RS (28.4%), protein (10.3%), TN (12.0%),TMA content (69.2%) 3. Considerable increase of ash (7%), fat (10.9%), NPN (130.8%), TVN (49.6%)	Faid et al., 1994
4.	Marine fish waste commercial fresh water fish waste and tilapia filleting residue	Acid digestion (formic acid and sulfuric acid) and anaerobic fermentation (*Lactobacillus plantarum*, sugarcane molasses)	Crude protein, amino acid composition	Saltwater (SW), commercial fresh water fish waste (FM), and tilapia filleting residue (TR) were mixed with 15% sugarcane molasses, 5% *Lactobacillus plantarum* and 0.25% sorbic acid (w/w) (fermented silage), while 2% formic acid (w/v) and 2% sulfuric acid (w/v) (acid silage) were added to the mixture	1. Crude protein: micro-Kjeldhal method according to AOAC-code 981.10 (1990) 2. Amino acid composition liquid chromotography, using a cationic exchange resin column and unhidrine post-column derivation in auto-analyzer	1. Acide silages: higher protein content 69.91% (SW), 44.38% (FW) and 39.59% (TR) compared to fermented silage 59.61% (SW) 42.09% (FW), and 35.84% (TR), respectively 2. Increase in histidine, threonine and serine levels for both processes and all three row meterials used 3. Decrease of valine iso-leucine, and leucine decreased in all products	Vidotti el al, 2003

Troell et al. (1997) evaluated the possibility of cultivating the red algae *Gracilaria chilensis* (Gracilariales, Rhodophyta) adjacent to salmon cages, so that the plants can utilise the released dissolved nutrients. At the first station (10m from the fish cages) the growth rate of *Gracilaria* was higher up to 40% (specific growth rate: 7%/d) compared to the second and third stations (150 and 1km distance), respectively. Close to the cages nitrogen and phosphorous contents were higher in algal tissues (1.9-2.1 mmol N/g (dry weight; dw) and 0.28-0.34 mmol P/g (dw) compared to the other distances. Furthermore, 1 ha cultivation of the algae, close to the fish cages, resulted in 5% removal of dissolved inorganic nitrogen and 27% of released dissolved phosphorous.

Rhodovulum sulfidophilum was grown in settled undiluted and nonsterilised sardine processing wastewater (SPW). Three levels of inoculum size (10, 20 and 30% v/v) were developed in glutamate-malate media (GMM) or settled and undiluted SPW. The results showed that highest biomass production was 4.8 g/l and reported after 96 h culture with 20% (v/v) inoculum size. The chemical oxygen demand (COD) decrease of SPW was greatest (85%) after 120 h culture with 30% (v/v) inoculum developed in GMM, whereas COD reduction of SPW was up to 79-83% when the inoculum was developed in SPW (Azad et al., 2003).

Bender et al. (2004) investigated the efficiency of a waste effluent treatment system for black sea bass (*Centropristis striata*) recycled-water mariculture. A biosolar filter system based on fluidised sand filters and microbial mats, which contained the cyanobacteria, *Oscillatoria* sp., on the upper surface and the purple non-sulphur bacterium, *Rhodopseudomonas* sp., below the surface, was constructed for fish waste bioremediation. When the microbial mat surfaces were covered with wastes, the fish wastewater was circulated under the mats, rather than on the surface. The results showed that the microbial mats effectively removed ammonia, due to cyanobacteria oxygen production which supported nitrification. In addition, oxygen concentrations ranged between 6 and 10 mg/L and total ammonia concentration remained below 1 mg/l (Figs. 9.4 and 9.5).

Fish-processing wastewater were used as a medium for cultivating the proteolytic yeast, *Candida rugopelliculosa*, as a diet for seed rotifer, *Brachionus plicatilis*. Fish processing wastewater were inoculated with a seed culture of *C. rugopelliculosa* and then fed into two continuous stirred tank reactors (CSTRs). On the other hand, the seed rotifer cultures were separately inoculated into two different media, which contained *C. rugopelliculosa* and *Saccharomyces cerevisiae* (used as commercial diet). Media containing *C. rugopelliculosa* was beneficial for rotifer growth, and increased cell population by 18.3% compared to the medium containing *S. cerevisiae*. Maximum growth of *C. rugopelliculosa* was $(6.09 \pm 0.04) \times 10^6$ cells/ml and reduction of influent soluble chemical oxygen demand, protein and phosphate concentration was up to 70.0%, 71.4% and 70.4%, respectively, at 6.3 h hydraulic retention time (HRT) (Fig. 9.6) (Lim et al., 2003).

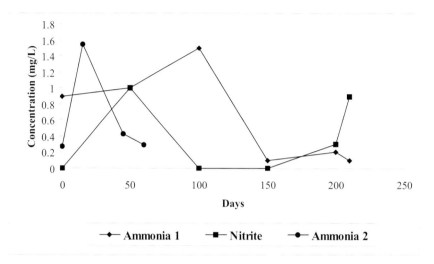

Fig. 9.4 Inorganic nirtrogen (ammonia and nitrite), and ammonia concentrations after 210 and 60 d of treatment, respectively (Ammonia 1: treatment with three denitrifying strains of bacteria; Ammonia 2: treatment with microbial mats) (Barak and van Rijn, 2000; Bender et al., 2004).

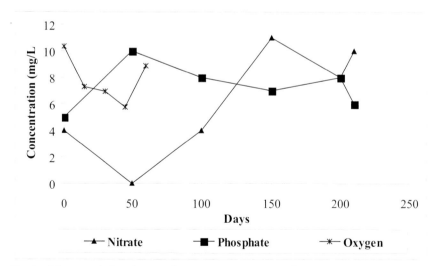

Fig. 9.5 Nitrate and phosphate, and oxygen concentrations after 210 and 60 d of treatment, respectively (Nitrate and phosphate: treatment with three denitrifying strains of bacteria; Oxygen: treatment with microbial mats) (Barak and van Rijn, 2000; Bender et al., 2004).

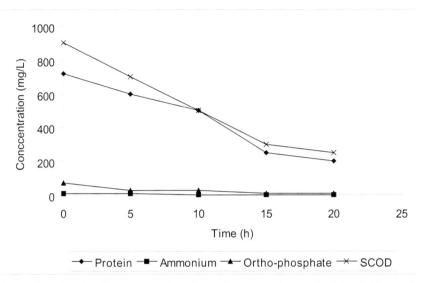

Fig. 9.6 Protein, phosphate, ammonia and SCOD concentrations at 6.3 h HRT (Lim et al., 2003).

Three strains of yeast, *Debaryomyces occidentalis* (P1), *Trichosporon ovoides* (P19) and a strain that could not identified (S27) were isolated from a drainage canal of a fish canning factory. The strain *Trichosporon ovoides* (P19) was immobilised and inoculated into fish processing wastewater, and cultured at 30°C. A total organic carbon decrease from 1.2×10^3 to 3.0×10^2 mg C/l per day was reported. In addition, a decrease in proteins (from 600 µg/ml to 250 µg/ml) and reducing sugars (from 100 µg/ml to 60 µg/ml) was also recorded (Urano et al., 2002).

A commercial bacterial inoculum, called Biostart, was applied to channel catfish (*Ictalurus punctatus*) ponds three times per week for 6 mon. There were few significant differences in concentrations of water quality variables, whereas there were no significant differences in bottom soil carbon and nitrogen between treated with bacteria and control ponds. Survival and net production of fish was significantly greater in ponds treated with the bacterial inoculum compared to controls (Queiroz and Boyd, 1998).

Four strains of *Rhodocyclus gelatinosus* (T6, R4, R5, R7) were isolated from the wastewater of seafood-processing plants and inoculated into tuna condensate (diluted and undiluted), shrimp-blanching water or effluent from a frozen-seafood plant under anaerobic conditions in light. Tuna condensate was the best medium for *Rhodocyclus gelatinosus* growth. Strain R7, gave the highest biomass production (4.0 g/l), cell yield (0.32 g cell/g COD), and COD removal (78%) in 1:10 (v/v) diluted tuna condensate. Shrimp-blanching water was also used to dilute tuna condensate, and an increase in growth rate (0.031-0.033 h^{-1}), biomass (4.1-4.6 g/L) and COD removal (77-80%) was reported. Furthermore, the addition of 1.0-5.0% (w/v) acetate, pyruvate, glucose, glutamate, propionate or malate to the tuna condensate did not increase cell yield, carotenoid or bacteriochlorophyll content or biomass protein. A maximum cell mass of 5.6 g/l (containing 50% protein) and 86% COD removal were observed after 5 d of incubation (Prasertsan et al., 1993).

Cultures of indigenous nitrifying bacteria (*Nitrobacter* and *Nitrosomonas* species) were immobilised onto porous clay pellets and tested for the TAN removal from prawn aquaculture water. Bacteria cultures were enriched from prawn pond water using open (continuous) and closed (batch) enrichment techniques. The results indicated that for both cultures TAN removal increased with increasing initial TAN concentrations (from 10 to 200 mg/l). Batch-enriched immobilised cultures were able to achieve maximum TAN removal (4.2-6.7 mg TAN/l per day) in pond water at density 1 pellet per 100 ml. At lowest pellet density (0.1 pellet /l) of pond water, after a 2-d lag phase, TAN concentration maintained at 0.5 mg/l under the fed-batch culture conditions of 3.2 mg TAN/l/ day (Shan and Obbard, 2001).

Tal et al. (2006) isolated and utilised the anaerobic ammonium-oxidising (Anammox) bacteria (*Planctomycetales* taxon), which were part of the bacterial community of the aerobic (nitrifying) and anaerobic (denitrifying) biofilter of the marine recirculating system, for inorganic nitrogen waste removal. Samples from both biofilters were cultured under anammox conditions, and ammonia and nitrite or hydrazine and nitrite oxidation rates were tested under anaerobic conditions. The removal rates recorded from the denitrifying biofilter were 1.08 µM nitrate N/bead/h under denitrifying conditions, and 0.098 µM ammonia N/bead/h and 0.16 µM hydrazine/ bead/ h under anammox conditions. On the other hand, in the nitrifying biofilter no significant hydrazine removal was reported.

Sponge clones of *Hymeniacidon perleve* were added to sterilised natural seawater (SNSW) supplemented seperately with *Escherichia coli* and *Vibrio anguillarum II* with defined density, and to natural seawater samples with significant bacteria in order to evaluate the ability of marine sponge *H. perleve* of removing the two pathogenic bacteria in aquaculture waters. In SNSW, *H. perleve* effectively removed 96% of *E. coli* within 10.5 h of treatment. For the tests on *V. anguillarum II* in SNSW, about 1.5 g fresh sponges kept pathogen growth at a lower initial density 3.6 x 10^4 cells/ ml of 200 ml water volume. In natural seawater concentrations of *E. coli*, *V. anguillarum II*, and total bacteria were significantly lower, at about 0.9%, 6.2%-34.5%, and 13.7%-22.5%, respectively, compared to untreated samples (Fu et al., 2006).

Three denitrifying strains of bacteria (*Pseudomonas aeruginosa, Paracoccus denitrificans*, and *Pseudomonas* sp.) isolated from a fluidised bed reactor were tested for their ability to remove nitrate and phosphate in a recirculating fish culture system. The treatment system consisted of a digestion basin, a denitrifying fluidised bed reactor and a nitrifying trickling filter. After 210 d of treatment inorganic nitrogen and phosphate concentrations (obtained from the trickling filter basin) did not accumulate (Figs. 9.4, 9.5). More than 90% of phosphorus was retained within the organic matter of the trickling filter. The organic material obtained from the fluidised bed reactor was rich in phosphorus up to 11.8% of dry weight, while in the trickling filter 1.9% of dry weight. In addition, nitrate was produced in the trickling filter, removed in the fluidised bed reactor and removed or produced in the digestion basin (Barak and van Rijn, 2000).

Treatment methods based on the use of various microorganisms for aquaculture waste remediation (parameters, quality control and results) are presented in Table 9.5.

Techniques Based on the Use of Microorganisms for Aquaculture Waste Remediation

Three ensiling experiments (two small silo and one large silo) of crab waste and wheat straw treated with different additives were conducted by Abazinge et al. (1993). In the first small silo experiment crab waste and wheat straw (1:1, wet basis) were ensiled with 0-20% dry molasses, 0-0.1% microbial inoculant, and 0-5.4% phosphoric acid (H_3PO_4), and 20% water (Experiment 1), while in the second small silo experiment 10-20% dry molasses, 0 or 0.1% microbial inoculant, and 0 or 20% water were added (Experiment 2). In large silo experiment crab waste and wheat straw (1:1) were mixed with 20% dry molasses, 20% dry molasses and 0.1% microbial inoculant and 16% acetic acid. In "Experiment 1" trimethylamine content (TMA) was lower up to 8.0-19.0% dry matter for molasses and inoculant treated silages at mixtures without added molasses or acid. In "Experiment 2" concentrations of volatile fatty acids (VFA), acetic, propionic, and isobutyric acid were higher in mixtures with added water and molasses, while addition of 10-20% dry molasses to crab-straw mixtures before ensiling resulted in substantial amounts of lactic acid. In large silo experiments treated silages with 20% molasses and 0.1% inoculant had higher lactic acid concentration (12.76% dry matter).

Effluent from a recirculation aquaculture system, inoculum and 0-8g/l sodium acetate (organic carbon supplementation) were added to two bacteria reactors, which were operated at different HRT (1-11 h). The results showed that 90% of inorganic N and 80% of ortho-phosphate-P were converted, whereas at 11, 6, 3 and 2h HRT, crude protein production was higher than volatile suspended solids (VSS) production. Furthermore, heterotrophic bacteria production on the solid waste obtained from the drum filter of the recirculating aquaculture system resulted in additional protein retention and nutrient discharge (Schneider, 2006a).

A shrimp hatchery effluent was fed in three glass reactors at 24-27°C, pH 7.8-8.0, oxygen concentration 8 mg/l, and hydraulic resistance time 5 d. Each reactor consisted of a retention chamber at the top for the water treatment, three stages with the constructed microbial mats overlaid at each stage, and several small holes serving as a sprinkling filter at the bottom. The effluent flowed throughout the surface of the first bed which contained mats, and then sprinkled to the second and third bed. The results indicated 97% ammonia N and 95% nitrate N concentration removal, while 80% biological oxygen demand (BOD) reduction was also reported (Paniagua-Michel and Garcia, 2003) (Fig. 9.7).

Effluent slurry produced in a recirculating aquaculture system with African catfish was used as substrate for growing heterotrophic bacteria. The effluent was pumped in a bacteria growth reactor through the drum filter of the recirculation system. The reactor was operated in a continuous flow and HRT of 6 h, while different levels of molasses (0-2.5 g C/l) were added to the effluent due to organic carbon deficiency of the fish waste. The results indicated inorganic N and orthophosphate-P reduction from the waste up to 90% and 98%, respectively. VSS concentration, Kjeldahl-N and TOC levels increased with increasing molasses concentration (Schneider, 2006b).

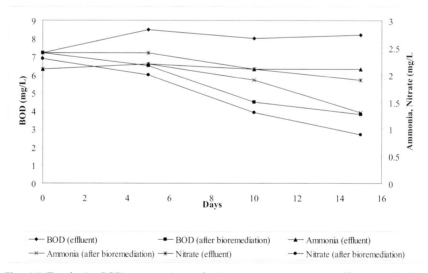

Fig. 9.7 Trends in BOD$_5$, ammonia, and nitrate concentration in effluent and after bioremediation with mats (Source: Paniagua-Michel and Garcia, 2003).

Table 9.5 Treatment methods based on the use of various microorganisms for aquaculture waste remediation (parameters, quality control and results)

No.	Kind of waste	Treatment	Parameters	Methodology	Quality control methods	Results	References
1.	Fisheries processing waste	Fermentation	Ash, carbohydrates, lipids, protein, biomass concentration	Compost extracts were used as a substrate for *Scytalidium acidophilum* ATCC 26774 (agitation speed of 150 rpm, temperature of 25.0°C, pH of 2.0 ± 0.1, and cultivatoin time 8–10 days)	1. Ash.: combustion in a muffle furnace at 600°C 2. Carbohydrates: anthrone reagent method presented by Morris (1948) 3. Lipids: method described by Folch et al. (1957) 4. Protein: multiplying the nitrogen, found by A.O.A.C. Micro-Kjeldahl Method 47.021 (Anon., 1980) by the factor 6.25 5. Biomass concentration: procedure of Martin et al. (1990)	1. Acid extracts of compost had higher concentrations of nutrients (lipids, carbohydrates, and protein) than compost 2. Substrate for the growth of acid-tolrant fungus *S. acidophilum* 3. Used for the biofiltration of liquid and gas effluents and in the growth of mushrooms	Martin, 1999
2.	Fish farm waste	Integrated marine cultivation of *Gracilaria chilensis* (Gracilariales, Rhodophyta)	Carbon, total nitrogen, total phosphorous, agar content	The algae were placed 10 m (1st station) and 150 m west of the cages (2nd station)and at 1 km south of the first farm (3rd station)	1. Carbon and total nitrogen: CHN elemental analyser 2. Total phosphorus: extraction using hydrocloric acid (Aspila et al.,1976) 3. Agar content: according to Cancino and Orellana (1987) and expressed at percentage of *Gracilhuria* dry weight	1. First station: 40% higher growth rate of *Gracilhuria* cultivated compared to the other stations 2. First station: higher algae nutrient content close to the cages (1.9-2.1 mmol N g^{-1} dw^{-1} and 0.28-0.34 mmol P g^{-1} dw^{-1}) compared to the other distances 3. 1 ha cultivation of the algae, close to the fish cages, resulted in 5% removel of dessolved inorganic nitrogen and 27% of released dissolved phosphorous	Troell et al., 1997

Table 9.5 contd....

3.	Sardine processing wastewater (SPW)	Treatment with *Rhodovulum sulfidophilum*	Total biomass, COD, specific growth rate	10, 20 and 30% v/v inoculum were devoloped gluta-matemalate media (GMM) or settled and undiluted SPW	1. Total biomass: drying the washed biomass at 102°C for 24 h (Sawada et al., 1977) 2. COD: according to APHA (1998) 3. Specific growth rate: according to Sasaki and Nagai (1979)	1. Highest biomass production 4.8 g/L (96h, 20% (v/v) inoculum size) 2. 85% COD decrease of SPW (120 h, 30%(v/v) innoculum development in GMM) 3. 79-83% COD reduction of SPW (innoculum was developed in SPW)	Azad et al., 2003
4.	Fish waste-water	Microbial mats combined with fludized sand filters	Oxyzen concentra-tions, pH, tem-perature ammonia concentrations	Biosolar filter system, based on fludized sand filters and microbial mats, which collained *Oscillatoria* sp., on the upper surface and *Rhodopseudomonas* sp., below the surface	1. Oxygen concentrations pH, and temprature: using a YSI computer-integrated probe system (YSI, Yellow Springs, OH) 2. Ammonia concentrations: using a kit from LaMotte Instruments (Chestertown, MD)	1. Ammonia removal 2. Total ammonia concentra-tion: <1mg/L 3. Oxygen concentration: 6-10mg/L	Bender et al., 2004
5.	Fish processing waste-water	Treatment by co-culture of *Candida rugopelliculosa* and *Brachionus plicatilis*	COD, fat, ammonia, organic nitrogen, total Kjeldahl ni-trogen, cations and anions, and heavy metals, cell numbers of yeasts and rotifers	Fish processing waste-water in-noculation with *C. rugopelliculosa* and then fed into two continuos stirred tank reactors (CSTRs). Seed toti-fer culture inoculat-ed (seperately) into two different media which contained *C. rugopelliculosa* and *Saccharomyces cerevistae*	1. COD: according Standard Methods (APHA, 1998) 2. Fat: according to the Bab-cock analysis (APHA, 1985) 3. Ammonia, organic nitrogen, total kjalahl nitrogen, according to the Kjeldahl method (ALPA,1998) 4. Cations and anions: by ion chromatograph 5. Heavy metals: by atomic ab-sorption spectrophotometer 6. Cell numbers of yeasts and rotifers: by haemacytometer	1. *C. rugopelliculosa* media: beneficial for rotifer growth, cell population in-crease by 18.3% compared to the medium contained *S. cerevisiae* 2. Maximum growth of *C. rugopelliculosa* (6.09±0.04) x10⁶ cells/mL 3. SCOD, protein and phos-phate reduction, 70.0%, 71.4%and 70.4% respec-tively (HRT 6.3h)	Lim et al.,2003

Table 9.5 contd.....

No.	Kind of waste	Treatment	Parameters	Methodology	Quality control methods	Results	References
6.	Fish processing waste-water	Bioremediation	Total organic carbon (TOC), inorganic carbon, proteins, peptides, redusing sugars	*Trichosporch ovoides* (p19) immobilization and inoculation into fish processing waste-water and cultivation at 30°C	Total organic carbon (TOC), inorganic carbon: by total carbon analyzer equipped with an infrared detector. Proteins peptides: according to the method of Smith et al. (1985) Reducing sugar: according to the method of Somogyi (1952)	1. TOC decrease: from 1.2×10^3 to 3.0×10^2 mg C/L per day 2. Reduction of proteins (from 600ppm to 250ppm) and reducing sugars (from 100ppm to 60ppm)	Urano et al., 2002
7.	Channel catfish ponds	Treatment with a bacteria inoculam (Biostart)	-	Application of a commercial becterial inoculam to the ponds (3 times/week for 6 months)	-	1. Significant differences in concentrations of water quality variables between treated with the bacteria and control ponds 2. No significant diffrences in bottom soil carbon and nitrogen between treated with bacteria and control ponds 3. Survival and net production of fish greater in ponds	Queiroz and Boyd, 1998

Filleting waste of silver carp (*Hypophthalmichthys molitrix*) was mixed with microbial transglutaminase (3 g/kg), dairy proteins, such as whey protein and sodium caseinate (10 g/kg), and NaCl (0, 10, 20 g/kg) for 2 min at 15°C. The restructured fish products obtained had different levels of salt (0, 1, and 2%). The results showed that the mechanical properties of fish gels grew with increasing the salt level of the samples containing dairy proteins. In addition, the mechanical properties of non-saltedd and low-salted fish gels increased, only when dairy proteins (sodium caseinate) were utilised for filleting waste treatment. Finally, microbial transglutaminase increased expressible water (Uresti et al., 2001).

The influence of various TAN concentrations (1-257 mg TAN/L) and nitrate-N concentrations (174 ± 29 mg/L) on nitrogen conversions in a recirculating aquaculture system was evaluated by Schneider (2006c). The effluent slurry was supplemented with 1.7g C/l sodium acetate, due to organic carbon deficiency, and was converted continuously in a suspended bacteria growth reactor with hydraulic retention time (HRT) of 6 h. The results showed that TAN was converted up to 100%, independently of TAN concentrations, compared to nitrate-N, while VSS production and yield did not alter with increasing TAN concentrations.

Table 9.6 provides a synoptical presentation of techniqes based on the use of microorganisms for fish waste remediation (parameters, quality control and results).

CONCLUSIONS

Aquaculture waste production can hardly be estimated, because of the high variety of aquaculture systems, such as ponds, flow through systems, cages, and recirculation aquaculture systems, and of fish species, such as herbivore, omnivore, carnivore, and of the different types of feed used, such as natural production, agriculture by-products, trash fish, high energy pellets, low protein feeds, and animal or human waste (Schneider, 2006d). The current approach to improving water quality in aquaculture is the application of microbes/enzymes to the ponds (Moriaty 1998). They result in a lower accumulation of slime or organic matter at the bottom of the pond , better penetration of oxygen into the sediment and a generally better environment for the farmed stock (Rao and Karunasagar 2000). Bioremediation is still considered to be a developing technology. One difficulty is that bioremediation is carried out in the natural environment, which contains diverse uncharacterised organisms. Another difficulty is that no two environmental problems occur under completely identical conditions; for example, variations occur in the types and amounts of pollutants, climate conditions and hydrogeodynamics (Watanabe, 2001). In the currect study the dominant microorganisms used for aquaculture waste remediation were *Aspergillus* sp., *Bacillus* sp. and *Lactobacillus* sp.

Table 9.6 Treatment based on the use of microorganisms for fish waste remediation (parameters, quality control and results)

No.	Kind of waste	Treatment	Parameters	Methdology	Quality control methods	Results	Refferences
1.	Crab waste	Ensiling of crab waste and wheat straw treatment with different additives	pH,Lactic acid, water-soluble carbohydrates(WSC), volatile fatty acids (VFA), terimithyl-amine, dry matter, Kjeldahl nitrogen, neutral detergent fiber (NDF), acid detergent fiber (ADF), lignin, cellulose, ash, hemicellulose	*1. small silo experiments:* Experiment 1. Crab waste and wheat straw (1:1,wet basis) were ensiled with 0–20% dry molasses, 0–0.1% microbial silage inoculant, and 0–5.4% phosphoric acid or 0–20% water Experiment 2. 10–20% dry molasses, 0 or 0.1% microbial silage innoculant, and 0 or 20% water were added to crab waste and wheat straw (1:1, wet basis) mixture *2. Large silo experiments:* Crab waste and wheat straw (1:1 proportions) were ensiled with 20% dry molasses, 20% dry molasses and 0.1% microbial silage inoculant, and 16% glacial acetic acid	1. pH: electrometically 2. Lactic acid: Banker and Summerson, 1941, as modified by Pennington and Sutherland (1956) 3. Water-soluble carbohydrates (WSC): Dubois et al, 1956 as adapted for corn plants by Johnson et al. (1966) 4. Volatile fatty acids (VFA): GC 5. Trimethyl-amine according to a colorimetric procedure (Dryer; 1959) after extraction in 7.5% trichloroacetic acid 6. Dry metter: by drying duplicate 200g samples in a forceddraft oven at a maximum of 60°C for 48 h. The sample were analyzed for DM and ash (AOAC, 1980) 7. Kjeldahl nitrogen (N): was determined on wet silage samples (AOAC, 1980) 8. Neutral detergent fiber (NDF) Van Soest and Wine, 1967 9. Acid detergent fiber (ADF): Van Soest, 1963 10. Lignin and Cellulose: Van Soest and Wine, 1968 11. Ash: AOAC, 1980 12. Hemicellulose: was determined by different (NDF–ADF)	*Small silo experiments:* 1. Lower TMA for molasses - and inoculant- treated silages upto 8.0–19.0% (Experiment 1) 2. Higher VFA, acetic, propionic, and isobutyric acid concentrations in mixtures with added water and molasses (Experiment 2). 3. 10–20% dry molasses to crab-straw mixtures before ensiling resulted in substentials amount of lactic acid (Experiment 2). *Large silo experiments:* Higher lactic acid concentration (12.76% dry matter) for ensiled mixture with added 20% molasses and 0.1% inculant, compared to molasses alone	Abazinge et al., 1993

No.	Source	Type	Parameters	Description	Methods	Results	Reference
2.	Effluent from a circulation aquaculture system	Remediation	Total solids (TS), total suspended solids (TSS), VSS, total ammonia nitrogen (TAN), nitrite-N, nitrate-N, orthophosphate-P, Kjeldahl nitrogen, total organic carbon (TOC)	Effluent, inoculum, and 0-8g/L sodium acetate (organic carbon supplementation) were added to two bacteria reactors (HRT 1-11h)	1. Total solids (TS): according to APHA -Method 2540 B 2. Total suspended solids (TSS): according to APHA-Method 2540D 3. VSS:according to APHA -Method 2540E 4. Total amonia nitrogen (TAN) nitrite-N,nitrate-N,ortho-phosphate-P, using the methods 155-006, 461-318,467-033, 503-317 from autoanalyser (SAN, Skalar, The Netherlands) 5. Kjeldahl nitrogen: using a Tecator 2020 Digestor and distillation by Tecator Kjeltec Autosampler system 1035 6. Total organic carbon (TOC) phato-metrically	1. Convertion of 90% of inorganic N and 80% of orthophos-phate-P 2. Crude protein production higher than volatile suspended solids (VSS) production 3. Heterotrophic bacteria production resulted in additional protein retention and nutient discharge	Schneider, 2006a
3.	Shrimp hatchery effluent	Ex-situ bioremediation	Total ammonia nitrogen (TAN), nitrate nitrogen, dissolved oxygen, BOD$_5$, pH	The effluent was fed in three glass reactors (24-27°C, pH 7.8-8.0, oxygen concentration 8 mg/L and HRT 5 days), each one of them consisted of a retention chamber, and three stages with the constructed microbial mats	1.Total ammonia nitrogen (TAN), nitrate nitrogen according to Standard Methods (APHA), 1995) 2. Dissolved oxygen: by modified Winkler method 4. pH by protentiomentry	1. 97% ammonia nitrogen and 95% nitrate nitrogen concentration removal 2. 80% BOD$_5$ reduction	Paniagua-Michel and Garcia, 2003
4.	Effluent slurry from a circulation aquaculture system	Remediation	Total solids (TS), total suspended solids (TSS), VSS, total ammonia nitrogen (TAN), nitrite-N, nitrate-	The effluent was pumped in a becteria growth reactor, which was operated in a continuous flow and HRT of 6 h, while different levels of molasses (0-2.5 g C/L) were added	1. Total solids (TS): according to APHA -Method 2540 B 2. Total suspended solids (TSS): according to APHA-Method 2540D 3. VSS: according to APHA -Method 2540E 4. Total ammonia nitrogen (TAN), nitrite-N, nitrate-N, ortho-phosphate-P: using the methods 155-006, 461-318	1. Inorganic N and orthophosphate-P readuction of 90% and 98% respectively 2. VSS Kjeldahl-N and TOC levels increased with increasing molasses concentration	Schneider, 2006b

In a lesser extent bacteria (cyanobacteria, nitrifying bacteria, ammonium-oxidizing bacteria, etc.), algae, yeasts, sponge, microbial inocula and microbial enzymes were also utilised for fish waste bioremediation.

ABBREVIATIONS

BOD_5	Biological Oxygen Demand
CFSR	Continuous Fixed-Slab Reactor
CSTRs	Continuous Stirred Tank Reactor
DM	Dry Matter
DOM	Dissolved Organic Matter
FW	Freshwater Fish
GMM	Glutamate-Malate Media
HRT	Hydraulic Retention Time
NPN	Non-Protein-Nitrogen
PM	Particulate Matter
RS	Reducing Sugars
SCOD	Soluble Chemical Oxygen Demand
SPW	Sardine Processing Wastewater
SW	Saltwater
TAN	Total Ammonia Nitrogen
TMA	Trimethylamine
TN	Total Nitrogen
TOC	Total Organic Carbon
TR	Tilapia Filleting Residue
TVN	Total Volatile Nitrogen
VBN	Volatile Basic Nitrogen
VFA	Volatile Fatty Acids
VSS	Volatile Suspended Solids

REFERENCES

Abazinge, M.D.A., Fontenot, J.P., Allen, V.G. and Flick, G.J. (1993). Ensiling characteristics of crab waste and wheat straw treated with different additives. J. Agric. Food Chem. 41: 657-661.

Ackefors, H. and Enell, M. (1990). Discharge of nutrients from Swedish fish farming to adjacent sea areas. Ambio. 19: 28-35.

Arvanitoyannis, I.S. and Kassaveti, A. (2008). Fish industry waste: treatments, environmental impacts, current and potential uses. Int. J. Food Sci. Technol. 43: 726-745.

Ayangbile, G.A., Fontenot, J.P., Kirk, D.J., Allen, V.G. and Flick, G.J. (1997). Effects of chemicals on preservation of crab-processing waste and fermentation characteristics of the waste-straw mixture. Food Chem. 45: 3622-3626.

Azad, S.A., Vikineswary, S., Chong, V.C. and Ramachandran, K.B. (2003). *Rhodovulum sulfidophilum* in the treatment and utilization of sardine processing wastewater. Lett. Appl. Microbiol. 38: 13-18.

Barak, Y. and van Rijn, J. (2000). Biological phosphate removal in a prototype recirculating aquaculture treatment system. Aquacult. Eng. 22: 121-136.

Basurco, B. and Lovatelli, A. (2005). The aquaculture situation in the Mediterranean Sea predictions for the future. [Online: http://www.iasonnet.gr/past_conf/abstracts/Basurco.pdf].

Bender, J., Lee, R., Sheppard, M., Brinkley, K., Phillips, P., Yeboah, Y. and Wah, R.C. (2004). A waste effluent treatment system based on microbial mats for black sea bass *Centropristis striata* recycled-water mariculture. Aquacult. Eng. 31: 73-82.

Bratvold, D., Browdy, C.L. and Hopkins, J.S. (1997). Microbial ecology of shrimp ponds: toward zero discharge. Oral presentation—World Aquaculture '97. Seattle, WA, USA.

Chen, C.-C. and Chen, S.-N. (2001). Water Quality Management with *Bacillus* spp. in the High-Density Culture of Red-Parrot Fish *Cichlasoma citrinellum* × *C. Synspilum*. N. Am. J. Aquacult. 63: 66-73.

Chen, S., Stechey, D. and Malone, R.F. (1994). Suspended solids control in recirculating aquaculture systems. In: Aquaculture Water Reuse Systems: Engineering Design and Management. Development in Aquaculture and Fisheries Sciences. M.B.Timmons and T. M. Losordo (eds). Elsevier, Amsterdam, The Netherlands pp. 61-100

Cho, C.Y., Hynes, J.D., Wood, K.R. and Yoshida, H.K. (1994). Development of high-nutrient-dense, low-pollution diets and prediction of aquaculture wastes using biological approaches. Aquaculture 124: 293-305.

Ervik, A., Samuelsen, O.B., Juell, J.E. and Sveier, H. (1994). Reduced environmental impact of antibacterial agents applied in fish farms using the liftup feed collector system or a hydroacoustic feed detector. Dis. Aquat. Org. 19: 101-104.

Faid, M., Karani, H., Elmarrakchi, A. and Achkari-Begdouri, A. (1994). A biotechnological process for the valorisation of fish waste. Biores. Technol. 49: 237-241.

FAO (2002). FAOSTAT, Aquaculture Production Quantities 1970-1999, FAO, Rome, Italy.

Fu, W., Sun, L., Zhang, X. and Zhang, W. (2006). Potential of the marine sponge *Hymeniacidon perleve* as a bioremediator of pathogenic bacteria in integrated aquaculture ecosystems. Biotechnol. Bioeng. 93: 1112-1122.

Gatlin, D.M. III, Li, P., Wang, X., Burr, G.S., Castille, F. and Lawrence, A.L. (2006). Potential application of prebiotics in aquaculture. Proceedings of the VIII International Symposium on Aquaculture Nutrition, 15-17 November, 2006, Mexico, I.E.C. Suarez, D.R. Marie, M.T. Salazar, M.G.N. Lopez, D.A.V. Cavazos and A.C.P.C.A.G.Ortega, pp. 371-376 [Online: http://w3.dsi.uanl.mx/publicaciones/maricultura/viii/pdf/22Gatlin.pdf].

Gowen, R.J. (1991). Aquaculture and the environment. In: The Environment. N. DePauw and J. Joyce (eds). Aquaculture European Aquaculture Society Special Publications No. 16, Ghent, Belgium.

Guerrero, R., Piqueras, Z.M. and Berlanga, Z.M. (2002). Microbial mats and the search for minimal ecosystems. Int. Microbiol. 5, 177-188.

Hammoumi, A., Faid, M., El Yachioui, M. and Amarouch, H. (1998). Characterisation of fermented fish waste used in feeding trials with broilers. Process Biochem. 33(4): 423-427.

http://nai.arc.nasa.gov/students/this_month/index.cfm (accessed 2008, January, 29)

http://www.bio.vu.nl/thb/users/vdberg/berg-mat.html (accessed 2008, January, 29)

http://www.microtack.com/html/microorganism.htm (accessed 2008, January, 29)

http://www.teicrete.gr/LEI/LAB/downloads/industrial_waste.pdf (accessed 2008, January, 29)

Hwang, S.-C., Lin, C.-S., Chen, I.-M., Chen, J.-M., Liu, L.-Y. and Dodds W.K. (2004). Removal of multiple nitrogenous wastes by *Aspergillus niger* in a continuous fixed-slab reactor. Biores. Technol. 93: 131-138.

Hwang, S.-C., Lin, C.-S., Chen, I-M. and Wuc, J.S. (2007). Removal of nitrogenous substances by *Aspergillus niger* in a continuous stirred tank reactor (CSTR) system. Aquacult. Eng. 36: 177-183.

Jones, A.B., Dennison, W.C. and Preston, N.P. (2001). Integrated treatment of shrimp effluent by sedimentation, oyster filtration and macroalgal absorption: a laboratory scale study. Aquaculture 193: 155-178.

Lim, J., Kim, T. and Hwang, S. (2003). Treatment of fish-processing wastewater by co-culture of *Candida rugopelliculosa* and *Brachionus plicatilis*. Water Res. 37: 2228-2232.

Liu, F. and Han, W. (2004). Reuse strategy of wastewater in prawn nursery by microbial remediation. Aquaculture 230: 281-296.

Maeda, M. (1994). Biocontrol of the larvae rearing biotope in aquaculture. Bull. Natl. Res. Inst. Aquacul. (Suppl.) 1: 71-74.

Martin, A.M. (1999). A low—energy process for the conversion of fisheries waste biomass. Renewable Energy 16: 1102-1105.

Moriarty, D.J.W. (1998). Control of luminous *Vibrio* sp. in penaeid aquaculture ponds. Aquaculture 164: 351-358.

Moriarty, D.J.W. (1999). Disease control in shrimp aquaculture with probiotic bacteria. In: Proceedings of the 8th International Symposium on Microbial Ecology. C.R. Bell, M. Brylinsky and P. Johnson-Green (eds). Atlantic Canada Society for Microbial Ecology, Halifax, Canada pp. 237-243.

Morimura, S., Nagata, H., Uemura, Y., Fahmi, A., Shigematsu, T. and Kida, K. (2002). Development of an effective process for utilisation of collagen from livestock and fish waste. Process Biochem. 37: 1403-1412.

Nisbet, E.G. and Fowler, C.M.R. (1999). Archaen metabolic evolution of microbial mats. Proc. Royal Soc. Biol. Sci. Ser. B 266: 2375-2382.

Organisation for Economic Co-operation and Development (OECD) (2000). Transition to responsible fisheries—post-harvesting practices and responsible fisheries: Case Studies. Directorate for Food Agriculture and Fisheries, Paris, France.

Paniagua-Michel, J. and Garcia, O. (2003). *Ex-situ* bioremediation of shrimp culture effluent using constructed microbial mats. Aquacult. Eng. 28: 131-139.

Pillay, T.V.R. (1991). Aquaculture and the environment. Blackwell Scientific, London, UK.

Prasertsan, P., Choorit, W. and Suwanno, S. (1993). Optimization for growth of *Rhodocyclus gelatinosus* in seafood processing effluents. World J. Microbiol. Biotechnol. 9: 593-596.

Queiroz, J.F. and Boyd, C.E. (1998). Effects of a bacterial inoculum in Channel Catfish ponds. J. World Aquacult. Soc. 29: 67-73.

Rao, S.P.S. and Karunasagar, I. (2000). Incidence of bacteria involved in nitrogen and sulphur cycles in tropical shrimp culture ponds. Aqua Int. 8: 463-472.

Reddington, C. (2003). Anaerobic digestion of fish waste: A novel solution to the problem? In: Proceedings of the First Joint Trans Atlantic Fisheries Technology Conference, The Icelandic Fisheries Laboratories, 10-14 June 2003 Reykjavik, Iceland. [Online: http://vefur.rf.is/TAFT2003/PDF/Reddington.pdf].

Saroglia, M. and Basurco, B. (2006). Mediterranean aquaculture. In: AQUA2006 including the 'Aquaculture Europe 2006' (Annual meeting of the European Aquaculture Society) and 'World Aquaculture 2006' (Annual meeting of the World Aquaculture Society). May 9-13, Firenze (Florence) Italy.

Sasikumar, C.S. and Papinazath, T. (2003). Environmental management: Bioremediation of polluted environment. In: Proceedings of the Third International Conference on Environment and Health. J. Martin, V. Bunch, Madha Suresh and T. Vasantha Kumaran (eds). Chennai, India, pp.15-17 December, 2003. Chennai: Department of Geography, University of Madras and Faculty of Environmental Studies, York University, pp. 465-469. [Online: http://www.yorku.ca/bunchmj/ICEH/proceedings/Sasikumar_CS_ICEH_papers_465to469.pdf].

Schneider, O. (2006a). Heterotrophic bacteria production utilizing the drum filter effluent of a RAS: influence of carbon supplementation and HRT. In: Fish Waste Management by Conversion into Heterotrophic Bacteria Biomass. Ph.D. Thesis, Wageningen University, The Netherlands, pp. 39-56 [Online: http://library.wur.nl/wda/dissertations/dis3962.pdf].

Schneider, O. (2006b). Molasses as C source for heterotrophic bacteria production on solid fish waste. In: Fish Waste Management by Conversion into Heterotrophic Bacteria Biomass.

PhD Thesis, Wageningen University, The Netherlands pp. 69-84 [Online: http://library.wur.nl/wda/dissertations/dis3962.pdf].

Schneider, O. (2006c). Introduction. In: Fish Waste Management by Conversion into Heterotrophic Bacteria Biomass. Ph.D. Thesis, Wageningen University, The Netherlands pp. 9-18 [Online: http://library.wur.nl/wda/dissertations/dis3962.pdf].

Schneider, O. (2006d). TAN and nitrate yield similar heterotrophic bacteria production on solid fish waste under practical RAS conditions. In: Fish Waste Management by Conversion into Heterotrophic Bacteria Biomass. Ph.D. Thesis, Wageningen University, The Netherlands pp. 57-68 [Online: http://library.wur.nl/wda/dissertations/dis3962.pdf].

Shan, H. and Obbard, J.P. (2001). Ammonia removal from prawn aquaculture water using immobilized nitrifying bacteria. Appl. Microbiol. Biotechnol. 57: 791-798.

Shih, I.-L., Chenb, L.-G., Yu, T.-S., Changb, W.-T. And Wang, S.-L. (2003). Microbial reclamation of fish processing wastes for the production of fish sauce. Enz. Microb. Technol. 33: 154-162.

Tacon, G.J. and Forster, I.P. (2001). Global trends and challenges to aquaculture and aquafeed development in the New Millennium. In: International Aqua Feed Directory and Buyer's Guide. Turret RAI PLC, Uxbridge pp. 4-24.

Tal, Y., Watts, J.E.M., and Schreier, H.J. (2006). Anaerobic ammonium-oxidizing (Anammox) bacteria and associated activity in fixed-film biofilters of a marine recirculating aquaculture system. Appl.Environ. Microbiol. 72: 2896-2904.

Thassitou, P.K., and Arvanitoyannis, I.S. (2001). Bioremediation: a novel approach to food waste management. Trends Food Sci. Technol.12: 185-196.

Troell, M., Halling, C., Nilsson, A., Buschmann, A.H., Kautsky, N. and Kautsky, L. (1997). Integrated marine cultivation of *Gracilaria chilensis* (Gracilariales, Rhodophyta) and salmon cages for reduced environmental impact and increased economic output. Aquaculture 156: 45-61.

Urano, N., Sasaki, E., Ueno, R., Namba, H. and Shida, Y. (2002). Bioremediation of fish cannery wastewater with yeasts isolated from a drainage canal. Mar. Biotechnol. 4: 559-564.

Uresti, R.M., Tellez-Luis, S.J., Ramirez, J.A. and Vazquez, M. (2001). Use of dairy proteins and microbial transglutaminase to obtain low-salt-fish products from filleting waste from silver carp (*Hypophthalmichthys molitrix*). Process Biochem. 36: 809-812.

Van Loosdrecht, M.C.M., Lyklema, J., Norde, W., Schraa, G. and Zehnder, A.J.B. (1987a). The role of bacterial cell wall hydrophobicity in adhesion. Appl. Environ. Microbiol. 53, 1893-1897.

Van Loosdrecht, M.C.M., Lyklema, J., Norde, W., Schraa, G. and Zehnder, A.J.B. (1987b). Electrophoretic mobility and hydrophobicity as a measure to predict the initial steps of bacterial adhesion. Appl Environ Microbiol. 53: 1898-1901.

Vidali, M. (2001). Bioremediation. An overview. Pure Appl. Chem. 73: 1163-1172.

Vidotti, R.M., Viegas, E.M.M., Dalton, V. and Carneiro, D.J. (2003). Amino acid composition of processed fish silage using different raw materials. Anim. Feed Sci. Technol. 105: 199-204.

Wang, Y.-B. and Han, J.-Z. (2007). The role of probiotic cell wall hydrophobicity in bioremediation of aquaculture. Aquaculture 269: 49-354.

Watanabe, K. (2001). Microorganisms relevant to bioremediation. Environ. Biotechnol. 12: 237-241.

Westerdahl, A., Olsson, C., Kjellerberg, S. and Conway, P. (1991). Isolation and characterization of turbot (*Scophthalmus maximus*) associated bacteria with inhibitory effects against *Vibrio anguillarum*. Appl. Environ. Microbiol. 57: 2223-2228.

Whiteley, A.S. and Bailey, M.J. (2000). Bacterial community structure and physiological state within an industrial phenol bioremediation system. Appl. Environ. Microbiol. 66: 2400-2407.

Yamamoto, M., Saleh, F., Ohtsuka, A. and Hayashi, K. (2005). New fermentation technique to process fish waste. Anim. Sci. J. 76: 245-248.

10

Microbial Reclamation of Fish Industry By-products

N.Bhaskar,* N.M. Sachindra, P.V. Suresh and *N.S. Mahendrakar*

INTRODUCTION

The world fish production has almost stagnated and presently stands at a little more than 14 million metric tonnes (MMT) (FAO, 2010). This can be seen by the considerably lower changes in production figures in the last 5 yr (Table 10.1). Fish processing in general is an economic activity that generates considerable foreign exchange to a maritime nation. For instance, Indian seafood exports in 2008 stood at about US $ 1908 billion with the total quantities of seafood exported being 602,835 MMT (MPEDA, 2010). Major aquatic products that dominate the export market come from the marine sector (both culture and capture). The major marine foods that are exported include shrimps, molluscs (squids, cuttle fishes, octopuses, oysters, mussels and clams) and fin fishes (tuna, seer, pomfret, etc.) mainly in frozen form. Apart from this processed marine products like *surimi*, canned products and ready-to-eat fish products are also exported.

Initially, fish sources appeared to be inexhaustible and by-products arising out of fish processing were seen as being worthless and discarded (Gildberg, 2004). This, apart from resulting in the loss of huge amounts of protein rich by-products also led to pollution problems. Although some by-products like shrimp waste are being utilized today, a huge amount is still being discarded creating both disposal and pollution problems. Annual discard from the fish industry is estimated to be 18-30 MMT around the world (Elvevoll, 2004), accounting for 25 to 30% of global fish production

Department of Meat, Fish and Poultry Technology, Central Food Technological Research Institute (CSIR), Mysore 570 020, India
*Corresponding author: E-mail: bhasg3@yahoo.co.in

(Rustad, 2003). The present status of disposition of global fish production indicates that >76% of fish produced is used for human consumption, while the rest goes for reduction or miscellaneous purposes (FAO, 2010). However, developed countries use fish in the processed form rather than being marketed fresh (Table 10.2) which would result in huge amounts of processing wastes at the points of production for such items. As far as global aquaculture production is concerned, China ranks first (contributing > 67% of world production) followed by India (~6% of world production); freshwater fish culture is the main contributor for aquaculture production with carps accounting for > 40% production (FAO, 2010).

Table 10.1 Global production (in metric tonnes) of different aquatic resources between 2003 and 2007

Year	Capture		Culture		*Total#*	*% change$*		
	Fishes*	Aq. plants	Fishes*	Aq. plants		Capture	Culture	Total
2003	90353972	1259379	42682153	12526482	133036125	-	-	-
2004	94363635	1366010	45924282	13930570	140287917	+4.43	+7.60	+5.45
2005	93253346	1305803	48149792	14789972	141403138	-1.18	+4.85	+0.79
2006	91994321	1143273	51653329	15075612	143647650	-1.37	+6.78	+1.56
2007	90063851	1104948	50329007	14858791	140392858	-2.14	-2.63	-2.32

*: Includes fishes, crustaceans, mollusks etc.
#: Total of capture and culture fish production—excludes aquatic plants.
$: % change calculated on year basis—only for fish category.
Source: FAO (2010).

The major by-products generated through marine products processing include visceral wastes, scales, waste (wash) water, filleting wastes (head, frame bones, skins and fins), air bladder, body/head shell waste, calcarious shells, etc. Similarly, freshwater fishes through the inland fishery activities mainly meet the domestic demand for fish and fish products. Most of the freshwater fishes find their way to consumer homes as table fishes. This effectively means that the major by-products generated in the domestic markets include scales, skins, visceral mass (viscera, air bladder, gonads and other organs), head and fins. Unlike the seafood processing sector, freshwater fish or the inland fisheries sector in general is unorganized and hence poses a different level of waste disposal problems. Further, most of these processing wastes are reported to be a good source of proteins (including enzymes) and fats [including omega (ω)-3 fatty acids and carotenoids]. If these biological compounds can be recovered, it would serve the dual purpose of recovery of these biomolecules and reducing the pollution problems associated therewith.

The present chapter excludes the general methods of preparation and applications of aquatic silage (including both fish and shell fish silages). For details on preparation and general uses (mainly as ingredients in

Table 10.2 Disposition (%) of global fish production between 2003 and 2007

	2003			2004			2005			2006			2007		
	Dvd.C	Dvn.C	World	Dvd.C	Dvn.C	World	Dvd.C	Dvn.C	World	Dvd.C	Dvn.C	World	Dvd.C	Dvn.C	World
a) For human consumption															
Fresh	5.4	50.6	39.6	5.2	47.4	37.9	4.2	47.8	38.3	4.1	48.3	39.0	3.9	47.9	38.9
Frozen	40.4	14.4	20.7	40.4	14.1	20.0	42.7	14.3	20.5	43.8	15.0	21.1	43.9	15.3	21.2
Curing	9.9	8.7	9.0	10.2	8.1	8.6	11.0	8.0	8.7	11.0	8.3	8.9	10.8	8.4	8.9
Canning	21.9	7.6	11.1	21.9	8.0	11.1	22.4	8.4	11.5	23.0	9.1	12.0	23.0	9.2	12.0
Total (a)	77.6	81.3	80.4	77.7	77.7	77.7	80.5	78.5	78.9	82.0	80.6	80.9	81.5	80.8	81.0
b) For other purposes															
Reduction	19.7	14.9	16.1	19.1	18.1	18.3	16.6	17.1	17.0	15.0	14.6	14.7	15.6	14.3	14.5
Miscellaneous	2.7	3.8	3.5	3.3	4.2	4.0	3.0	4.4	4.1	3.0	4.8	4.0	2.9	4.9	4.5
Total (b)	22.4	18.7	19.6	22.3	22.3	22.3	19.5	21.5	21.1	18.0	19.4	19.1	18.5	19.2	19.0

Dvd.C: Developed countries; Dvn.C: Developing countries; Reduction mainly includes fish meal and oil production.
Source: FAO (2010).

livestock/aquaculture feeds) readers may refer to comprehensive reviews by other authors (Raa and Gildberg, 1982; Gildberg, 2004; Arruda et al., 2007; Bechtel, 2007). In the context of this review, microbial reclamation of fish industry waste especially with reference to fermentation techniques and application of proteases of microbial origin only are considered. This chapter reviews and discusses microbial reclamation from the view points of preservation as well as recovery of valuable components like enzymes, carotenoids, polyunsaturated fatty acids (PUFA), etc. from fish industry wastes [both finfish and shell fish (mainly crustacean shell fish)].

QUALITY AND QUANTITY OF FISH INDUSTRY WASTES

Considering the world fish production and fish production in India in particular, it is obvious that large quantities of by-products are generated through processing. For instance, as can be seen from the production figures from both capture and culture fisheries in India between 2003 and 2007 (Table 10.3A), India alone generates > 3 MMT of by-products annually due to fish processing activities. If one considers only visceral waste and shrimp shell waste , Indian fish processing industry generates >700,000 tonnes annually; while global production of fish viscera and shrimp shells would amount to 11.4 MMT (Table 10.3B).

The major by-products arising out of fish processing include viscera, skin, scales, bones and bone frames (in case of *surimi* production). These by-products are rich in protein and fat which make them more perishable. Reports on the yield of by-products arising from processing operations in terms of weight percentage of live fish/shellfish are very limited. A generalized schematic representation of different by-products and their yield is presented in Fig. 10.1. There are only a few detailed reports with regards to yield of processing by products. For instance, frame bones resulting from filleting operations account for 15% of live weight in case of cod and salmon with more than 60% live weight of fish being fillets (Gildberg et al., 2002; Liaset et al., 2003). Similarly, shell waste including body carapace and head in case of some Indian shrimp varieties account for 48 to 56% of live weight (Sachindra et al., 2005a) and meat content of some of the Indian crabs (freshwater and marine) varies from 28 to 30% of live weight (Sachindra et al., 2005b). Some of the available data regarding yield of by-products in different finfish and shell fish species is presented in Table 10.4.

As mentioned earlier fish processing wastes/by-products are good raw materials for fishmeal and oil production (Rustad, 2003; Gbogouri et al., 2006; Guerard, 2006). These by-products can be categorized into material used for non-food purposes [as feed ingredients/fertilizers/

leather making etc. (e.g., silage, composting)], food purposes (e.g., fish sauce) and for specialty products (mainly bioactive molecules; e.g. carotenoids, collagen, peptides, enzymes, etc.). There is normally a far better profitability in making products for human consumption while the highest profitability is in producing bioactive compounds that have biotechnological/pharmaceutical applications (Rustad, 2003). Several methods have been reported for recovering proteins and lipids from these processing wastes. These include recovering proteins/lipid and related products through various physical, chemical and biological treatments or their combinations. The possible methods of recovering protein/lipid and related molecules from fin fish and shell fish processing waste is summarized in Fig. 10.2. As can be seen fermentation or hydrolysis using enzymes results in simultaneous recovery of proteins, fats and related compounds.

Table 10.3A Fish production (in metric tonnes) in India in the last five years and estimate of by-products generated there from

Year	Capture	Culture	Total	Waste generated[Est.]
2003	3712149	2312971	6025120	2711304
2004	3391009	2794636	6185645	2783540
2005	3481136	2837751	6318887	2843499
2006	3844837	3169303	7004140	3151863
2007	3953476	3354754	7308230	3288704

[Est.]: Estimated considering the average waste on live weight basis to be 45%
Source: FAO (2010); All the values are weights in tonnes.

Table 10.3B Global and Indian fish production and current scenario of estimated waste generation (in million metric tonnes) (FAO, 2010)

	Global	India
Capture	90.10	3.95
Culture	50.33	3.35
Total	140.43	7.30
Estimated waste generation*	63.2	3.3
Visceral waste alone**	8.4	0.4
Shrimps / Prawns only	6.1	0.3
Estimated waste generation***	3.0	> 0.1

*: Estimated considering 45% of live weight to be waste.
**: Estimated considering an average of 6% to be contributed by viscera.
***: Estimated considering that head and body carapace constitute 50% of live weight.
Source : FAO (2010)

① Head (9-12%)
  Food, meal, oil
② Meat portion (30-40%)
  Food (fillet / mince / steak), meal, oil
③ Visceral mass (5-15%)
  Fish meal, food (fish sauce), oil, enzymes
④ Skin & Scales (3-8%)
  Collagen, gelatin, leather
⑤ Bones (10-20%)
  Meal, collagen, gelatin
⑥ Fins (3-6%)
  Meal, oil, collagen, gelatin

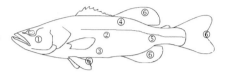

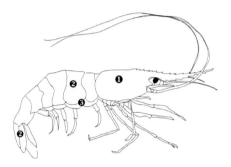

❶ Head shell waste (35-40%)
  Chitin, chitosan carotenoids, enzymes
❷ Meat portion (40-45%)
  Food (fillet / mince / steak), meal, oil
❸ Body carapace (5-10%)
  Chitin, chitosan, carotenoids
❹ Skin & Scales (3-8%)
  Collagen, gelatin, leather
❺ Ovaries

① Shell (30-40%)
  Chitin, calcium, carotenoid
② Meat portion (20-35%)
  Food
③ Claw (5-15%)
  Food, chitin, calcium, carotenoid
④ Ovaries (5-15%)
  Food, protein, enzymes, carotenoids

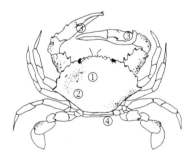

Fig. 10.1 Finfish and crustacean shellfish by-products and their possible uses. (*Values shown in the figure are compiled from various authors mentioned elsewhere in the text*).

CHARACTERISTICS OF FISH INDUSTRY WASTES WITH REFERENCE TO BIOACTIVE MOLECULES

Fish wastes remain a source of bioactive materials that include proteinaceous, lipid based and other bioactive compounds. These include peptides/amino acids, carotenoids, vitamins, antioxidants and minerals (Hurst, 2006). Some of the important bioactive materials that can be derived from fish processing wastes are discussed in brief in this section. For detailed information on these bioactive materials of aquatic origin, readers can refer to reviews by several authors (Kristinsson and Rasco, 2000c; Matusno, 2001; Aneiros and Garateix, 2004; Waldron et al., 2004; Kim and Mendis, 2006; Bhaskar et al., 2006a, b; Hurst, 2006; Kristinsson, 2007; Arvanitoyannis and Kassaveti, 2008; Chandini et al., 2008; Hayes et al., 2008a, b; Kim et al., 2008; Kristinsson, 2008; Miyashita and Hosokawa, 2008).

Table 10.4 Composition (%) of different by-products resulting from the processing of various fin fish (fresh water carps) and shell fishes

(A) Freshwater carps[a]				
	Common carp	Catla	Rohu	Mrigal
Meat	55.9	63.4	69.6	54.9
Viscera	13.9	3.9	4.9	11.0
Fins	3.4	2.9	3.1	3.6
Scale	3.7	3.6	3.1	4.3
Skin	5.7	4.4	3.9	8.0
Bone	6.1	4.5	3.9	8.1
Head	8.5	14.4	10.0	8.8
Gills	2.9	2.9	2.7	1.5
(B) Crustacean shell fishes–Shrimps				
	Penaeus mondon[b]	*Solonocera indica*[c]	*Aresteus alcocki*[c]	*Pandalus borealis*[d]
Meat	51.3	34.4	37.4	48.1
Head	34.4	53.4	47.5	38.9
Carapace	14.3	12.2	15.1	13.0
(C) Crustacean shell fishes - Crabs[p]				
	Potamon potamon		*Charbdys cruciata*	
Meat	28.8		29.7	
Shell	35.7		34.4	
Rest[#]	35.5		35.9	
(D) Molluscan shell fishes–Squids[q]				
	Squid–Plant A		Squid–Plant B	
Head & arms	42.8		52.4	
Fins	25.3		-	
Funnel	12.9		29.5	
Tube	8.0		-	
Viscera	6.1		10.8	
Pens	1.6		0.3	

a: Sachindra et al., 2004; Common carp (*Cyprinus carpio*), Catla (*Catla catla*), Rohu (*Labeo rohita*) and Mrigal (*Cirrhinus mrigala*) are all major freshwater carps available in India.
b: Sachindra et al., 2005a; c: 2006a; d: Heu et al., 2003.
p: Sachindra et al., 2005b; *P. potamon* is a freshwater crab while *C. cruciata* is a marine crab available in India.
#: Includes parts other than meat and shell (e.g., gills, viscera etc.) that were discarded.
q: Lian et al., 2005; Squid–*Loligo pealei*.

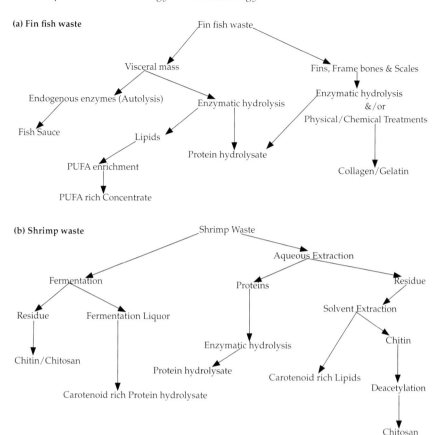

Fig. 10.2 Possible methods of utilization of fish and shrimp industry wastes along with their resultant products.

Fish processing by-products are generally rich in proteins and lipid based compounds. In addition, crustacean shellfish processing wastes contain pigments that contribute for their characteristic colour on cooking. Proximate composition of different fish processing industry by-products is presented in Table 10.5. As can be seen from the table, protein contents in by-products of marine species vary from 7.9% (shrimp carapace) to as high as 25% (Pollock skin) of fresh weights whereas it ranges from 8.5 to ~14% in freshwater fish by-products. Apparently, fish wastes are also rich in lipids that can be recovered for further applications. In case of crustacean shell fishes, recoverable components include chitin (~5% of dry weight), protein (~10%) and carotenoids (~60 to 185 µg/g fresh waste) (Table 10.5). In the ensuing paragraphs, general characteristics of fish proteins and lipids will not be discussed rather specific lipid/protein components that have potential applications will be described.

Table 10.5 Proximate composition of by-products resulting from the processing of different finfish and shell fish species

Fish	Moisture*	Protein*	Fat*	Ash*	Carotenoid**	Chitin	Source
Herring	68.6	11.7	16.2	2.9	-	-	Aidos et al., 2001
Hake	77.5	16.5	2.0	3.5	-	-	Batista, 1999
Monk fish	80.8	14.5	0.7	3.7	-	-	
Pollock Viscera	63.5	15.2	19.1	1.6	-	-	
Pollock frames	80.9	16.3	0.9	3.4	-	-	Bechtel, 2003
Pollock heads	79.9	15.2	1.2	4.6	-	-	
Pollock skin	78.2	25.0	0.4	0.7	-	-	
Tuna fins	37.2	24.4	0.4	40.0	-	-	Aewsiri et al., 2008
Tuna viscera	79.2	16.0	1.6	3.6	-	-	Cha and Cadwallader, 1998
Cod by pdts.#	70.5	16.1	9.6	3.6	-	-	Dauksas et al., 2005
Salmon-head	66.0	11.3	16.7	6.0	-	-	Ramirez, 2007
Salmon-BB	64.3	14.1	15.2	6.4	-	-	
Catla viscera	76.3	8.5	12.5	2.5	-	-	Bhaskar et al., 2008
VW-FW carps	58.9	11.9	27.8	2.5	-	-	Ganesan et al., 2009
Squid by pdts.	81.2	14.1	2.9	1.8	-	-	Lian et al., 2005
Shrimp—*Crangon crangon*	71.1	40.6	10.0	27.5	-	17.8	Synowiecki and Al-Khateeb, 2003
Shrimp A—Head	77.2	8.6	8.1	4.0	185.3	3.3	Sachindra et al., 2006b
Shrimp A—Carapace	79.8	7.9	3.0	4.9	117.4	4.4	
Prawn—Head	72.6	19.3	0.9	7.1	-	-	Balogun and Akegbejo-Samsons, 1992
Shrimp Waste	81.4	28.4$	17.9$	19.2$	59.8	26.3$	Bhaskar et al., 2007c

#: Viscera+Backbone; BB–back bone; *: g per 100g wet waste, unless otherwise mentioned; **: μg per gram of wet waste; $: g per 100g dry weight.
Shrimp A: *Aresteus alcocki*; Prawn: *Macrobrachium rosenbergii*; Shrimp waste–Mix of head and carapace from *Penaeus monodon*; VW-FW carps: Mix of visceral waste from major fresh water carps (Catla, Rohu & Common Carp).

Protein Related Compounds

Collagen and Gelatin

By-products from fish processing are a potential source of collagen. It is the main component in the skin (Norland, 1989; Sikorski and Borderias, 1994) that can be collected separately from other by-products. Skin and bones constitute about 30% of total fish weight (Muyonga et al., 2004; Aewsiri et al., 2008).

Several authors have studied isolation and characterization of collagen from fish wastes (Nomura et al., 1997; Nagai and Suzuki, 2000; Nagai, 2004; Mizuta et al., 2005; Arnesen and Gildberg, 2006). Collagen in the purified form has a number of applications in pharmaceutical and cosmetic industry (Borderias et al., 1994). Similarly gelatin, the hydrolyzed form of collagen, is an ingredient in the food industry. Gelatin is used as a food additive to increase the texture, the water holding capacity and stability of several food products (Borderias et al., 1994). The uniqueness of fish collagen from cold water fish lies in the lower content of proline and hydroxy proline (Haard et al., 1994). Although fish gelatin does not form particularly strong gels, it is well suited for certain industrial application; i.e., micro-encapsulation, light sensitive coatings and low set time glue. There is also a market for non-gelling gelatin which has a potential in the cosmetic industry as an active ingredient. Fish collagen and gelatin generates new applications as a food ingredient because it has properties different from mammalian collagen (Rustad, 2003).

Protamine and Taurine

Protamine is a basic peptide containing more than 80% arginine. It has been found in the testicles of more than 50 fish species. Protamine has the ability to prevent the growth of *Bacillus* spores, being used as an antimicrobial agent in food processing and preservation (Rustad, 2003). Taurine is a sulphur containing amino acid most abundant in any tissue and is unique due to the fact that it does not form a constituent of any proteins nor is it linked to any protein by a peptide bond. Aquatic foods including fish processing by-products are rich sources of taurine (Sakaguchi et al., 1982; Zhao et al., 1998; Gormley et al., 2007). It has a hypolipedemic effect and is known to stimulate bile acid synthesis (Ogawa, 1996). The physiological benefits of taurine is a subject of several research works (di Wu et al., 1999; Liu and Li, 2000; Militante and Lombardini, 2002; Berger and Mustafa, 2003; Fennessey et al., 2003). Fish processing wastes are potential sources from which taurine can be recovered for further use in both foods as well as biomedical applications.

Protein Hydrolysates

Being rich in protein, fish processing by-products hold a potential for preparation of protein hydrolysates. Fish protein hydrolysates have application mainly in nutritional management of individuals who can not digest whole protein. The most prevalent application of protein hydrolysate has been for feeding infants with food hypersensitivity (Silvestre, 1997). The majority of works on preparation of protein hydrolysates from fish processing wastes reveal that the hydrolysates possess anti-oxidative properties (Amarowicz and Shahidi, 1997; Batista, 1999; Kim et al., 2001; Mendis et al., 2005; Je et al., 2007). The preparation of protein hydrolysates, their functional and physiological properties have been reviewed by several authors (Kristinson and Rasco, 2000c; Guerard, 2006; Kristinsson, 2007, 2008; Fujita and Yoshikawa, 2008; Okada et al., 2008).

Fish Peptone

Fish peptone is a protein, which has been digested to peptides and free amino acids by proteolytic enzymes. The soluble protein fraction of fish viscera silage is rich in essential amino acids and has a high nutritional value (Strom and Eggum, 1981; Wright, 2004). Due to low average molecular weight, it is suitable as an easily digestible protein supplement in the feed for juvenile fish and domestic animals (Ragunath and Gopakumar, 2002). Fish viscera peptone is an excellent nitrogen source in microbial growth media. It has been used successfully in cultivating fish pathogens like *Vibrio salmonicida* for vaccine production and has been shown that it provides better growth than high quality peptones like Bacto tryptone and Bacto peptone obtained from microorganisms (Clausen et al., 1985; Almas, 1990; Vecht-Lifshitz et al., 1990).

Enzymes

Apart from the above, fish processing by-products are also rich sources of diverse enzymes. Several digestive proteases, chitinases, lipases, transglutaminases, collagenases, etc. are the major enzymes that can be recovered from fish processing wastes, mainly visceral wastes (Morrissey and Okada, 2007). The enzymes available are so diverse in nature that the topic has been the subject of several reviews (Haard, 1998; Gildberg et al., 2000; Shahidi and Janak-Kamil 2001) and publication as a text by Haard and Simpson (2000). Most of these enzymes of aquatic origin find application in preparation of autolytic silage, sauce preparation, fermented product preparation etc.

Lipid Related Compounds Including Carotenoids and Antioxidants

Fish Oil

Lipids play a major role in human nutrition and health while accounting for about 40% of the total calories in most of the industrialized countries (Gordon and Ratlif, 1992). Fish oils, especially of marine origin, are unique compared to other lipids. Fish oils or lipids (mainly marine lipids) have been considered as an available source of omega-3 fatty acids, especially docosahexaenoic acid (DHA) and eiocsapentaenoic acid (EPA) (Aidos et al., 2001; Gbogouri et al., 2006). The various mechanisms by which EPA and DHA prevent/modify cardiovascular diseases have been reviewed (Gordon and Ratlif, 1992; Connor, 2000; Holub, 2002; Tapiero et al., 2002; Li et al., 2003; Calder, 2003; Bhaskar et al., 2006a).

Carotenoids

Apart from the PUFA, marine by-products are also rich sources of carotenoids, especially processing wastes of crustacean shellfishes. Crustacean wastes are one of the major sources of recoverable natural carotenoids, especially astaxanthin (Shahidi et al., 1998; Armenta-Lopez et al., 2002; Sachindra et al., 2005a, b, 2006a, b; Simpson, 2007). The subject of physiological benefits of astaxanthin has been reviewed (Sachindra and Mahendrakar 2005; Sachindra et al 2006b; Olaizola 2008).

Chitin and Chitosan

Chitin is the second most abundant biopolymer that is composed of N-acetyl-D-glucosamine units and gives structural strength to insects and crustacean shells. Native chitin is insoluble and has limited applications ranging from large scale technical applications in functional membranes, food technology and protein precipitation, to very sophisticated applications on medicine and cosmetics (Tsugitta, 1990; Knorr, 1991; Taranathan and Kittur, 2003; Hayes, 2008b). Shellfish like crab, shrimp, prawn, lobster and cray fish contain 14-35% chitin on dry matter basis (Tsugita, 1990; Rao et al., 2000; Healy et al., 2003; Bhaskar et al., 2007a). Chitosan can be best defined as a family of polymers with varying deacetylation levels, size and charge distribution (Tharanathan and Kittur, 2003). Preparation of chitiosan with the aid of enzymes and its characterization (Vishnukumar et al., 2004a, b; Kittur et al., 2005), chitosan films for food packaging (Taranathan, 2003; Srinivasa et al., 2004b), properties of dried chitosan films (Tharanathan, 2003; Srinivasa et al., 2004a), biodegradation of chitosan (Harish Prashanth et al., 2005), safety of chitosan coating (Ramesh et al., 2004) and characterization of depolymerized products of chitosan (Harish Prashanth and Taranathan, 2005) have been recently reported.

MICROBIAL RECLAMATION OF FISH INDUSTRY WASTE

Fin Fish Wastes

The major fish wastes which have been studied for microbial reclamation is the visceral mass followed by the collagenous components like skin, bone and fins. Fish visceral wastes contain several spoilage as well as lactic acid bacteria (LAB) (Sudeepa et al., 2007a; Bhaskar et al., 2007b). In addition, fish processing wastes including viscera have been reported to be a good source of proteins including enzymes and fats (Gildberg, 2004; Dauksas et al., 2005; Bhaskar et al., 2007a; Morrissey and Okada, 2007) and also, good substrates for lactic acid fermentation (Gao et al., 2006).

Lactic Fermentation

Fermentation is the major technique that is being used for stabilizing the nutrient rich fish waste, especially visceral mass, for further processing and recovery. Conversion of fish processing wastes into ensilage upgrades the biowastes and this approach is eco-friendly, safe, technologically flexible and economically viable (Mathew and Nair, 2006). Ensilation of this waste can be accomplished either by direct addition of mineral or organic acid (acid silage) or biologically by fermentation with LAB. The usefulness of fermentation preservation is based on the fact that it is eco-friendly in nature compared to acid/alkali preservation (Hall et al., 1997). Lactic acid bacteria and its metabolic products during fermentation prevent the growth of unwanted spoilage microorganisms. Researchers working on fermentation have used varying levels of sugar cane molasses (10 to >35%) and different fermentation time (2 to 6 d) for ensiling of fish wastes (Ahmed and Mahendrakar 1995, Lallo et al., 1997; Zahar et al., 2002; Mathew and Nair, 2006; Ganesan et al., 2009).

The fermentation process can be effectively used for simultaneous recovery of protein hydrolysate as well as lipids available in the waste materials (Fig. 10.2). Recently fermentative reclamation of fish visceral waste proteins by preparing fish sauce has been reported by Shih et al. (2003). They found that any alterations in treatment conditions resulted in alterations in volatile compounds as well as content of amino acids, although not much change was observed in the over all quality of fish sauces prepared from Bonito visceral waste. It was observed that the aroma compounds found in the fermented sauces prepared from fish visceral wastes mainly resulted from the lipid components of the material, although other components like sugars and proteins also contributed to the flavour (Cha and Cadwallader, 1998). In another report, viscera of one species of fish that is rich in endogenous enzymes and *in situ* LAB have been used in preparation of sauce from whole sardines (Chrost, 1990; Kim et al., 2002; Klomklao et al., 2006). Chitin recovery was also facilitated from shrimp waste by lactic acid fermentation (Hall and Silva, 1992; Rao et al., 2000).

Apart from fermentation, several enzymes of microbial origin have also been employed for the production of protein hydrolysates from fish industry wastes (Benjakul and Morrissey, 1997; Kim et al., 2001; Lian et al., 2005; Nilsang et al., 2005; Ramirez, 2007; Bhaskar et al., 2008; Bhaskar and Mahendrakar, 2008). Traditional methods for preparation of autolytic hydrolysate like fish silage exploit the endogenous enzymes and it is rather difficult to control the autolysis by endogenous enzymes due to several factors including fish species and seasonality as well as the type and amount of enzymes (Sikoroski and Naczk, 1981). Alternatively, enzymatic proteolysis can be employed to recover biomass from fish visceral mass and results in a soluble product generally referred to as fish protein hydrolysate (Guerard et al., 2002). The benefit of hydrolyzing fish proteins, to make functional protein ingredients and nutritional supplement is a more recent technology, although the first commercially available protein hydrolysate appeared in the late 1940s. In spite of the production being massive world wide, proper control of the process and the exact mechanism behind protein hydrolysis is not fully understood (Kristinnson and Rasco, 2000c).

Application of Exogenous Enzymes

Addition of exogenous enzymes could make the hydrolytic process more controllable, apart from hastening it, thereby making it reproducible. Several factors such as pH, incubation period, enzyme: substrate ratio and temperature influence enzymatic activity (Viera et al., 1995; Liaset et al., 2000; Guerard et al., 2002). Proteolytic enzymes from plants and microorganisms have been found to be suitable to produce fish protein hydrolysate (Benjakul and Morrisey, 1997; Guerard et al., 2001; Nilsang et al., 2005). Acid proteases, even though is better for microbial growth prevention, have only low protein yield and thus, milder enzymes at neutral and slightly alkaline conditions have been used more frequently (Kristinsson and Rasco, 2000c). Enzymes used to produce fish protein hydrolysate have at least one common characteristic; they should be food grade and if they are of microbial origin, the producing organism has to be non-pathogenic. The choice of substrate, protease employed and the degree to which the protein gets hydrolyzed generally affect the physico-chemical properties of the resulting hydrolysates (Mullaly et al., 1995). Alcalase—an alkaline bacterial protease produced from *Bacillus licheniformis*, has been proven to be one of the best enzymes used in the preparation of fish protein hydrolysate (Hoyle and Merritt, 1994; Shahidi et al., 1995; Benjakul and Morrisey, 1997; Dufosse et al., 1997; Kristinsson and Rasco, 2000a, b; Guerard et al., 2001). Further, it has been reported that fish protein hydrolysates prepared using alcalase had less bitter principles as compared to those made with plant proteases like papain (Hoyle and Merritt, 1994; Kristinsson and Rasco, 2000b). Recently hydrolysis conditions for preparation of protein hydrolysates from freshwater fish processing wastes using alcalase have been optimized (Bhaskar et al., 2008). The flow-chart for preparation of fish protein hydrolysate using commercial proteases is outlined in Fig. 10.3.

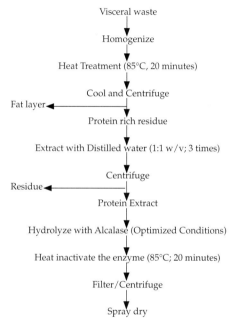

Visceral waste
↓
Homogenize
↓
Heat Treatment (85°C, 20 minutes)
↓
Cool and Centrifuge
↓
Fat layer ← | Protein rich residue
↓
Extract with Distilled water (1:1 w/v; 3 times)
↓
Centrifuge
↓
Residue ← | Protein Extract
↓
Hydrolyze with Alcalase (Optimized Conditions)
↓
Heat inactivate the enzyme (85°C; 20 minutes)
↓
Filter/Centrifuge
↓
Spray dry

Fig. 10.3 Flow sheet for the production of protein hydrolysate from the visceral wastes of Catla (*Catla catla*) using alcalase under optimized conditions. (from Bhaskar et al., 2008).

Although, alcalase is favoured by several researchers in the preparation of fish protein hydrolysate (Benjakul and Morrisey, 1997; Kristinsson and Rasco, 2000b; Bhaskar et al., 2007a, c), it requires a suitable pH for its activity. However, pH of fish visceral wastes usually varies between 5.85 and 6.25 (Bhaskar and Mahendrakar, 2007; Bhaskar et al., 2007a, c). Adjustment of pH prior and/or during hydrolysis, although is common, is not industrially desirable as the added acid/alkali results in unwanted inorganic mass (salt) that may result in undesirable effects and/or may be difficult to remove later in the hydrolysate recovery process (Aspmo et al., 2005). In this regard, neutral proteases of microbial origin look promising as natural (*in situ*) pH of the visceral waste itself can be utilized to prepare protein hydrolysates (Guerard et al., 2002; Dauksas et al., 2005; Nilsang et al., 2005; Dumay et al., 2006; Bhaskar and Mahendrakar, 2008).

The amino acid composition of several protein hydrolysates prepared using different fish industry wastes is presented in Table 10.6. Such hydrolysates prepared from low quality materials like visceral waste has the potential for application as ingredients in aquaculture feeds (Vidotti et al., 2003; Nilsang et al., 2005) and as a source of nitrogen in microbial growth media (Dufosse et al., 1997; Guerard et al., 2001).

Attempts to recover collagen from fish industry waste using enzymes from microbial sources are very scanty. However, enzymatic hydrolysis of other proteins present in the fish industry waste would leave a collagen rich mass that can be further chemically/physically treated to obtain collagen and gelatin (Fig. 10.2).

Shellfish Industry Wastes

Crustacean wastes are also important sources of chitin, with chitin content ranging from 13 to 42% depending on the species (Johnson and Peniston, 1982; Balogun and Akegbejo-Samsons, 1992; Bechtel, 2003; Heu et al., 2003). Deproteinization, demineralization and decolourization are the major steps involved in production of chitin from crustacean wastes. Strong alkali is used in the traditional process of deproteinization at high temperatures and prolonged periods, which results in depolymerization and deacetylation of chitin (Synowiecki and Al-Khateeb, 2003). Further, the disposal of spent alkali poses environmental problems. Enzymatic and fermentation techniques are developed to overcome these problems. In addition these bioprocessing techniques allow recovery of other valuable products such as protein and carotenoids (Healy et al., 1994, 2003). Stabilization of shrimp waste by lactic acid fermentation and evaluation of resulting silage as a feed ingredient has been attempted by many workers. The studies on fermentation of shrimp wastes included the effect of different carbohydrate sources, starter cultures on fermentation and quality of fermented silage (Fagbenro, 1996; Fagbenro and Bello-Olusoji, 1997; Cha and Cadwallader, 1998; Shirai et al., 2001; Cira et al., 2002). Sachindra et al. (1994) reported a method for fermentation of shrimp waste proteins after removal of the chitinous shell. Fagberono (1996) fermented raw heads from freshwater prawn with 5% *Lactobacillus plantarum* as starter culture and 15% molasses as the carbohydrate source and achieved a pH of 4.5 in 7 d. Such fermented silage was found to be a good protein supplement in the diets for cat fish (Fagbenro and Bello-Olusoji, 1997). Shirai et al. (2001) demonstrated that initial glucose concentration and level of LAB are the critical factors in shrimp waste ensilation. The conditions for fermentation of Indian shrimp waste using *Lb. plantarum* have been optimized (Sachindra et al., 2007). Further different lactic cultures were evaluated for their efficiency to ferment shrimp waste and *Pedicoccus acidolactici* was found to be the best starter culture (Bhaskar et al., 2007).

Table 10.6 Amino acid composition of protein hydrolysates obtained from processing wastes of different fish species

	Pacific whiting (A)	Alaska Pollack (B)	Squid (C)	Fish soluble (D)	Salmon (E)	Catla (F)	Shrimp (G)
Essential Amino acids							
Histidine	2.10	0.00	1.45	2.06	1.67	2.06	2.40
Isoleucine	4.30	1.01	3.87	0.35	4.00	3.60	5.75
Leucine	7.16	1.98	7.47	0.77	7.67	7.17	9.84
Lysine	8.33	1.55	0.66	0.54	7.67	7.07	8.19
Methionine	3.02	1.72	3.32	0.48	3.00	2.02	3.13
Phenyl alanine	3.80	2.05	3.73	0.64	3.67	3.53	4.55
Tyrosine	3.50	0.00	3.11	0.40	7.00*	2.57	0.54
Threonine	5.12	2.45	2.97	0.74	3.67	4.02	2.58
Tryptophan	0.14	-	-	-	1.33	-	-
Arginine	7.29	6.38	10.02	2.76	6.33	10.82	5.93
Valine	4.72	1.90	4.08	0.56	5.00	4.79	7.42
Non Essential Amino acids							
Aspargine/Aspartate	10.10	3.05	9.46	0.67	-	8.50	5.48
Glutamine/Glutamate	13.80	5.13	12.63	1.52	-	15.01	11.26
Serine	5.33	5.78	4.01	0.67	-	4.34	1.99
Glycine	7.88	41.47	4.01	2.98	-	10.99	8.35
Alanine	6.53	9.47	4.96	1.33	-	7.04	18.47
Proline/Hydroxy proline	6.00	9.73	4.29	1.45	-	6.24	4.05
Cysteine/cystine	0.92	-	0.49	0.05	-	0.23	0.09

*: Tyrosine + Phenylalanine; A : Benjakul and Morrissey, 1997; B–Kim et al., 2001; C–Lian et al., 2005; D–Nilsang et al., 2005; E–Wright, 2004; F–Bhaskar et al., 2008; G–Bhaskar et al., 2010.

Fermented shrimp wastes were found to contain all the essential amino acids (Lopez-Cervantes et al., 2006). One of the integrated approaches using fermentation to recover carotenoids and chitin simultaneously is presented in Fig. 10.2 (Bhaskar et al., 2008). The fermented liquor is reported to have very good antioxidant potential (Sachindra and Bhaskar, 2008) as well as a balanced amino acid composition (Table 10.6) (Bhaskar et al., 2008). In addition, fermentation ensilaging was found to be a better option for stabilizing carotenoids in shrimp waste without affecting its recovery (Sachindra et al., 2007; Bhaskar and Mahendrakar, 2007).

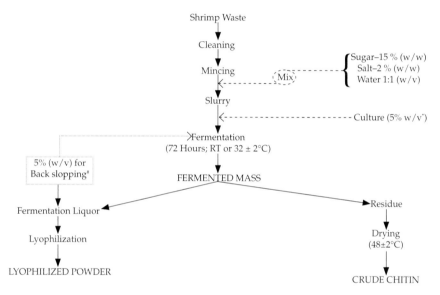

*: *Pediococcus acidolactici* CFR 2182; 8 to 9 log cfu/ml
#: Can be used from the 2nd batch onwards, if not contaminated; has 8 to 9 log cfu/ml

Fig. 10.4 Process flow chart for simultaneous recovery of carotenoids and chitin by fermentation (Bhaskar et al., 2010).

Protease producing microorganisms have been successfully used for deproteinization of crustacean wastes (Oh et al., 2000; Wang and Chio, 1998; Yang et al., 2000). Wang and Chio (1998) demonstrated 81% deproteinization of crab shell by fermentation with protease producing *Pseudomonas aeruginosa* K-187. Other proteolytic organisms used for deproteinization of crustacean shell wastes were *Pseudomonas maltophilia* LC-102 (Shimahara et al., 1984), and *Bacillus subtilis* (Yang et al., 2000). The deproteinization process with demineralization also takes place from the crustacean shells during fermentation (Shirai et al., 2001). Fermentation of shrimp shell using *Bacillus subtilis* resulted in production of sufficient quantities of acid to remove the minerals from the shell and the protease produced by the organism was responsible for the deproteinization of the shell (Sini et al., 2007).

There is a growing interest in the derivatives obtained from the chitin hydrolysis, chitoligomers, N-acetylglucosamine and glucosamine (Shaikh and Deshpande, 1993). Ramirez-Coutino et al. (2006) used fungal chitinase for production of oligosaccharides from the chitin obtained from shrimp wastes. Crustacean wastes have been used as a substrate for chitinase production. Suresh and Chandrashekaran (1998) used prawn waste for production of chitinase by marine fungus, *Beauveria bassiana* through solid state fermentation. Shellfish waste was used as a substrate for solid-state cultivation of *Aspergillus* sp. and chitinase production (Rattanakit et al., 2002). Bioconversion of shellfish chitin wastes for the production of *Bacillus subtilis* W-118 chitinase was reported by Wang et al. (2006). Shrimp waste was used as proteo-chitinaceous substrate for production of proteases and chitinases by *Bacillus thuringensis* (Rojas-Avelizapa et al., 1999).

Chitin is prepared from crustacean wastes by a two-step process that involves removal of constituent proteins including carotene-proteins followed by the removal of ash content. Traditionally chitin preparation from shrimp wastes involves the use of alkalis (usually 4% sodium hydroxide) for deproteinization and acids (e.g., 4% hydrochloric acid) for demineralization, making this process ecologically aggressive and a source of pollution (Hall et al., 1997; Rao et al., 2000). Several researchers have suggested fermentation of shrimp biowaste using LAB as an alternative to the chemical process of chitin production (Rao et al., 2000; Bautista et al., 2001; Healy et al., 2003; Rao and Stevens, 2006).

Carotenoids extracted from crustacean wastes would find a potential use as a source of pigmentation in cultured fish and shrimp. Aquatic animals, which are cultured, do not show the same colouration as that of their wild counter parts (Spinelli and Mahnken, 1978). Pigmentation of cultured species like salmonids and crustacean is done through dietary manipulation (Shahidi et al., 1998). Feeding pigment-enriched diet is regarded as one of the most important management practices for marketing farmed salmon (Moe, 1990).

CONCLUSIONS

Fish processing industrial wastes offer a huge wealth of biomolecules that have potential uses. Biotechnological processes like fermentation and enzymatic hydrolysis can be effectively integrated with the processing operations to reclaim the nutrient mass, which apart from being lost can otherwise create an environmental hazard. Further, the recovered bioactive molecules like enzymes, protein hydrolysates, peptides, carotenoids can be effectively utilized in aquaculture feeds, biomedical industries, as flavour precursors and colourants.

ACKNOWLEDGEMENT

The authors gratefully acknowledge partial funding of their research by ICAR, DBT, DST, MoEF (all of which are Govt. of India organizations) and JSPS, Govt. of Japan through various project grants. The authors also thank the Director, CFTRI, for permission to publish this work.

ABBREVIATIONS

DHA   Docosahexaenoic acid
EPA   Eiocsapentaenoic acid
LAB   Lactic acid bacteria
PUFA   Polyunsaturated fatty Acids

REFERENCES

Aewsiri, T., Benjaku, S., Visessanguan, W. and Tanaka, M. (2008). Chemical compositions and functional properties of gelatin from pre-cooked tuna fin. Int. J. Food Sci. Technol. 43: 685-693.

Ahmed, J., and Mahendrakar, N.S. (1995). Effect of different levels of molasses and salt on acid production and volume of fermenting mass during ensilage of tropical freshwater fish viscera. J. Food Sci. Technol. 32: 115-118

Aidos, I., Van-der-Padt, A., Boom, R.M. and Luten, J.B. (2001). Upgrading of maatjes herring by-products: production of crude fish oil. J. Agric. Food Chem. 49: 3697-3704.

Almas, K.A. (1990). Utilization of marine biomass for production of microbial growth media and biochemicals. In: Advances in Fishery Technology and Biotechnology for Increased Profitability. M.N. Voigt and J. R. Botta (eds). Technomic Publ.Co., Inc., Basel, Switzerland pp. 361-372.

Amarowicz, R. and Shahidi, F. (1997). Antioxidant activity of peptide fractions of capelin protein hydrolysates. Food Chem. 58: 355-359.

Aneiros, A. and Garateix, A. (2004). Bioactive peptides from marine sources: pharmacological properties and isolation procedures. J. Chromat. B 803: 41-53.

Armenta-Lopez, R., Guerrero, L.I. and Huerta, S. (2002). Astaxanthin extraction from shrimp waste by lactic fermentation and enzymatic hydrolysis of the carotenoprotein complex. J. Food Sci. 67: 1002-1006.

Arnesen, J.A. and Gildberg, A. (2006). Extraction of muscle proteins and gelatin from cod head. Process Biochem. 41(3): 697-700.

Arruda, L.F. de, Van der Padt, A. and Boom, R.M. (2007). Use of fish silage—a review. Braz. Arch. Biol. Technol. 50: 879-886.

Arvanitoyannis, I.S. and Kasaveti, A. (2008). Fish industry waste: treatments, environmental impacts, current and potential uses. Int. J. Food Sci. Technol. 43: 726-745.

Aspmo, S.I, Horn, S.J and Eijsink, V.G.H. (2005). Enzymatic hydrolysis of Atlantic cod (*Gadus morhua* L.) viscera. Process Biochem. 40: 1957-1966.

Balogun, M.A. and Akegbejo-Samsons, Y. (1992). Waste yield, proximate and mineral composition of shrimp resources of Nigeria's coastal waters. Biores. Technol. 40: 157-161.

Batista, I. (1999). Recovery of proteins from fish waste products by alkaline extraction. Eur. Food Res. Technol. 210: 84-89.

Bautista, J., Jover, M., Gutierrez, J.F., Corpas, R., Cremades, O., Fontiveros, E.E., Iglesias, F. and Vega, J. (2001). Preparation of crawfish chitin by *in situ* lactic acid production. Process Biochem. 37: 229-234.

Bechtel, P.J. (2003). Properties of different fish processing by-products from pollock, cod and salmon. J. Food Process. Preserv. 27: 101-116.

Bechtel, P.J. (2007). By-products from seafood processing for aquaculture and animal feeds. In: Maximizing the Value of Marine By-products. F. Shahidi (ed). CRC Press, Boca Raton, FL, USA pp. 435-449.

Benjakul, S. and Morrisey, M.T. (1997). Protein hydrolysate from Pacific Whiting solid waste. J. Agric. Food Chem. 61(1/2): 131-138.

Berger, M.M. and Mustafa, I. (2003). Metabolic and nutritional support in acute cardiac failure. Curr. Opin. Clin. Nutr. Metab. Care 6: 195-201.

Bhaskar, N. and Mahendrakar, N.S. (2007). Chemical and microbiological changes in acid ensiled visceral waste of Indian major carp *Catla catla* (Hamilton) with emphasis on proteases. Ind. J. Fish 54 (2): 217-225.

Bhaskar, N. and Mahendrakar, N.S. (2008). Protein hydrolysate from visceral waste proteins of Catla (*Catla catla*): Optimization of hydrolysis conditions for a commercial neutral protease. Biores. Technol. 99: 4105-4111.

Bhaskar, N., Hosokawa, M. and Miyashita, K. (2006a). Physiological effects of eicosapentaenoic acid (EPA) and docosahexenoic acid (DHA)—a review. Food Rev. Intl. 22: 291-307.

Bhaskar, N., Hosokawa, M. and Miyashita, K. (2006b). Conjugated fatty acids occurring in aquatic and terrestrial plants and their physiological effects: a review. In: Nutraceutical Lipids and Their Co-products. Nutraceutical Food Science and Technology, Volume 5. F. Shahidi (ed). CRC Press, Boca Raton, Fl, USA pp. 221-236.

Bhaskar, N., Sudeepa, E.S., Rashmi, H.N. and Tamilselvi, A. (2007a). Partial purification and characterization of protease of *Bacillus proteolyticus* CFR3001 isolated from fish processing waste and its antibacterial activities. Biores. Technol. 98: 2758-2764.

Bhaskar, N., Sathisha, A.D., Sachindra, N.M., Sakhare, P.Z. and Mahendrakar, N.S. (2007b). Effect of acid ensiling on the stability of visceral waste proteases of Indian Major Carp, *Labeo rohita*. J. Aquatic Food Prod. Technol. 16: 73-86.

Bhaskar, N., Suresh, P.V., Sakhare, P.Z. and Sachindra, N.M. (2007c). Shrimp biowaste fermentation with *Pediococcus acidolactici* CFR2182: Optimization of fermentation conditions by response surface methodology and effect of optimized conditions on deproteination/demineralization and carotenoid recovery. Enz. Microb. Technol. 40: 1427-1434.

Bhaskar, N., Benila, T., Radha, C. and Lalitha, R.G. (2008). Optimization of enzymatic hydrolysis of visceral waste proteins of catla (*Catla catla*) for preparing protein hydrolysate using a commercial protease. Biores. Technol. 99: 35-343.

Bhaskar, N., Suresh, P.V., Sakhare, P.Z., Lalitha, R.G. and Sachindra, N.M. (2010). Yield and chemical composition of fractions from fermented shrimp biowaste. Waste Manag. Res. 28: 64-70.

Borderias, J., Angeles M. and Montero, P. (1994). Use of image analysis to determine fat and connective tissue in salmon muscle. Z. Lebensm Unters Forsch. 199: 255-261.

Calder, P.C. (2003). Long chain n-3 fatty acids and inflammation: potential application in surgical and trauma patients. Braz. J. Med. Biol. Res. 36: 433-466.

Cha, Y.J. and Cadwallader, K.R. (1998). Aroma-active compounds in skipjack tuna sauce. J. Agric. Food Chem. 46: 1123-1128.

Chandini, S.K., Ganesan, P., Suresh, P.V. and Bhaskar, N. (2008). Seaweeds as sources of nutritionally beneficial compounds—a review. J. Food Sci. Technol. 45: 1-13.

Chrost, R.J. (1990). Microbial ectoenzymes in aquatic environments. In: Aquatic Microbial Ecology. J. Overbeck and R.J. Chrost (eds). Springer, New York, USA pp. 47-78.

Cira, L.A., Huerta, S., Hall, G.M. and Shirai, K. (2002). Pilot lactic acid fermentation of shrimp wastes for chitin recovery. Process Biochem. 37: 1359-1366.

Clausen, E., Gildberg, A. and Raa, J. (1985). Preparation and testing of an autolysate of fish viscera as growth substrate for bacteria. Appl. Environ. Microbiol. 50: 1556-1557.

Connor, W.E. (2000). Importance of n-3 fatty acids in health and diseases. Am. J. Clin. Nutr. 71 (suppl.): 171S-175S.

Dauksas, E., Falch, E., Slizyte, R.S. and Rustad, T. (2005). Composition of fatty acids and lipid classes in bulk products generated during enzymatic hydrolysis of cod (*Gadus morhua*) by-products. Process. Biochem. 40: 2659-2670.

di-Wu, Q., Wang, J.H., Fennessy, F., Redomond, H.P. and Bouchier-Hayes, D. (1999). Taurine prevents high-glucose-induced human vascular endothelial cell apoptosis. Am. J. Physiol. Cell Physiol. 277: C1229-C1238.

Dufosse, L., De La Broise, D. and Guerard, F. (1997). Review: Fish protein hydrolysates as nitrogen sources for microbial growth and metabolite production. In: Recent Research Developments in Microbiology, Volume 1. S.G. Pandalai (ed). Research Sign Post Publ., Trivandrum, India, pp. 365-381.

Dumay, J., Donnay-Moreno, C., Barnathan, G., Jaouen, P. and Berge, J.P. (2006). Improvement of lipid and phospholipid recoveries from sardine (*Sardina pilchardus*) viscera using industrial proteases. Process. Biochem. 41: 2327-2332.

Elvevoll, E. (2004). Fish waste and functional Foods. In: Total Food.. K.Waldron, C. Faulds and A. Smith (eds). Norwich, England, www.totalfood2004.com/elvevoll_files/frame. htm

Fagbenro, O.A. (1996). Preparation, properties and preservation of lactic acid fermented shrimp heads. Food Res. Int. 29: 595-599.

Fagbenro, O.A. and Bello-Olusoji, O.A. (1997). Preparation, nutrient composition and digestibility of fermented shrimp head silage. Food Chem. 60: 489-493.

FAO (2010). Year Book of Fishery Statistics—Latest summary of tables. (ftp://ftp.fao.org/ FI/STAT/SUMM_TAB.HTM). (Available at http://www.fao.org/fishery/statistics/en; last accessed on March 2, 2010).

Fennessy, F.M., Moneley, D.S., Wang, J.H., Kelly, C.J. and Bouchier-Hayes, D. (2003). Taurine and vitamin C modify monocyte and endothelial dysfunction in young smokers. Circulation 107(3): 410-415.

Fujita, H. and Yoshikawa, M. (2008). Marine derived protein hydrolysates, their biological activities and potential as functional food ingredients: ACE inhibitory peptides derived from bonito. In: Marine Nutraceuticals and Functional Foods. Nutraceutical Food Science and Technology, Volume 7. C. Barrow and F. Shahidi, (eds). CRC Press, Boca Raton, Fl., USA pp. 247-248.

Ganesan, P., Pradeep, M.G., Sakhare, P.Z., Suresh, P.V. and Bhaskar, N. (2009). Optimization of conditions for natural fermentation of freshwater fish processing waste using sugar cane molasses. J. Food Sci. Technol. 45 (in press).

Gao, M.T., Hirata, M., Toorisaka, E. and Hano, T. (2006). Acid hydrolysis of fish wastes for lactic acid fermentation. Biores. Technol. 97: 2414-2420.

Gbogouri, G.A., Linder, M., Fanni, J. and Parmentier, M. (2006). Analysis of lipids extracted from salmon (*Salmo salar*) heads by commercial proteolytic enzymes. Eur. J. Lipid Sci. Technol. 108: 766-775.

Gildberg, A. (2004). Enzymes and bioactive peptides from fish waste related to fish silage, fish feed and fish sauce production. J. Aquat. Food Prod. Technol. 13(2): 3-11.

Gildberg, A., Simpson, B.K. and Haard, N.F. (2000). Use of enzymes from marine organisms. In: Seafood Enzymes. N.F. Haard and B.K. Simpson (eds). Marcel Dekker, New York, USA pp. 619-639.

Gildberg, A., Arnesen, J.A. and Carlehog, M. (2002). Utilization of cod backbone by biochemical fractionation. Process Biochem. 38: 475-480.

Gordon, D.T. and Ratliff, V. (1992). The implications of omega-3-fatty acids in human health. In: Advances in Seafood Biochemistry. G. J. Flick and R. E. Martin (eds). Technomic Publ. Co. Inc., Lancaster, Basel, Switzerland, pp 69-98.

Gormley, T.R., Neumann, T., Fagan, J.D. and Brunton, N.P. (2007). Taurine content of raw and processed fish fillets. Eur. Food Res. Technol. 225: 837-842.

Guerard, F. (2006). Enzymatic methods for marine by-products recovery. In: Maximizing the Value of Marine By-products. F. Shahidi (ed). CRC Press, Boca Raton, Fl, USA pp. 107-143.

Guerard, F., Duffose, L., De La Broise, D. and Binet, A. (2001). Enzymatic hydrolysis of proteins from yellow fin tuna (*Thunnus albacares)* wastes using alcalase. J. Mol. Catalysis B: Enzymatic 11: 1051-1059.

Guerard, F., Guimas, L. and Binet, A. (2002). Production of tuna waste hydrolysates by a commercial neutral protease preparation. J. Mol. Cat. B: Enz. 19/20: 489-498.

Haard, N.F. (1998). Specialty enzymes from marine organisms. Food Technol. 53(7): 64-67.

Haard, N.F. and Simpson, B.K. (2000) Seafood Enzymes. Marcel Dekker, New York, USA.

Haard, N.F., Simpson, B.K. and Sikorski, Z.E. (1994). Biotechnological applications of seafood proteins and other nitrogenous compounds. In: Seafood Proteins. Z.E. Sikorski, B.S. Pan and F. Shahidi (eds). Chapman and Hall, New York, U.S.A, pp. 195-216.

Hall, G.M. and Silva, S. (1992). Lactic acid fermentation of shrimp (*Penaeus monodon*) waste for chitin recovery. In: Advances in Chitin and Chitosan. C. J. Brine, P. A. Sandford and J. P. Zikakis (eds). Elsevier Applied Science, London, UK pp. 633-638.

Hall, G.M., Zakaria, Z. and Reid, C.I. (1997). Recovery of value added products from shellfish waste by a lactic acid fermentation process. In: Seafood from Producers to Consumers, Integrated Approach to Quality. J.B. Luton and T. Bonneson (eds). Elsevier Applied Science, London, UK pp. 97-102.

Harish Prashanth, K.V. and Taranathan, R.N. (2005). Deploymerized products of chitosan as potent inhibitors of tumor-induced angiogenesis. Biochim. Biophys. Acta 1722: 22-29.

Harish Prashanth, K.V., Lakshman, K., Shamala, T.R. and Taranathan, R.N. (2005). Biodegradation of chitosan-graft-polymethylmethacrylate films. Int. Biodeter. Biodegrad. 56: 115-120.

Hayes, M., Carney, B., Slater, J. and Bruck, W. (2008a). Mining marine shellfish wastes for bioactive molecules: Chitin and Chitosan–Part A: Extraction methods. Biotechnol. J. 3: 871-877.

Hayes, M., Carney, B., Slater, J. and Bruck, W. (2008b). Mining marine shellfish wastes for bioactive molecules: Chitin and chitosan–Part B: Application. Biotechnol. J. 3: 878-889.

Healy, M.G., Romo, C.R. and Bustos, R. (1994). Bioconversion of marine crustacean shell waste. Resour. Conserv. Recycl. 11: 139-147.

Healy, M., Green, M. and Healy, A. (2003). Bioprocessing of marine crustacean shell waste. Acta Biotechnol. 23: 151-160.

Heu, M.S., Kim, J.S. and Shahidi, F. (2003). Components and nutritional quality of shrimp processing by-products. Food Chem. 82: 235-242.

Holub, B.J. (2002). Clinical nutrition: 4. Omega-3 fatty acids in cardiovascular care. Can. Med. Assoc. J. 166: 608-615.

Hoyle, N.T. and Merritt, J.H. (1994). Quality of fish protein hydrolysate from Herring (*Clupea harengus*). J. Food Sci. 59: 76-79 & 129.

Hurst, D. (2006). Marine functional foods and functional ingredients—a briefing document. Marine Foresight Series 5. Available as PDF document at www.marine.ie

Je, J-Y., Qian, Z-J., Byun, H-G. and Kim, S.K. (2007). Purification and characterization of an antioxidant peptide from tuna backbone protein by enzymatic hydrolysis. Process. Biochem. 42: 840-846.

Johnson, E.L. and Peniston, Q.P. (1982). Utilization of shellfish waste for chitin and chitosan production. In: Chemistry and Biochemistry of Marine Food Products. R.E. Martin, G. J. Flick, C. E. Hobard and D. R. Ward (eds). AVI Publication Co, Westport, CA, USA pp. 415-419.

Kim, S.K. and Mendis, E. (2006). Bioactive compounds from marine processing by-products —a review. Food Res. Int. 39: 383-393.

Kim, S.K., Kim, Y.T., Byun, H.G., Nam, K.S., Joo, D.S. and Shahidi, F. (2001). Isolation and characterization of antioxidative peptides from gelatin hydrolysate of Alaska Pollack skin. J. Agric. Food Chem. 49: 1984-1989.

Kim, S.K., Mendis, E. and Shahidi, F. (2008). Marine fisheries by-products as potential nutraceuticals: an overview. In: Marine Nutraceuticals and Functional Foods. Nutraceutical Food Science and Technology, Volume 7. C. Barrow and F. Shahidi (eds). CRC Press, Boca Raton, Fl., USA pp 1-22.

Kim, Y.K., Bae, J.H., Oh, B.K., Lee, W.H. and Choi, J.W. (2002). Enhancement of proteolytic enzyme activity excreted from *Bacillus stearothermophilus* for a thermophilic aerobic digestion process. Biores. Technol. 82: 157–164.

Kittur, F.S., Vishnukumar, A.B., Varadaraj, M.C. and Taranathan, R.N. (2005). Chitooligosaccharides–preparation with the aid of pectinase isozyme from *Aspergillus niger* and their antibacterial activity. Carbohyd. Res. 340: 1239-1245.

Knorr, D. (1991). Recovery and utilization of chitin and chitosan in food processing waste management. Food Technol. 45: 114-120.

Klomklao, S., Benjakul, S., Visessanguan, W., Kishimura, H. and Simpson, B.K. (2006). Effects of addition of spleen of skipjack tuna (*Katsuwonus pelamis*) on the liquefaction and characteristics of fish sauce made from sardine (*Sardinella gibbosa*). Food Chem. 98: 440-452.

Kristinsson, H.G. (2007). Aquatic food protein hydrolysates. In: Maximizing the Value of Marine By-Products. F. Shahidi (ed). CRC Press, Boca Raton, Fl., USA pp. 229-248.

Kristinsson, H.G. (2008). Functional and bioactive peptides from hydrolyzed aquatic food proteins. In: Marine Nutraceuticals and Functional Foods. Nutraceutical Food Science and Technology, Volume 7. C. Barrow and F. Shahidi (eds). CRC Press, Boca Raton, Fl, USA pp. 229-246.

Kristinsson, H.G. and Rasco, B.A. (2000a). Biochemical and functional properties of Atlantic salmon (*Salmo salar*) muscle proteins hydrolysed with various alkaline proteases. J. Agri. Food Chem. 48: 657-666.

Kristinsson, H.G. and Rasco, B.A. (2000b). Kinetics of the hydrolysis of Atlantic salmon (*Salmo salar*) muscle proteins by alkaline protease and a visceral serine protease mixture. Process Biochem. 36: 131-139.

Kristinsson, H.G. and Rasco, B.A. (2000c). Fish protein hydrolysates: production, biochemical and functional properties. Crit. Rev. Food Sci. Nutr. 40: 43-81.

Lallo, C.H.O., Singh, R., Donawa, A.A. and Madoo, G. (1997). The ensiling of poultry offal with sugar cane molasses and lactobacillus culture for feeding to growing/finishing pigs under tropical conditions. Animal Feed Sci. Technol. 67: 213-222.

Li, D., Bode, O., Drummond, H. and Sinclair, A.J. (2003). Omega-3 (n-3) fatty acids. In: Lipids for Functional Foods and Nutraceuticals. F. D. Gunstone (eds). The Oily Press, New York, USA pp. 225-262.

Lian, P.Z., Lee, C.M. and Park, E. (2005). Characterization of squid-processing by-product hydrolysate and its potential as aquaculture feed ingredient. J. Agric. Food Chem. 53: 5587-5592.

Liaset, B., Lied, E. and Espe, M. (2000). Enzymatic hydrolysis of by-products from the fish-filleting industry; chemical characterization and nutritional evaluation. J. Sci. Food Agric. 80: 581-589.

Liaset, B., Julshman, K. and Espe, M. (2003). Chemical composition and theoretical nutritional evaluation of the produced fractions from enzymatic hydrolysis of salmon frames with protamex. Process Biochem. 38: 1747-1759.

Liu, X.Q. and Li, Y.H. (2000). Epidemiological and nutritional research on prevention of cardiovascular disease in China. Br. J. Nutr. 84-S2: 199-203.

Lopez-Cervantes, J., Sanchez-Machado, D.I. and Rosas-Rodrigues, J.A. (2006). Analysis of free amino acids in fermented shrimp waste by high-performance liquid chromatography. J. Chromatography A 1105: 106-110.

Mathew, P. and Nair, K.G.R. (2006). Ensilation of shrimp waste by *Lactobacillus fermentum*. Fish. Technol. 43: 59-64.

Matsuno, T. (2001). Aquatic animal carotenoids. Fish. Sci. 67: 771-783.

Mendis, E., Rajapakse, N. and Kim, S.K. (2005). Antioxidant properties of a radical-scavenging peptide purified from enzymatically prepared fish skin gelatin hydrolysate. J. Agric. Food Chem. 53: 581-587.

Militante, J.D. and Lombardini, J.B. (2002). Treatment of hypertension with oral taurine: experimental and clinical studies. Amino Acids 23: 381-393.

Miyashita, K. and Hosokawa, M. (2008). Beneficial health effects of marine carotenoid, fucoxanthin. In: Marine Nutraceuticals and Functional Foods. Nutraceutical Food Science and Technology, Volume 7. C. Barrow and F. Shahidi (eds). CRC Press, Boca Raton, Fl, USA pp. 297-320.

Mizuta, S., Fujisawa, S., Nishimoto, M., and Yoshinaka, R. (2005). Biochemical and immunochemical detection of types I and V collagens in tiger puffer Takifugu rubripes. Food Chem. 89(3): 373-377.

Moe, N.H. (1990). Key factors in marketing farmed salmon. Proc. Nutr. Soc. New Zealand, 15: 16-22.

Morrissey, M.T. and Okada, T. (2007). Marine enzymes from seafood by-products. In: Maximizing the Value of Marine By-Products. F. Shahidi (ed). CRC Press, Boca Raton, Fl., USA pp. 374-396.

MPEDA (2010). Marine products export performance review; http://www.mpeda.com; last accessed on March 2, 2010.

Mullaly, M.M., O'Callaghan, D.M., Fitzgerald, R.J., Donnelly, W.J. and Dalton, J.P. (1995). Zymogen activation in pancreatic endoproteolytic preparations and influence on some whey protein characteristics. J. Food Sci. 60(2): 227-233.

Muyonga, J.H., Cole, C.G.B. and Duodu, K.G. (2004). Characterization of acid soluble collagen from skins of young and adult Nile perch (*Lates niloticus*). Food Chem. 85(1): 81-89.

Nagai, T. (2004). Characterization of collagen from Japanese sea bass caudal fin as waste material. Eur. Food Res. Technol. 218(5): 424-427.

Nagai, T. and Suzuki, N. (2000). Isolation of collagen from fish waste material—skin, bone and fins. Food Chem. 68(3): 277-281.

Nilsang, S., Lertsiri, S., Suphantharika, M. and Assavanig, A. (2005). Optimization of hydrolysis of fish soluble concentrate by commercial proteases. J. Food Eng. 70: 571-578.

Nomura, Y., Yamano, M., Hayakawa, C., Ishii, Y. and Shirai, K. (1997). Structural property and *in vitro* self-assembly of shark type I collagen. Biosci. Biotechnol. Biochem. 61(11): 1919-1923.

Norland, P.E. (1989). Fish gelatine. In: Advances in Fisheries Technology and Biotechnology for Increased Profitability. M.N. Voigt and J. R. Botta (eds). Technomic Publishing Co. Inc., Lancaster, Pennsylvania, USA pp. 325-333.

Ogawa, H. (1996). Effect of dietary taurine on lipid metabolism in normcholesterolemic and hypercholesterolemic stroke prone spontaneously hypertensive rats. In: Advances in Experimental Biology, Volume 403. Taurine 2 Basic and Clinical Aspects. S. Schafer, J.B. Lombardini and R.J. Huxtable (eds). Plenum Press, New York, USA pp. 107-115.

Oh, Y., Shih, I., Tzeng, Y. and Wang, S. (2000). Protease produced by *Pseudomonas aeruginosa* K-187 and its application in the deproteinization of shrimp and crab shell wastes. Enz. Microb. Technol. 27: 3-10.

Okada, T., Hosokawa, M., Ono, S. and Miyashita, K. (2008). Enzymatic production of marine derived protein hydrolysates and their bioactive peptides for use in foods and nutraceuticals. In: Biocatalysis and Bioenergy. C.T. Hou and J.R. Shaw (eds). John Wiley & Sons, New York, USA pp. 491-520.

Olaizola, M. (2008). The production and health benefits of astaxanthin. In: Marine Nutraceuticals and Functional Foods. Nutraceutical Food Science and Technology, Volume 7. C. Barrow and F. Shahidi (eds). CRC Press, Boca Raton, Fl., USA pp. 321-344.

Raa, J. and Gildberg, A. (1982). Fish Silage. CRC Crit. Rev. Food Sci .Nutr. 16(4): 383-491.

Ragunath, M.R. and Gopakumar, K. (2002). Trends in production and utilization of fish silage. J. Food Sci. Tech. 39: 03-110.

Ramesh, H.P., Viswanatha, S.and Taranathan, R.N. (2004). Safety evaluation of formulations containing carboxymethyl derivatives of starch and chitosan in albino rats. Carbohyd. Polym. 58: 435-441.

Ramirez, A. (2007). Salmon by-product proteins. FAO Fish. Circ. No. 1027, Food and Agric. Org., Rome, Italy.

Ramirez-Coutino L, Carmen M, Cervantes M, Huerta S, Revah S and Shirai K. (2006). Enzymatic hydrolysis of chitin in the production of oligosaccharides using *Lecanicillium fungicola* chitinases. Process Biochem. 41: 1106-1110.

Rao, M.S. and Stevens, W.F. (2005). Chitin production by *Lactobacillus* fermentation of shrimp biowaste in a drum reactor and its chemical conversion to chitosan. J. Chem. Technol. Biotechnol. 80: 1080-1087.

Rao, M.S., Munoz, J. and Stevens, W.F. (2000). Critical factors in chitin production by fermentation of shrimp biowaste. Appl. Microbiol. Biotechnol. 54: 808-813.

Rattanakit, N., Plikomol, A., Yano, S., Wakayama, M. and Takashi Tachiki, T. (2002). Utilization of shrimp shellfish waste as a substrate for solid-state cultivation of *Aspergillus* sp. S 1-13: Evaluation of a culture based on chitinase formation which is necessary for chitin-assimilation. J. Biosci. Bioeng. 93: 550-556.

Rojas Avelizapa, L.I., CruzCamarillo, R., Guerrero, M.I., RodriguezVazquez, R. and Ibarra, J.E. (1999). Selection and characterization of a proteo-chitinolytic strain of *Bacillus thuringiensis*, able to grow in shrimp waste media. World J. Microbiol. Biotechnol. 15: 261-268.

Rustad, T. (2003). Utilization of marine by-products. Electronic J. Environ. Agri. Food Chem. 2 (4): 458-463.

Sachindra, N.M. and Mahendrakar, N.S. (2005). Extractability of carotenoids from shrimp waste in vegetable oils and process optimization. Biores. Technol. 96: 1195-1200.

Sachindra, N.M. and Bhaskar, N. (2008). *In-vitro* antioxidant activity of liquor from fermented shrimp biowaste. Bires. Technol. 99: 9013-9016.

Sachindra, N.M., Rao, Lalitha S. and Sripathy, N.V. (1994). Ensilaging of prawn carapace by lactic fermentation. Seafood Exp. J. 25(13): 13-16.

Sachindra, N.M., Bhaskar, N., Sathisha, A.D. and Sakhare, P.Z. (2004). Collagen in different body components of Indian Major Carps (IMCs) and common carp (*Cyprinus carpio*). Presented at the 16th Indian Convention of Food Scientists and Technologists, December 10-11, 2004, Mysore, India.

Sachindra, N.M., Bhaskar, N. and Mahendrakar, N.S. (2005a). Carotenoids in different body components of Indian shrimps. J. Sci. Food Agric. 85: 165-172.

Sachindra, N.M., Bhaskar, N. and Mahendrakar, N.S. (2005b). Carotenoids in crabs of marine and freshwaters of India. LWT Food Sci. Technol. 38: 221-225.

Sachindra, N.M., Bhaskar, N. and Mahendrakar, N.S. (2006a). Carotenoids in *Solonocera indica* and *Aristeus alcocki*, deep-sea shrimp from Indian waters. J. Aquat. Food Prod. Technol. 15(2): 5-16.

Sachindra, N.M., Bhaskar, N. and Mahendrakar, N.S. (2006b). Recovery of carotenoids from shrimp waste in organic solvents. Waste Manage. 26: 1092-1098.

Sachindra, N.M., Bhaskar, N., Siddegowda, G.S., Sathisha, A.D. and. Suresh, P.V. (2007). Recovery of carotenoids from ensiled shrimp waste. Biores. Technol. 98: 1642-1646.

Sakaguchi, M., Murata, M. and Kawai, A. (1982). Taurine levels in some tissues of yellowtail (*Seriola quinqueradiatd*) and mackerel (*Scomber japonicus*). Agric. Biol. Chem. 46: 2857-2858.

Shahidi, F. and Janak-Kamil, Y.V.A. (2001). Enzymes from fish and aquatic invertebrates and their application in food industry. Trends Food Sci. Technol. 12: 435-464.

Shahidi, F., Metusalach and Brown, J.A. (1998). Carotenoid pigments in seafoods and aquaculture. CRC Crit. Rev. Food Sci Nutr. 38: 1-67.

Shaikh, S.A. and Deshpande, M.V. (1993). Chitinolytic enzymes: their contribution to basic and applied research. World J. Microbiol. Biotechnol. 9: 468–475.

Shih, I.L., Chen, L.G., Yu, T.S., Chang, W.T. and Wang, S.L. (2003). Microbial reclamation of fish processing wastes for the production of fish sauce. Enz. Microb. Technol. 33: 154-162.

Shimahara, K., Yasuyuki, T., Kazuhiro, O., Kazunori, K. and Osamu, O. (1984). Chemical composition and some properties of crustacean chitin prepared by use of proteolytic activity of *Pseudomonas maltophilia* LC102. In: Chitin, Chitosan and Related Enzymes. J. P. Zikakis (ed). Academic Press, Orlando, Fl., USA pp. 239–255.

Shirai, K., Guerrero, I., Huerta, S., Saucedo, G., Castillo, A., Gonzalez, R.O. and Hall, G.M. (2001). Effect of initial glucose concentration and inoculation level of lactic acid bacteria in shrimp waste ensilation. Enz. Microb. Technol. 28: 446–452.

Sikoroski, Z.E. and Naczk, M. (1981). Modification of technological properties of fish protein concentrate. CRC Crit. Rev. Food Sci. Nutr. 13: 201-230.

Sikorski. Z.E. and Borderias, J.A. (1994). Collagen in the muscle and skin of marine animals. In: Seafood Proteins. Z. E. Sikorski, B.S. Pan and F. Shahidi (eds). Chapman and Hall, New York, USA pp. 58-70.

Silvestre, M.P.C. (1997). Review of methods for the analysis of protein hydrolysates. Food Chem.60: 263-271.

Simpson, B.K. (2007). Pigments from by-products of seafood processing. In: Maximizing the Value of Marine By-Products. F. Shahidi (ed). CRC Press, Boca Raton, Fl., USA pp. 413-434.

Sini, T.K., Santhosh, S. and Mathew, P.T. (2007). Study on the production of chitin and chitosan from shrimp shell by using *Bacillus subtilis* fermentation. Carbohyd. Res. 342: 2423-2429.

Spinelli, J. and Mahnken, C. (1978). Carotenoid deposition in pen reared salmonids fed diets containing oil extracts of red crab (*Pleuronnocodes planipes*). Aquaculture13: 213-216.

Srinivasa, P.C., Ramesh, M.N., Kumar, K.R. and Taranathan, R.N. (2004a). Properties of chitosan films prepared under different drying conditions. J. Food Eng. 63: 79-85.

Srinivasa, P.C., Susheelamma, N.S., Ravi, R. and Taranathan, R.N. (2004b). Quality of mango fruits during storage: Effect of synthetic and eco-friendly films. J. Sci. Food Agric. 84: 818-824.

Strom, T. and Eggum, B.O. (1981). Nutritional value of fish viscera silage. J. Sci. Food Agric. 32: 115-120.

Sudeepa, E.S., Rashmi, H.N., Tamilselvi, A. and Bhaskar, N. (2007a). Proteolytic bacteria associated with fish processing wastes: Isolation and characterization. J. Food Sci. Technol. 44: 281-284.

Sudeepa, E.S., Tamilselvi, A. and Bhaskar, N. (2007b). Antimicrobial activity of a lipolytic and hydrocarbon degrading lactic acid bacterium isolated from fish processing waste. J. Food Sci. Technol. 44: 417-421.

Suresh, P.V. and Chandrasekaran, M. (1998). Utilization of prawn waste for chitinase production by the marine fungus *Beauveria bassiana* by solid state fermentation. World J. Microbiol. Biotechnol. 14: 655-660.

Synowiecki J. and Al-Khateeb N.A.A.Q. (2003). Production, properties and some new applications of chitin and its derivatives. Crit. Rev. Food Sci. Nutr. 43: 145-171.

Tapiero, H., Ba, G.N., Couvreur, P. and Tew, K.D. (2002). Polyunsaturated fatty acids (PUFA) and eicosanoids in human health and pathologies. Biomed. Pharmacother. 56: 215-222.

Taranathan, R.N. and Kittur, F.S. (2003). Chitin-the undisputed biomolecule of great potential. Crit.. Rev. Food Sci. Nutr. 43(1): 61-87.

Tharanathan, R.N. (2003). Biodegradable films and composite coatings: past, present and future. Trends Food Sci. Technol. 14: 71-78.

Tsugita, T. (1990). Chitin/Chitosan and their applications. In: Advances in Fisheries Technology and Biotechnology for Increased Profitability. M.N. Voigt and J.R. Botta (eds). Technomic Publ. Co. Inc., Basel, Switzerland pp. 287-298.

Vecht-lifshitz, S.E., Almas, K.A., and Zomer, E. (1990). Microbial growth on peptones from fish industrial waste. Lett. Appl. Microbiol. 10: 183-186.

Vidotti, R.M., Viegas, E.M.M. and Careiro, D.J. (2003). Amino acid composition of processed fish silage using different raw materials. Anim. Feed Sci. Technol. 105: 199-204.

Viera, G.H.F., Martin, A.M., Sampaiao, S.S., Omar, S. and Gonsalves, R.C.F. (1995). Studies on the enzymatic hydrolysis of Brazilian lobster (*Panulirus* spp) processing wastes. J. Sci. Food Agric. 69: 61-65.

Vishnukumar, A.B. and Taranathan, R.N. (2004). A comparative study on depolymerization of chitosan by proteolytic enzymes. Carbohyd. Polym. 58: 275-283.

Vishnukumar, A.B., Lalitha, R.G. and Taranathan, R.N. (2004a). Non-specific depolymerization of chitosan by pronase and characterization of the resultant products. Eur. J. Biochem. 271: 713-723.

Vishnukumar, A.B., Varadaraj, M.C., Lalitha, R.G. and Taranathan, R.N. (2004b). Low molecular weight chitosans: Preparation with the aid of papain and characterization. Biochim. Biophys. Acta 1670: 137-146.

Waldron, K., Faulds, C. and Smith, A. (eds). (2004). Total Food. Norwich, England, www.totalfood2004.com/elvevoll_files/frame.htm

Wang, S.L. and Chio, S.H. (1998). Deproteinization of shrimp and crab shell with the protease of *Pseudomonas aeruginosa* K-187. Enz. Microb. Technol. 22: 629-633.

Wang, S., Lin, T., Yen, Y., Liao, H. and Chen, Y. (2006). Bioconversion of shell fish chitin wastes for the production of *Bacillus subtilis* W-118 chitinase. Carbohyd. Res. 341: 2507-2515.

Wright, I. (2004). Salmon by-products. Aqua Feeds: Formulation and Beyond. 1: 10-12.

Yang, J.K., Shih, I.L., Tzeng, Y.M. and Wang, S.L. (2000). Production and purification of protease from a *Bacillus subtilis* that can deprotenize crustacean wastes. Enz. Microb. Technol. 26: 406-413.

Zahar, M., Benkerroum, N., Guerouali, A., Laraki, Y., El-Yakoubi, K. (2002). Effect of temperature, anaerobiosis, stirring and salt addition on natural fermentation silage of sardine and sardine wastes in sugarcane molasses. Biores. Technol. 82: 171-176.

Zhao, X., Jia, J. and Lin, Y. (1998). Taurine content in Chinese food and daily taurine intake of Chinese men. In: Advances in Experimental Biology, Volume 442. Taurine 3, Cellular and Regulatory Mechanisms. S. Schafer, J.B. Lombardini and R.J. Huxtable (eds). Plenum Press, New York, USA pp. 501-505.

Color Plate Section

Chapter 4

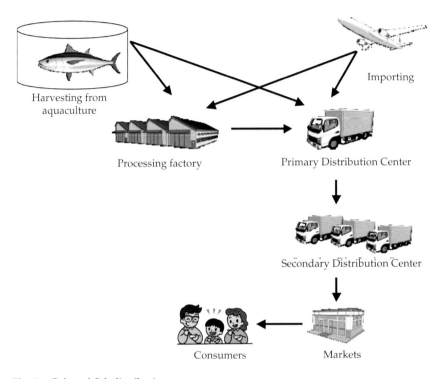

Fig. 4.1 Cultured fish distribution route.

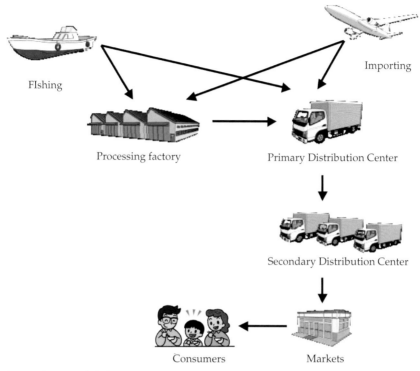

Fig. 4.2 Captured fish distribution route.

Chapter 6

A. Plara

B. Som fug (Plaa som)

C. Som fug (Plaa som)

D. Hoi Dorang (Mussel)

E. Tilapia (Som-fug)

Fig. 6.1 Traditional fermented fish products of Thailand.

Chapter 8

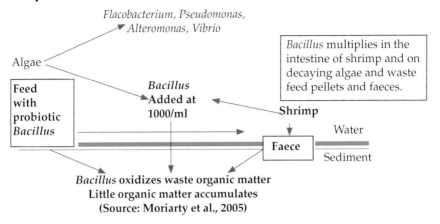

Fig. 8.1 Effect of *Bacillus* at high population density in the pond for organic matter from algae, feed and animals. Specially selected *Bacillus* displaces pathogenic *Vibrio* (Moriarty et al., 2005).

Index